RAYMOND McFARLAND

A HISTORY OF THE NEW ENGLAND FISHERIES

WITH MAPS

Elibron Classics
www.elibron.com

Elibron Classics series.

ISBN 1-4021-9706-3 (paperback)
ISBN 1-4021-3051-1 (hardcover)

This Elibron Classics Replica Edition is an unabridged facsimile of the edition published in 1911 by University of Pennsylvania; D. Appleton and Co., New York.

A HISTORY OF THE NEW ENGLAND FISHERIES

WITH MAPS

BY

RAYMOND McFARLAND, A.M.
ASSISTANT PROFESSOR IN MIDDLEBURY COLLEGE

UNIVERSITY OF PENNSYLVANIA
D. APPLETON AND COMPANY, AGENTS, NEW YORK
1911

A HISTORY OF THE NEW ENGLAND FISHERIES

WITH MAPS

BY

RAYMOND McFARLAND, A.M.
ASSISTANT PROFESSOR IN MIDDLEBURY COLLEGE

UNIVERSITY OF PENNSYLVANIA
D. APPLETON AND COMPANY, AGENTS, NEW YORK
1911

J. F. TAPLEY CO.
NEW YORK

DEDICATED TO

CAPTAIN JOHN W. McFARLAND

MASTER MARINER, OF GLOUCESTER

PREFACE

There is no industry in America that antedates the fisheries. Fishermen from France, Spain, Portugal and England frequented the banks of Newfoundland before any trade relations between the Old and the New World came into existence. The codfishery of the New England coast was carried on many years previous to the establishment of a permanent English colony in our country. Voyages of exploration and settlement to New England were prompted by reports of the valuable fishing grounds lying between Cape Cod and Cape Sable. With the permanent settlement of New England the fisheries became the chief industry of the people, and continued down to the Revolution to be a main source of wealth, and to supply the principal article of trade between the merchants of New England and those of the West Indies and Europe.

The fisheries of New England were of especial importance during the colonial period and still continue to be of greater economic importance than at any previous period of our history; yet the industry, with the exception of the whale fishery, has no literature that adequately sets forth its history and value. There are several good accounts of the whale fishery; but the story of the cod and mackerel and inshore fisheries does not appear to have been an attractive field to writers. The accounts we have of the history and of the methods of the fisheries have been prompted largely by government inquiry and exist as government reports.

The present volume aims to show the development and importance of the New England fisheries from pre-colonial days to the present time. Little attention is paid to the whale fishery, as that industry, now almost passed away, has received proper attention in the literature of the fisheries.[1] A brief discussion of the early fisheries in Newfoundland waters has been introduced in order to place the New England fisheries in their proper setting relative to the fisheries of Europe. The fisherman as an individual has not received the attention that he deserves; but this is a study in industrial rather than social history. The Story of the Fisherman has yet to be written, and it can be made a volume of keen interest, a narrative of heroic daring.

The information presented in the work has been derived from a wide range of sources, from personal inquiry in the principal fishing towns from New York to Newfoundland, and from experience in the fishing industry itself. The author is indebted to the Carnegie Institution of Washington, D. C., for assistance received in preparing this volume. Acknowledgments also are made of the courtesy and kindly assistance received from the officers of the Worcester Public Library, and the library of the American Antiquarian Society, and to Professor Emory R. Johnson, of the University of Pennsylvania, who has kindly read the manuscript and offered valuable suggestions and criticisms during the progress of the work.

RAYMOND McFARLAND.

Middlebury College, Vt.

1 Cf. W. S. Tower, History of American Whale Fishery. Publications of the University of Pennsylvania, Philadelphia, 1907.

CONTENTS

CONTENTS

LIST OF MAPS

NEW ENGLAND FISHERIES

CHAPTER I

THE FISHING GROUNDS OF THE NORTH ATLANTIC

The consideration of the fishing grounds embraces the geographical range of important food-fish, the kinds of fish found in a particular locality, the nature of the ocean bottom, whether muddy, sandy or rocky, and the depth and temperature of the water. Fishing grounds are inshore or offshore, according as they are adjacent to the coast or lie at considerable distances from the mainland. Inshore grounds are creeks, rivers, bays, harbors and nearby shoals. Offshore grounds are banks that may be elevated portions of coasts that have been submerged, like the small banks of the Gulf of Maine; or a part of the continental shelf, of which George's Bank off the east coast of Massachusetts is an example; or submarine plateaus, as the Grand Bank of Newfoundland.

For purposes of this work the fishing grounds of the North Atlantic may be divided into three parts: the northernmost part, including Iceland, extends from Davis Strait southward along the east coast of Labrador to Cape Race in Newfoundland; the second part is more irregular in form, and is included between Cape Race and Montauk Point, Long Island; a third part occupies an unbroken stretch from Long Island to Cape Florida. The grounds extend over forty degrees of latitude from 65° to 25° north latitude. The distance from Holsteinborg in Greenland to Cape Race is 1,600 miles, and from Cape Race southwest to Cape Florida is 2,400 miles.

DAVIS STRAIT AND ICELAND.

In Davis Strait the fishing grounds are less clearly defined than in any other part with the possible exception of the Icelandic grounds. The principal edible fish taken is the halibut; others are the lump-fish, the Norway haddock and the capelin. Cod are so scarce that it is not profitable to catch them for market. The banks of the Strait extend from latitude 68° to 63°, they are from 20 to 40 miles from the mainland and slope abruptly on their inner or eastern edges. Their surface is of varied character, though generally rocky with scattered sandy and muddy spots. They range in depth from 20 to 50 fathoms, a condition that makes fishing easier than on the Grand Bank of Newfoundland where the fish are found in deeper water. During the months of July and August, the season is most favorable for fishing both on account of the temperature, which ranges from 40° to 45°, and the greater abundance of fish that move to and from the banks during these months. Owing to the peculiarities of the tidal currents in Davis Strait fishing is rendered impossible for five or six hours at a time; as the nights in this latitude are light enough to permit fishing at all hours in July, it is possible for the fishermen to select the most convenient time for their work, and to take their rest when the tides are running. The principal harbors for fishermen on the Greenland shore are Sukkertoppen, Goothaab Fiord and Holsteinborg.

The first American vessel to visit Iceland for fish was the *Membrino Chief*, of Gloucester, which went on a salt halibut voyage in 1873.[1] The failure of the venture in-

[1] Fishermen's Memorial and Record Book, p. 96.

fluenced other American vessels to avoid the region for a decade. In 1884 three American vessels made successful voyages and brought back valuable information concerning the fishing grounds of Iceland and the abundance of cod and halibut in the vicinity. The natives fish invariably within the three-mile limit. The American vessels found better fishing from 5 to 25 miles offshore. The character of the bottom was described as rough, broken and "catchy." The currents are very irregular and uncertain, usually not faster than one mile an hour, the general trend being back and forth along the coast. Very rough weather is experienced on these grounds during some seasons, and at times fishing operations are suspended by the floe-ice that is driven on the coast. The halibut are much larger than those caught at Greenland; and are principally of the white variety, while the Greenland halibut are nearly all gray.

LABRADOR AND EASTERN NEWFOUNDLAND.

As early as 1758 a description of the fishing grounds of the coast of Labrador[1] was given by the captain of an American fishing vessel that visited the coast during the previous summer. The account was as follows:

"The coast is full of islands, many of them large, capable of great improvement as they have more or less good harbors, abounding in fish and seal, water and land fowls, good land covered with woods, in which are great numbers of fur beasts of the best kind. Along the coast are many excellent harbors, very safe from storms; in some are islands, with sufficient depth of water for the largest vessels to ride between, full of codfish, and rivers with plenty of salmon, trout and other fish."

More than a century elapsed before a more complete account of the coast was obtained by a representative of

[1] Massachusetts Historical Society Collections for 1792.

the Newfoundland government; in recent years, the work and writings of Dr. Grenfell have added greatly to our knowledge of the Labrador coast. A succession of deep, narrow fiords along the coast has for an outer fringe a vast multitude of islands about twenty-five miles in stretch from the mouths of the fiords seaward. About fifteen miles outside these fringing islands are numerous banks and shoals, the feeding ground of cod from the middle of June to October, while outside the shoals there appears to be a second range of banks, where, probably, the cod feed in winter. This island-studded area forms an immense codfishing ground, which has been estimated at 5,200 square miles in area.

The mean length of the fishing season for cod over four degrees of latitude on the northeastern coast of Newfoundland has been estimated to be 142 days, for southern Labrador over three degrees of latitude it is 87 days, and for northern Labrador over three and one-half degrees of latitude it is 52 days. The cod does not travel far in its annual migration on this coast. After the spawning season the fish retires to deeper water offshore. Each year the cod returns to its birthplace with the school, and haunts the same neighborhood the short season of its inshore life. The school of cod arrives on the coast about a week later for every degree of latitude farther north. For a period of about forty days the codfishing goes on simultaneously during August and September throughout the length of a coast line, extending from latitude 47° to latitude 58° 30′, or more than 700 miles.[1]

No fishing banks are found off the east coast of Newfoundland, but the coast furnishes a vast area for boat-fishing for cod. Squid, capelin and herring abound and are taken mainly for use as bait to be sold to French and

[1] Grenfell, Labrador, pp. 297–327; Goode, The Fishery Industries of the United States, Sec. III, pp. 12–13.

American bank-fishermen. In this region the fishing is carried on from near the middle of June to the first week in November. On the Newfoundland coast as well as on the Labrador fishing grounds the fishermen employ the cod trap for catching the fish. The contrivance is a large room with floor and walls of twine arranged with an opening on the landward side through which schools of cod may enter but can not pass out again. As many as one hundred quintals of cod have been caught at one haul on many occasions; on the other hand, a whole fishing season may pass without a school of cod entering the trap.

CAPE RACE TO MONTAUK POINT.

From Flemish Cap, a fishing ground, 8° E.N.E. of Cape Race, to Montauk Point, are found the most famous and valuable fishing grounds of American waters. The shore line of this section of the coast is exceedingly irregular, deeply indented with gulfs and bays, fiords, harbors and inlets. It is a rock-bound coast, beautiful throughout its extent, formerly occupied exclusively by the humble cottages of the fishermen, now fast becoming dotted with substantial summer residences of wealthy American families. The drumlin-islands of Boston harbor and the fiord-harbors of southern Maine still attest the influence of the glacial epoch and the subsequent sinking of the coast. A strong current from the Arctic Ocean encounters the Gulf Stream off the southeast coast of Newfoundland. It is checked in its course, its bergs are devoured by the warm stream from the South, and it is forced to the westward. The cold current follows along the south coast of Newfoundland until it reaches Cape Breton Island where it divides, sending one branch into the Gulf of Saint Lawrence and another along the Nova Scotia coast until it is finally lost in the Gulf of Maine.

The fishing grounds of the section include those of the Gulf of Saint Lawrence, the Gulf of Maine, southern New England, and the offshore banks that extend from George's Bank on the southwest to Flemish Cap on the east, forming "an almost continuous series of broad, submarine elevations, stretching a distance of 1,100 miles, and with a varying width of 50 to 250 miles." Within this area are found cod, hake, haddock, pollock, halibut, cusk, bluefish, salmon, mackerel, herring, lobsters, clams and oysters. The bold fishermen of France, Portugal and England first came to the rich fishing grounds of this region; here American fisheries began, and still persist after four centuries of fishing from the waters of this part of the North Atlantic. It is a region of historical interest, as well as of economic importance. No part of the banks is without its tragedy, there is no point of land that has not broken strong hulls and compassed the death of fathers and sons. On the other hand, the fishermen of New England mark every headland by some noteworthy catch of fish and cover their charts with dots and lines that tell of the rich harvest gathered from the sea.

THE GULF OF SAINT LAWRENCE.

The Gulf of Saint Lawrence is of great importance and value to the fisheries. It lies between the parallels of 45° and 50°, is irregular in shape, shallow throughout most of its area and contains only a few good harbors. The bottom is generally rocky and diversified with areas of greater or less extent of sand, gravel and mud. The western coast of Newfoundland and the region between Cape Breton Island, Prince Edward Island and Cape Gaspé are of the most importance. A cod-fishing ground, known as the Cape North Ground, lies off the northern end of Cape Breton. The shore here is high and steep and the depth of the water rapidly reaches 60 to 100 fathoms. Cod are more plentiful in May and June when moving in towards the shallow grounds of the Gulf of Saint Lawrence. Near this point mackerel are last seen in June on their annual migration from the waters off Cape Hatteras to their summer home in the Gulf.

Fifty miles northwest of the Cape North Ground are the Magdalen Islands,—rocky, exposed and irregular in shape, inhabited by French-speaking fisherfolk. During the winter season the islands are isolated from the rest of the world by the encompassing ice-floes, but they are frequented in summer by fishermen from New England and the Maritime Provinces in search of herring, cod and mackerel. Pleasant Bay, in the south part of the Islands, is widely famous for the abundance of herring that frequent its waters.

The Bradelle Bank is a stony patch lying about fifty miles west of the Magdalen Islands. It is thirty miles in extent north and south, and twenty miles wide, with an average depth of twenty-five fathoms. Orphan Bank, a smaller fishing ground, lies north of Bradelle Bank and off the mouth of the Bay of Chaleur. Closer inshore towards the New Brunswick coast are Miscou Flat and Pigeon-hill Ground. Codfishing is carried on at all these banks during the summer season.

Prince Edward Island, lying between the meridians of 62° and 64°, has excellent fishing grounds off its northeastern coast, especially for mackerel. The principal harbors are Tignish on the north and Souris on the southeast point of the island; these harbors, with Port Hood on the west coast of Cape Breton Island, are generally the headquarters of American vessels sojourning in the Gulf.

NOVA SCOTIA COAST.

Passing to the outer coast of Newfoundland and the coast Provinces, we find extensive inshore fishing grounds along the entire region. The average width of this coast belt is about eighteen miles, and it ranges in depth from 10 to 50 fathoms. Cod and halibut are scarce, but were formerly taken by American vessels off Scatari and Flint Island. The south coast of Nova Scotia is deeply serrated, affording excellent harbors for boat-fishing. At numerous places along the coast herring are abundant, and are taken for bait to supply American and Provincial bank-fishermen. Mackerel make their appearance on the southwest coast of Nova Scotia the last of May. They follow the coast eastward to Canso, where they turn north to enter the Gulf either through the Strait of Canso or by Scatari and the eastern coast of Cape Breton Island.

BAY OF FUNDY.

In the Bay of Fundy are extensive herring grounds. Mackerel sometimes enter the bay, but few American vessels now visit the region for them. The herring approach Grand Manan in July and remain in the vicinity until September. Later in the fall and winter other schools of herring swarm in to the mainland about Campobello Island and the waters between Point Lepreau, New Brunswick, and Eastport, Maine. The fish are used for the sardine trade, and for bait for bank fishing. The sardine canning industry of this region is the most extensive and valuable in the Western Hemisphere. The Pollock Grounds to the north and west of Campobello Island, the Mud Hake Grounds lying to the east, and the Wolves Haddock Grounds farther to the northeast, are visited by fishermen from eastern Maine and western New Brunswick for pollock, hake and haddock.

NEW ENGLAND SHORE.

The principal inshore fishing grounds of the coast of Maine, within the limits of the sixty-fathom line, lie at an average distance of twelve to fifteen miles from the shore. They are generally rocky or gravelly patches abounding in cod, haddock and pollock; hake are found on the muddy bottoms between the harder patches. Herring and mackerel are abundant in their season, although the latter in the last twenty years have been less abundant than previously. Soft clams also abound and form the winter business of many of the citizens who ordinarily do not engage in fishing. At no other place in the United States are lobsters found in such numbers as on the Maine coast.

From Grand Manan to Isle au Haute is a succession of detached rocks, ridges, ledges and grounds more or less famous for their fisheries. Among the most valuable are Outer Schoodic Ridge, 22 miles southeast from Baker's Island, one of the best shore-fishing grounds of Maine; Mount Desert Rock, famous, no less, for its being a well-known landmark for seamen than for the fish obtained around it; the Grumpy, 10 miles southeast from the western head of Isle au Haute, excelling in hake in summer, in haddock in winter, and in cod the entire year; and Hatchell Ground, southeast from the Grumpy, and 9½ miles from Isle au Haute, next to the Grumpy considered one of the best grounds inside Mount Desert Rock.

The islands of Matinicus, Monhegan and Seguin are well-known centers for important fishing grounds grouped about them. Only a few of the groups may be mentioned. Matinicus Sou'sou'west Ground, 6 miles distant from Matinicus Rock, is triangular in shape, about 9 miles in extent on each side. Here are good fishing-grounds for cod and haddock. Monhegan Island is surrounded by patches of fishing grounds, Old Jeffrey's to the west being the best. This group of grounds is the resort of trawlers and hand-liners. Seguin Ground, which lies S. W. by S. from Seguin Island, and Kettle Bottom, 10 miles south from the island, are the best grounds in the vicinity. Kettle Bottom is circular in shape, about 10 miles across, and is surpassed by no other ground on the Maine coast in the annual yield of fish.

A series of inshore grounds stretches from the West Cod Ledge off Casco Bay to the Isle of Shoals, the principal ones being Cape Porpoise Peaks, Tanter, the Inner and the Outer Bumbo, the Nubble, Boone Island Ground, Ten Acres, and Blue Clay.

Stillwagen's, or Middle Bank, is a submarine elevation connecting Cape Ann with Cape Cod and forming the boundary limit between the Gulf of Maine and Massachusetts Bay. The shallower portion of the bank, 9½ to 19 fathoms, has a sandy bottom; the deeper part of the bank is of gravel and pebbles. From Eastern Point, Cape Ann, to the northern end of the bank, the region is dotted with ledges, ridges and small banks, the best known of which are Old Man's Pasture, Western Part Ridge, Hart's Ridge, Eagle Ridge, Brown's Ledge, Kettle Island Ledge, Spot of Rocks, and Saturday Night's Ledge. A stretch of rocky ledges extends from Boston Harbor to Plymouth Harbor; this region is frequented mainly by local fishermen who take their catch to the Boston markets.

The bottom off the east and southeast coast of Massachusetts is sandy with a gradual slope offshore. Morris Ledge, lying to the eastward from Chatham, Outer Crab Ledge and Pollock Rip Grounds, to the southeast of Cape Cod, are fished upon for cod chiefly by fishermen of the Cape towns. The island of Nantucket is the center of important fishing grounds. To the southeast are the Nan-

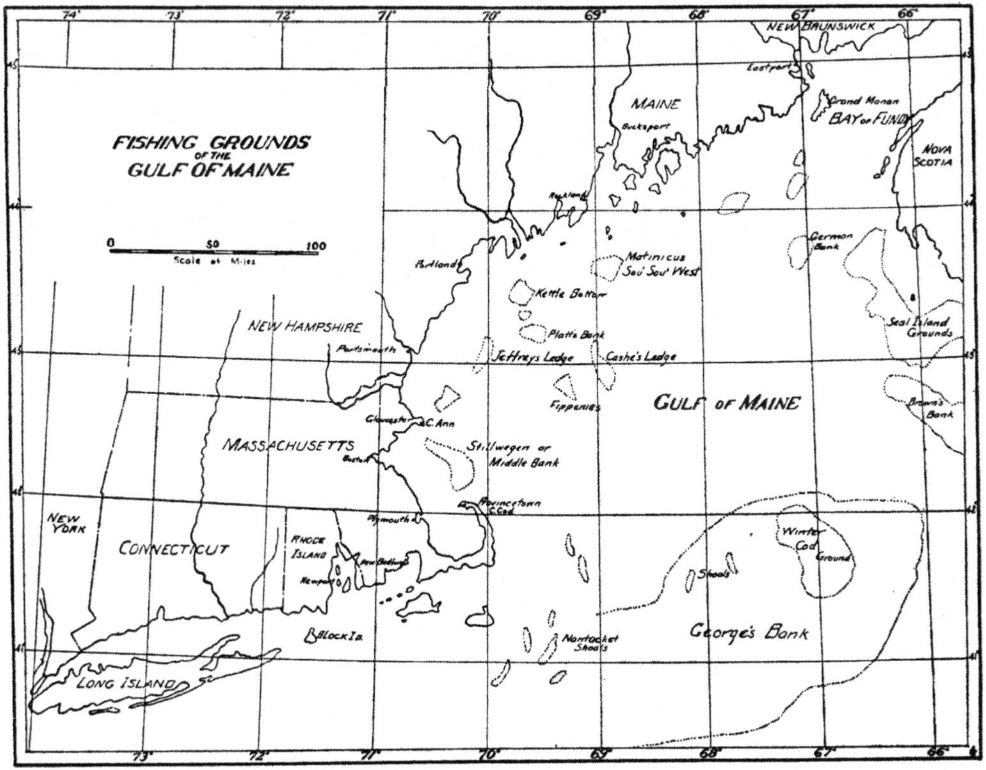
FISHING GROUNDS OF THE GULF OF MAINE
0 50 100
Scale of Miles
NEW BRUNSWICK
Eastport
MAINE
Bucksport
Grand Manan
BAY of FUNDY
NOVA SCOTIA
Rockland
German Bank
Portland
Matinicus Sou' Sou' West
Kettle Bottom
Seal Island Grounds
NEW HAMPSHIRE
Platts Bank
Portsmouth
Jeffreys Ledge
Cashe's Ledge
Fippenies
GULF OF MAINE
Brown's Bank
Gloucester
C. Ann
MASSACHUSETTS
Stillwegen or Middle Bank
Boston
Provincetown
C. Cod
NEW YORK
Plymouth
Winter Cod Ground
CONNECTICUT
RHODE ISLAND
New Bedford
Shoals
Newport
Block Id.
Nantucket Shoals
George's Bank
LONG ISLAND
74°
73°
72°
71°
70°
69°
68°
67°
66°

tucket Shoals, shallow and dangerous, with strong flood tides running to the northeast and ebb tides to the southwest. The Shoals include Davis Shoal, 13 miles from Sankaty Head Light, Fishing Rip, 29 miles to the southeast, and Phelp's Bank, 38 miles to the southeast of the Light. An extension of this shoal three degrees to the E.N.E. forms the famous George's Bank. Along the east and south coasts of Nantucket there are grounds within two miles of the shore which are fished on for cod and bluefish. The principal fishing grounds in Nantucket Sound are Bishop and Clark's Ledge, South West Ground, Off Hyannis, Off Cotuit, and Mutton Shoal Ground west of Nantucket.

The region off Vineyard Sound and Gay Head furnishes excellent grounds for sea bass, tautog and lobsters. The first two kinds are found in less frequent numbers in Buzzards Bay. Cod are caught on Brown's Ledge, west from Gay Head, on Southwest Ledge, 13 miles S.W. by W. from Gay Head, and on Cox's Ledge, ten miles farther out and in the vicinity of No Man's Land.

Near Block Island are cod and haddock grounds at Shark's Ledge, which lies about nine miles to the southeast, the South West Ledge, and the North Ground between the island and the mainland. On these grounds, as well as those of No Man's Land, important catches of mackerel are made in their season.

GULF OF MAINE.

The Gulf of Maine is rectangular in shape, 215 miles in length from Cape Cod to Cape Sable, with an average width of 80 miles. It is of the greatest importance to the fisheries, being unsurpassed in America for the abundance and variety of the catch of fish as well as for its proximity to Gloucester, Boston, Provincetown, Portland and other fishing ports. The banks of the Gulf of Maine are resorted to by the New England fishing fleet. Cod, hake, cusk, haddock, pollock and some halibut are taken. The mackerel fishery is of especial importance, the depth of the water furnishing excellent opportunity for the use of purse-seines; occasionally several hundred barrels of mackerel are taken at a single setting of the seine.

In the eastern part of the Gulf are Grand Manan Bank, German Bank, and Marblehead Bank, each on the meridian of 67° and south from Eastport, Maine. Cashe's Ledge lies 76 miles east from Cape Ann on the 69th meridian, and is 22 miles long north and south by 17 miles wide. On this ledge are three small shoals, one being only four fathoms deep, on which the water breaks in severe storms. Fippenies Bank lies about twelve miles west of the southern part of Cashe's Ledge. Platt's Bank, or New Ledge, lying 53 miles E. by N. ½ N. from Thatcher's Island, is considered one of the best cod and haddock grounds in the Gulf of Maine. Ten miles north of New Ledge is Mistaken Ledge.

Extending northeast from Cape Ann for 42 miles is a nearly continuous ridge of fishing grounds, the part nearer the Cape being known as the East Shoal Water of Cape Ann, about 18 miles in length, while the part farther off-shore is called Jeffrey's Ledge, one of the famous fishing grounds of this region. The bottom consists of rocks, pebbles and coarse gravel, a favorite feeding ground for ground-fish except hake. Tillie's Bank, eighteen miles east by south from Thatcher's Island, was formerly one of the best fishing grounds off Cape Ann.

GEORGE'S BANK.

While the banks of the Gulf of Maine are offshore banks, they lie within and are smaller than the banks that extend in a more or less continuous line from George's Bank to the Grand Bank of Newfoundland. George's Bank is a northeast extension of Nantucket Shoals from which it is separated, at the meridian of 69° west longitude, by the South Channel with a depth of less than 50 fathoms. It extends eastward to the meridian of 66°, and its width is embraced between the parallels of 40° 30′ and 42° 8′. The bottom is principally of sand, with occasional patches of gravel, pebbles and rocks. Its depth is from two to fifty fathoms. Several shoals, of which Southwest Shoal and Cultivator Shoal are well-known, are found on the western part of the bank with depths ranging from two to fifteen fathoms. The tides of George's are strong, sweeping in a circular motion over the bank. For many years it was the firm conviction of fishermen that the tides of George's would under-run and sink any vessel that should anchor on the grounds, so that it was not until 1821 that the crew of a Gloucester schooner had the courage to drop anchor on the bank and prove the falsity of the belief. On the shoals the tides cause rips, while in stormy weather the sea breaks, making navigation extremely dangerous.

East of the shoals is the "winter fishing ground" for cod, where the fish appear in large numbers in February, March and April. During the roughest season of the year fishermen frequent that part of George's which, during easterly storms, gives a lee shore on the dangerous shoals. No place is more to be dreaded than this, few places have caused more shipwrecks and greater loss of life. Cod, haddock and halibut are the great staples of George's Bank, although within a few years extensive catches of mackerel have been made in July and August.

NOVA SCOTIA OFFSHORE BANKS.

Fifteen miles to the northeast from George's Bank and south from Cape Sable, Nova Scotia, is Brown's Bank with an extreme length of 63 miles and a width of 43. Between this bank and Cape Sable are the Seal Island Grounds, while to the northeast are Roseway Bank and Le Have Bank, the latter being a favorite fishing ground for American fishermen.

The parallel of 44° and the meridian of 60° pass through Sable Island. This island, narrow and crescent-shaped, is in the eastern part of the extensive Sable Island Bank, which is elliptical in form with a length of 156 miles and a width of 56. To the northeast of the Island is the Gully—famous for its halibut—a deep passageway between Sable Island and Banquereau, another important fishing bank with an area of about 2,800 square miles. To the west of Sable Island Bank are the Owl and Doubtful Bank, while on the northern shoal is another called the Middle Ground. North of Banquereau is an irregular stretch of grounds termed Misane Bank, flanked on the west by Canso Bank and on the east by Artimon Bank.

NEWFOUNDLAND BANKS.

The Bank of Saint Pierre, situated off the southern coast of Newfoundland, extends in a northwest and southeast direction. On its northern edge it approaches to within ten miles of the French islands of Saint Pierre and Miquelon. Cod and halibut are the principal fish taken on this bank. On its eastern edge a gully, attaining a depth of 100 fathoms in places, separates the Bank of Saint Pierre from Green Bank by a distance of fifteen miles. Halibut are taken on the latter bank.

Seven miles to the east of Green Bank lies the Grand Bank of Newfoundland. It is roughly triangular in shape, one side facing N.N.E., another S.W., and the third about E. by S. North and south it extends from below the parallel of 43° to beyond that of 47°; its width is between the

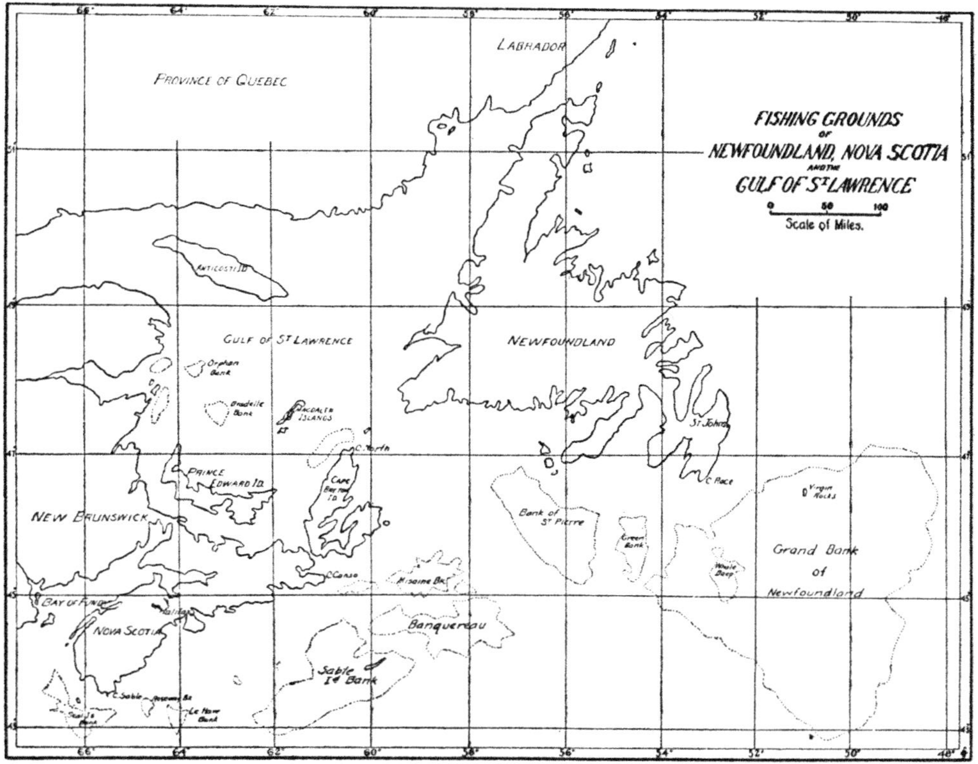
FISHING GROUNDS
OF
NEWFOUNDLAND, NOVA SCOTIA
AND THE
GULF OF S^T LAWRENCE
0 50 100
Scale of Miles.
PROVINCE OF QUEBEC
LABRADOR
ANTICOSTI ID
GULF OF S^T LAWRENCE
NEWFOUNDLAND
Orphan Bank
Bradelle Bank
MAGDALEN ISLANDS
St. John's
C. North
PRINCE EDWARD ID
Cape Breton Id
C. Race
Virgin Rocks
NEW BRUNSWICK
Bank of St Pierre
Green Bank
Grand Bank of Newfoundland
Whale Deep
C. Canso
Misaine Bk
BAY OF FUNDY
NOVA SCOTIA
Banquereau
Sable Id Bank
C. Sable
Le Have Bank
66° 64° 62° 60° 58° 56° 54° 52° 50° 48°
51° 49° 47° 45° 43°

meridians of 48° and 54°, giving it an area of 37,000 square miles, or more than that of the state of Indiana. A channel separates the bank from Cape Race, which is 35 miles distant. The polar current sweeps across the bank from the north to the southwest, bearing icebergs to the northern and eastern parts of the bank, causing dense fogs, preventing fishing much of the season each year, and imperiling, not only fishermen, but trans-Atlantic liners as well. In summer, the Gulf Stream sweeps over the southern part of the bank; in winter, it flows past a few miles to the south.

Shoals are found at Virgin Rocks and Eastern Shoals, located in the northern part of the bank. The Virgin Rocks consist of a group of small rocky shoals, the more important being Main Ledge, with three to 19 fathoms' depth, Briar Shoal, Southwest Rock, and Bucksport Shoal with a depth of 4¾ to 11 fathoms. The Eastern Shoals lie 15 miles eastward of the Virgin Rocks, and have a range in depth from 7 to 27 fathoms. The average depth of the bank as a whole is between 35 and 45 fathoms.

The southern part of the bank has a bottom of fine sand of varying color; a middle section consists of sand, gravel and pebbles over certain areas; the northern part is of gravel, pebbles and rocks. The eastern edge of the bank descends rapidly into deep water. The southwest edge of the bank is the best region for halibut. Cod are found principally on the southern half of the bank and on the central and northern part. Other fish that are caught are bank quahogs, bank clams, clams, periwinkles, sea anemones, sea pumpkins, sea lemons, crabs, shrimp, starfish, specimens of octopus, squid and capelin.

The fishing season lasts from April to October. Early in the season the fishing is carried on at the southern part of the bank. As the season advances the fleet moves northward to the vicinity of the Virgin Rocks. In June capelin make their appearance, and the shoal of cod found following them has been termed the "capelin school" by fishermen. In July the body of cod on the ground is called the "squid school," since the fish no longer take the herring and capelin offered them for bait, but are attracted by the quantities of squid that arrive at this time. Hand-liners report seeing the cod in large numbers lazily swimming along in the clear shoal water of Virgin Rocks and refusing to touch hooks baited with herring.

Before any permanent French or English settlements had been made in America the Grand Bank of Newfoundland was frequented by fishing vessels from France, Portugal, Spain and England. France led in these Newfoundland fisheries, making regular voyages as early as 1504. The first accounts of Portugal, Spain and England are in 1517, when there were fifty vessels of all sorts at Newfoundland. Sixty years later the number of vessels had increased to 315. For more than four centuries, during which time the fishing has been well nigh continuous in its season, the Grand Bank of Newfoundland has been the principal source of supply of the codfish to American and European consumers. Since the close of the Civil War the halibut fishery, also, has been carried on successfully by American fishermen. It would be difficult to find in the whole world an equivalent area of the ocean bed which equals in historic interest and economic importance the Grand Bank of Newfoundland.

LONG ISLAND TO FLORIDA.

The coast of the Atlantic from Montauk Point, Long Island, to Cape Florida, is in striking contrast to that described above. North of the parallel of 41° the coast is irregular and rocky, scarred by the ice of the glacial epoch; to the south it is unbroken and sandy. In one section are to be found many gulfs and bays, the nursery of good harbors; in the other few bays occur, and, consequently, there are few harbors. The polar current washes the northern coasts with its cold waters, while the Gulf Stream sweeps in undisputed sway off the coast of the United States, tempering the adjacent waters and bringing its northern rival to a full check in the Gulf of Maine. More edible fish, and consequently more valuable ones, abound in the colder waters; but a greater variety of species is found on the warmer shores. Immense areas of submarine banks, the spawning and feeding grounds of edible fish, occur between Newfoundland and Nantucket Island, but grounds of a similar kind are small and scattered to the southward.

The good harbors of New England and the Provinces have led to a monopoly, in the North, of the shipbuilding of the country, an industry that is of especial value to the successful prosecution of deep-sea fisheries. Nature has designed New England to be the center of the Atlantic fisheries of North America. To the natural influences that have helped to crown her work in this direction with success, one must not fail to add the indirect influences,—the barren condition of the soil of New England and the severity of its winters,—which led the early settler upon these shores to go down to the sea in ships to eke out a scanty living. And this tendency to resort to the abundance of the ocean's storehouse for food has become well-nigh an hereditary trait with the coast inhabitant of New England.

Along the southern shore of Long Island Sound, good sea bass grounds occur; scup and "snappers" are caught in most of the bays and harbors of the sound. South of the island the principal regions for fish are Shinnicook Ground, Offshore Fire Island Ground, and Dillberry Ground, all visited by New York smacks fishing for codfish. The same smacks also visit the grounds off the New Jersey coast, principally in search of cod in winter and bluefish in summer. The grounds are Rocky Spots, southeast of Sandy Hook, Shark River Ground, Squan Ground, East of Barnegat Light Ground, Egg Harbor Ground, Absecum Ground and Five Fathom Bank, east of Cape May.

From Cape Henlopen to Cape Charles there are few offshore grounds, but inshore seining for menhaden, bluefish and sea mullet is carried on. Fifteen miles southeast of Cape Henlopen is Old Ground, one of the largest and oldest-known grounds on this coast for cod in winter and sea bass in summer.

From Cape Charles to southern Florida there are long stretches of sandy beaches. Scattered fishing grounds occur along almost the entire coast, but most of the fish are taken on the sandy shores. An abundance of bluefish is found from Albemarle Sound to Cape Hatteras, although the fish range from Florida to Massachusetts. The fishing grounds south of Long Island are of interest to New England fishermen almost wholly in connection with the southern spring mackerel catch; the fish are taken offshore from Cape Hatteras north to the Long Island grounds in the months of April and May.

The range of fish varies with the species. Chesapeake Bay is the center of the oyster fishery, now one of the most important branches of the fisheries of the country; oysters and squeteague are taken along the entire coast from Cape Cod to the Gulf of Mexico. Clams are taken from Chesapeake Bay to Nova Scotia; lobsters from Delaware Bay to Newfoundland; alewives from Georgia to the Gulf of Saint Lawrence; shad from Florida to Maine; mullet from North Carolina to Mexico; herring from Martha's Vineyard to Newfoundland; mackerel from Cape Hatteras to the Gulf of Saint Lawrence; cod and haddock from New Jersey to Newfoundland.

CHAPTER II

THE BACKGROUND OF NEW ENGLAND COLONIZATION

The fisheries of the North Atlantic have always occupied an important place in the industrial development of New England. They were well established along the coast several years previous to the settlement of Plymouth. During the period of settlement the fisheries were often the only source of food supply for the colonists, and in several instances they were largely instrumental in preventing settlements from being broken up. With the growth of colonial trade and the expansion of commerce the fisheries early became the principal industry of Massachusetts, and their development was so rapid that they were recognized as "the cornerstone of New England prosperity." For two centuries the fisheries were regarded as a prominent factor in her industrial organism.

They have been honored and honorable in our national history. Fishermen formed the nucleus of our first navy, and in four of the country's wars, they have furnished officers and men who were leaders in their important calling. For three-quarters of a century, the deep-sea fisheries were benefited by a system of bounties granted by the government, and whenever our fishing rights upon the high-seas have been in peril the country has been a unit in supporting the contentions of the fishermen of New England. Before tracing the development of the fisheries of New England during the periods of colonial and national life, it will be well to inquire into the importance of the fisheries before any English settlements had been made in this region, and to note the influences that these fisheries had, both direct and indirect, in shaping the colonization of New England.

The importance of the fisheries carried on by European countries in American waters previous to the settlement at Plymouth is not generally recognized. In 1497, John Cabot announced to Europe the abundance of codfish at Newfoundland. Within seven years, at least, the fishermen of France had taken advantage of Cabot's discovery by sending vessels to the fishing grounds of America, and every year for more than four centuries the flag of France has been carried by French fishermen to the banks of Newfoundland. Nearly a score of years before De Soto first saw the Mississippi River, houses for the accommodation of fishermen had been erected in Newfoundland. The year that Drake sailed from England on his memorable voyage around the world witnessed also the quiet departure from the harbors of Europe of over 300 ships to fish in American waters. Before the pioneer voyages of Gosnold and Pring and De Monts had been made and recorded, the hardy fishermen of western Europe had made thousands of voyages across the Atlantic with scarce a thought of the hardships of their trips, and scarce a word written to chronicle their deeds.

The influence of these voyages was felt on both sides of the Atlantic. The Parliament of England passed legislation favorable to the development of the fisheries. In France, the Newfoundland codfishery was placed under the protection of the government. Disputes and bloodshed occurred on the fishing grounds between the fishermen of France and England. Both countries were quickened in their attempts at colonization in America. The coasts of New England were explored, bays and harbors were named, and the severity of the winter climate was tried by Europeans who were eager to develop this industry of the sea, many years before the Pilgrims started from Leyden. The Pilgrims settled at a place that had been visited previously by three different European explorers, and as many times given a different name. The welcome accorded to them by the chieftain Samoset was in English that he had learned from fishermen along the coast. When their food supply became exhausted and starvation was near at hand, these men from Leyden secured their living from fishing, as they had intended before leaving the shores of Europe.

While Spain was centering her attention, with good returns, on the southern part of the new world, England, France and Portugal were not idle in the northern part. To England and France, in particular, belongs the credit of discovering and exploring the coasts of this country and Canada, of making attempts at settlement, of permanent colonization, of developing the trans-Atlantic trade in fish and furs, and of transplanting an old civilization to a new world. John Cabot, born in Italy, but living in Bristol, England, was the first to lead the way in 1497. Sailing under a commission from Henry VII, he visited the eastern coast of North America, and sailed along the shores of Labrador, Newfoundland, and, possibly, Cape Breton Island. He returned to England in August after an absence of three months.[1]

Immediately after his return, Cabot made a chart and globe on which he described his voyage. In conversation with a writer of the day he stated that "the sea is covered with fishes, which are caught not only with the net, but in baskets, a stone being tied to them in order that the baskets may sink in the water;" and further, "his comrades say that they will bring so many fishes that this kingdom will no longer have need of Iceland, from which country there comes a great store of fish which are called

[1] Calendar of State Papers, Venetian, 1202–1509, p. 262.

stock-fish."[1] A later account of Cabot's voyages[2] describes these regions as yielding "plenty of fish, and those very great, as seales, and those which commonly we call salmons: there are soles also above a yard in length: but especially there is great abundance of that kinde which the Savages call Baccalaos."

The methods of the Indian fishermen and the Icelandic supply of fish are interesting facts. The announcement to the old world of this great supply of fish across the Atlantic was a matter of economic importance to the Europeans, who were great consumers of fish. To encourage the fisheries, the law makers compelled people to eat fish by creating fast days, and at this time every third day was observed as such.[3] The most perplexing fact in the Cabot narrative is the statement that the natives called the codfish *Baccalaos*, a name that was applied by the seamen of the Bay of Biscay to that fish long before Columbus sailed on his voyage of discovery.

The conclusion has been made that the fishermen of France must have visited these regions long before the voyage of Cabot. Local traditions have laid claim to the discovery of the banks of Newfoundland by the fishermen of Normandy and Brittany before 1492. Those coasts were peopled by a race of adventurous mariners, and such ports as Dieppe, Saint Marlo, Honfleur, and others, had already furnished men and leaders for voyages of exploration and discovery. The fishing population of the French provinces were accustomed to voyages of considerable length, as they had already visited the Canaries and the coast of Africa. There is no reason why fishing voyages might not have been made to the banks of New-

[1] Winsor, Narrative and Critical History of the United States, III, p. 54.

[2] Hakluyt, Principall Navigations, III, p. 27.

[3] Winsor, IV, p. 3.

foundland before 1497, but there is no positive evidence that such voyages were made. "If in the original Basque, *baccalaos* is the word for codfish, and if Cabot found it in use among the inhabitants of Newfoundland, it is hard to escape the conclusion that Basques had been there before him."[1] Granting, then, that voyages were made before Columbus, no practical good that we are aware of ever came from them, either in connection with European commerce or with the development of the fisheries.

In 1500, Gaspar Cortereal, in the service of the king of Spain, made a voyage to the northeast coast of America, and again in 1501, he made a more extensive one with his brother, Miguel. In each of the two following years, expeditions were sent out under the direction of the Cortereal family, but they failed to find either the Northwest passage, or the lost members of previous expeditions.

In these voyages the Cortereals had been unsuccessful in their attempt to reach the Indies by a new route; but they inflamed their countrymen with zeal to prosecute the fisheries in American waters, in consequence of which the business grew to large proportions and produced considerable revenue for the State.[2] The influence of the early Portuguese voyages is seen in the names of Cape Race, Fortune Bay, and Island of Codfish (*Y dos Bocalhas*),—these with other names appearing in the chart made by the Portuguese pilot Reinal, and ascribed to the year 1503.[3]

The fishermen of Dieppe, Saint Marlo and other ports of France were quick to follow in the track of the Cabots and Cortereals. Very soon they became skillful and powerful in the American fisheries. The first authentic record we have of the ships of Brittany at Newfoundland is as early as 1504. The French found the fishing on the

1 Parkman, Pioneers of France in the New World, pp. 171–172.

2 Winsor, IV, p. 4.

3 Winsor, From Cartier to Frontenac, p. 5.

banks so good that from that time forward they never missed a year. They have left the name of Cape Breton Island as an enduring trace of their early voyages.[1] In 1506, Jean Denys, of Honfleur, is said to have visited the Gulf of Saint Lawrence, and to have made a chart of it. Two years later, Thomas Aubert, from Dieppe, is credited with having sailed up the Saint Lawrence River eighty leagues. In 1518, the Baron De Lery attempted a French settlement in the new country at Sable Island, but his endeavors ended in failure. By 1522, there had been constructed at Newfoundland between forty and fifty houses for the accommodation of fishermen.[2] Thus, it appears, that within a quarter of a century after the discovery of the mainland by Cabot, the French were well established in the prosecution of the Newfoundland fisheries, and had made considerable headway toward securing a permanent foothold on land.

The English seemed to have played a minor part in these fisheries. In a letter written by John Rut to Henry VIII in 1527, and dated from Saint John's, Newfoundland, the writer states that there were eleven sail of Normans, one Brittain, and two Portuguese in the harbor, while he knew of but one English vessel along the coast at the time. In another report for the same year, it is stated that ships from Spain, England, France and Portugal to the number of fifty were employed in fishing at Newfoundland that year. One writer places the date ten years earlier, but this is a mistake.[3] The English made an attempt to establish a colony in Newfoundland in 1536. The affair was at the expense of Mr. Hoare, a wealthy London merchant. A company of one hundred and twenty persons, of whom

[1] Fiske, New France and New England, p. 4.

[2] Sabine, Report on the Principal Fisheries of the American Seas, p. 36.

[3] Winsor, IV, p. 68.

thirty were gentlemen of refinement, was formed, but the venture was a failure. Many of the colonists died of starvation, while the miserable remainder, subsisting for a time on the bodies of their dead comrades, seized a French ship in Newfoundland waters and at last reached England.[1]

After France became consolidated under Francis I it was in a position to profit by the skill of her fishermen in navigation. Accordingly Chabot, admiral of France, entrusted to Jacques Cartier, a native of Saint Marlo, the command of an expedition in 1534. Cartier explored the Gulf of Saint Lawrence but his attempt at colonization was unsuccessful, as was another expedition under his command the following year. The French appear to have had establishments on the shore at Newfoundland for their fishery in 1540, but the information concerning them is uncertain. Under the influence of Sully, Henry IV placed the Newfoundland fishery under the protection of the French government.[2]

The first measure of encouragement to English fisheries was enacted by Parliament in 1548, when heavy fines were imposed on all persons who should eat flesh on fish days. Another act passed the same year forbade admiralty officers to make exactions for the privilege of carrying on the fisheries. This applied in particular to Newfoundland with the result that from this time forward the Newfoundland fishery became free for every inhabitant in the realm. In 1563, Parliament declared it to be unlawful to eat flesh on Wednesdays and Saturdays, affixing a penalty of three pounds, except by special license purchased from the government.[3] Again, in 1571, fishermen were encouraged by being permitted to export their sea-fish free of customs; while, by another act, foreign fishermen, anchoring on the

[1] Sabine, p. 36.

[2] Isham, The Fishery Question, p. 3.

[3] McPherson, Annals of Commerce, pp. 102, 138.

coast or interfering in waters where nets were spread, were liable to seizure and confiscation.[1]

During these years the Newfoundland fishing business prospered and increased with all nations engaged in it. The profits already were so manifest that the English early desired a monopoly of the privilege of making and curing the fish on the island. Merchants had begun to quarrel among themselves for the advantage of shore locations. After the accession of Elizabeth to the throne there was greater activity in her seafaring subjects.[2] In 1577, there was a total of 315 sails at Newfoundland, some authorities placing the number as high as four hundred. France was still in the lead with 150 ships. Spain had about 100, Portugal had 50, while England had but fifteen sails. England still sent fishing vessels to Iceland, and this fact may account for the small number sent to Newfoundland. Sabine states that the English ships were superior to those of the other nations, that they gave protection to the others, and exacted tribute or payment for the service.

Nearly half a century had now passed since the English made their first attempt to establish a permanent settlement in America. The merchants of the day realized the disadvantage of carrying on the fisheries at so great a distance from the base of supply; they saw that a colony in the New World would facilitate the prosecution of the Newfoundland fisheries and would lead to the development of other kinds of business. Further attempts at colonization, therefore, were quickened by the desire to place the American fishing interests on a more economical basis.

In 1578, Sir Humphrey Gilbert had secured letters patent to discover, settle and regulate these remote countries. Previous to his departure from England the government sent Sir Thomas Hampshire to the fishing grounds

[1] Sabine, p. 38.

[2] Isham, p. 7.

of Newfoundland for the purpose of settling disputes about the preëmption of shore stations. Gilbert arrived at Newfoundland in 1583 with a charter for colonization, the first of its kind that was issued by the English government. He took possession of the island and all surrounding lands within a circumference of two hundred leagues in the name of Queen Elizabeth. His right was acknowledged by the fishermen of the vessels at Saint John's at the time of his arrival, who paid him quit-rent for privileges that he granted to them.

England's claim to Newfoundland and its fishing grounds rested on the results of the discovery by Cabot in 1497. Subsequently, her rights to this claim were emphasized by Gilbert's acts at this time. Apparently the English held the right to exclude the vessels of all other nations from the fishing grounds; possibly she compelled the acknowledgment of this right, for in 1585 Sir Bernard Drake with a fleet of ships captured and sent back to England several vessels of the fishing fleet laden with furs and fish.

Although the voyage of Gilbert was disastrous to himself and his colonists, yet the enterprise was helpful in awakening greater interest in American affairs. During the following decade rapid progress was made by the English in developing the Newfoundland fishing industry. According to a statement made by Sir Walter Raleigh in Parliament in 1593 these fisheries were the "stay and support" of the west counties of England. The interest that the public had in the fisheries was due largely to the great demand for products of the sea at that period. By the statute-books there were 143 days in the year when the subjects of the realm were required to abstain from eating flesh and to live on fish.[1]

The prosperous condition of the English fisheries at New-

[1] Sabine, p. 39.

foundland at the opening of the new century can be seen from the estimate that at this time there were about 200 ships that went annually to Newfoundland, and that employment was furnished on shore and aboard the vessels for ten thousand men and boys. It was the custom then, as it later became the custom of the fishermen of New England, to sail for the fishing grounds in the early spring with the vessel fitted out with a supply of provisions, salt and fishing tackle sufficient to meet the needs of a voyage lasting for several months. The return home was made in September. As fast as the fish were caught during the summer months they were carried ashore and there dressed, salted, cured and pressed by the shore crew. When the vessel sailed from Newfoundland it was with a cargo of fish ready for the market, and sometimes the vessel would sail to a market port for the purpose of selling and discharging the cargo before returning to the home port. The fishermen spent their winters at home, idly spending their hard earned "share-money."

Sir William Monson, writing in 1610 of the value of the discovery of North America to the English people, said, "England may boast that the discovery, from the year aforesaid (1497) to this very day, hath afforded the subject, annually, one hundred and twenty thousand pounds, and increased the number of many a good ship, and mariners, as our western parts can witness by their fishing in Newfoundland."[1]

The French were still regarded as formidable rivals of the English in the fisheries, although there seems to have been a decline in the number of French vessels engaged in the enterprise during the early part of the seventeenth century. There must have been more than a hundred of their vessels frequenting the fishing grounds of Newfound-

[1] Sir William Monson. Naval Tracts, Book VI, in Churchill's Collection of Voyages and Travels.

land at this time. One old French fisherman, Scavelet, is credited with forty voyages to these grounds previous to 1609.[1]

De Monts, in 1604, came to take possession of Acadia which had been granted to him by the king, Henry IV, of France, together with a monopoly of that region. The first attempt at settlement was made at the mouth of the Saint Croix River. A more attractive place across the bay led them to settle Port Royal, afterwards called Annapolis. While the colonists were building a storehouse and forts, one of their number, Champlain, was engaged in exploring and delineating the New England coast. Two summers were spent in this manner, the explorers getting a good knowledge of the harbors and shore-line as far as Hyannis on the southern coast of Cape Cod. No harbor was found with advantages that were preferable to those of Port Royal, and the colony, after undergoing great trials and hardships, became a permanent station on the American coast.

In a few years the French pushed still farther west. Under the Jesuits a station was founded by Fathers Biard and Masse at a place called Saint Saviour on Mount Desert Island, where they were soon joined by Father Du Thet. The settlement was short-lived. With a commission from the governor of Virginia, Sir Samuel Argall, who was fishing in the vicinity, unexpectedly fell upon the French, scattered the people and destroyed the settlement. A little later another Jesuit settlement made by La Saussaye on the western shore of Somes Sound, in the same locality, met with a similar fate. This was the beginning of the active hostility between the English and French in America. Numerous quarrels between the fishermen of these two nations had already taken place on these shores. After the French had pillaged one of the English ships, the crews

[1] Sabine, p. 8.

of the latter vessels went armed to the banks. In 1594, the English seized a hostile French ship and sent it back to England.[1] The French continued to frequent the coast of New England for several years after their encounters with Captain Argall. One of their ships was lost at Cape Cod and another in Boston Harbor, in 1619, their crews being killed.[2]

These hostile relations were not confined to encounters between the Europeans. At the close of the sixteenth century the fishermen began to commit outrages on the natives of Newfoundland. The Indians were driven from the fishing waters, they were hounded and slaughtered on the land without discrimination of age or sex, and finally they were exterminated from their native soil. It is not a pleasant task to record such deeds against the early fisher-folk of England; but it is only a part of the history of many others of a like kind. Similar conditions prevailed on the coast of Maine when the fishermen began to have dealings with the natives. The Indians were taught "drunkenness, wickedness and lewdness," and other abuses were committed against them,—in the language of the day, "to the overthrow of our trade and the dishonour of the government."

The habits of the fishermen who first frequented American waters were loose and immoral. This was due in part to the low moral tone of England at the time, to the particular nature of the fisherman's calling, and to the circumstances in which they were placed. The condition of society in Elizabeth's day was far below the standard of the civilization of to-day. With social irregularities existing at the court of the realm it is natural to expect worse conditions to prevail among the humbler class of subjects. The fisherman's lot was cast among unattractive surround-

[1] Isham, p. 8.
[2] Winsor, IV, p. 110.

ings; his food was coarse, his clothing rough and heavy, his bed a bundle of straw. He lived in the midst of tubs, of hooks and lines, of salt, bait and fish. He bathed only as the downpour of rain fell upon him at work or as the splash of the breaking waves soaked through his garments. His only preparation for sleep was to divest himself of his heavy boots and oil clothes; if he "turned in" wet to the skin it was with the expectation that the heat from his body would dry his wet clothing during the night. During his sojourn in the New World he was separated for months at a time from his home and family and from whatever refining influences his humble village life afforded. His calling offered few relaxations, no amusements. Whatever diversions he found ashore in the wilderness of America seem to have been used to the extreme. The results were what naturally would follow under such circumstances,—disregard for the rights of native men and women, and lasting dishonor to himself. The pity of it is that men possessing so many admirable traits of character, ready to separate themselves from their families for months, patient to endure the monotonous fare, the peril and the hard work to be found on a codfisherman of the Grand Bank of Newfoundland of that day, did not practice that nobler virtue of self-restraint in the presence of lawlessness and license. Instead they left, in their dealings with the Indians, a record of cruelty, of debauchery and of injustice that time will not efface, nor their good deeds obliterate, nor the circumstances of the period or the conditions of their calling excuse.

While the French were engaged in colonizing in Acadia and in exploring the adjacent coasts, the English, too, were making explorations and discoveries for themselves on the New England coast. Bartholomew Gosnold, in March, 1602, made the first direct voyage across the Atlantic to New England and commenced the fisheries on these shores.

He gave to Cape Cod its present name on account of the multitude of fish that he took near it. Continuing his way southward he built a storehouse and fort on Elizabeth Island, intending to establish a permanent settlement there. But when the time came for his ship to return, the intended colonists to a man embarked aboard the vessel and returned with Gosnold to England. Gosnold spread most favorable reports of the new country, declaring among other things that the fishing was superior to that on the banks of Newfoundland.

Several merchants of Bristol became interested in such enterprises and sent out Martin Pring in 1603 with two vessels. Pring reached the coast of Maine near Penobscot Bay, discovered and named Fox Island from several silver-gray foxes that he saw there, and coasted as far south as Martha's Vineyard. He entered Plymouth Harbor where he remained six weeks, exploring the region, becoming acquainted with the natives, gathering cargoes of sassafras, and even planting seeds to test the soil. He gave the name of Saint John's Harbor to the place, and, after loading his vessels, sailed for England. He confirmed Gosnold's statement that the fisheries of New England were superior to those of Newfoundland. "Thus two years before Champlain explored Plymouth Harbor, naming it Port of Cape Saint Louis, ten years before the Dutch visited the place, calling it Crane Bay, and seventeen years before the arrival of the Leyden Pilgrims, Englishmen became familiar with the whole region, and loaded their ships with the fragrant products of the neighboring woods."[1]

In 1605, George Waymouth, under the patronage of several Englishmen of rank, came to the Maine coast in the vicinity of Penobscot Bay. Waymouth, like Gosnold and Pring, considered the fishing prospects of the coast of Maine greater than those of Newfoundland. An account of

[1] Winsor, III, p. 174.

the voyage published in London the same year relates that as they were setting sail for England they caught very large fish, so that there was good ground to warrant "in a short voyage, with good fishers, to make a more profitable return from hence than from Newfoundland; the fish being so much greater, better fed," and more to the same purpose.

The effect of these voyages and reports was to arouse the zeal of other Englishmen to undertake colonization in this quarter of the New World. The narrative of Waymouth's voyage was published at once, and attracted the notice of Sir John Popham, chief-justice of England. Popham, with his brother George, Sir Ferdinando Gorges, and other men of influence, became interested in a project to found a colony near the fishing grounds of New England. Accordingly, in 1607, a vessel was sent out, which reached the mouth of the Kennebec River in August. At a place called Saint George the colonists built cabins, a storehouse and some slight fortifications; they also built a pinnace called the "Virginia," the first vessel built in New England, which crossed the Atlantic several times. Under the command of their president, George Popham, forty-five emigrants were left on the shore for that winter. The following year a ship returned to announce the death of the chief-justice; the colonists became disheartened at the prospects of the enterprise, and the settlement was abandoned.

This attempt to establish a colony on the coast of Maine sprang from a desire on the part of English merchants to have a permanent base in New England for the better prosecution of the shore fisheries. Great credit is due to Sir Ferdinando Gorges for his interest in American colonization. He was an Episcopalian, a firm friend of the Stuarts, and a member of the Plymouth Council. His immediate objects were to establish fisheries, to erect sawmills, and to open communication with the Indians. "It may be admitted that his purposes were entirely personal, and that he aimed solely to acquire wealth; but still, whatever were his motives, the voyage of Challon, in 1605; the enterprise of the Pophams and the Gilberts to the Kennebec, the following year, in which he had an interest; the voyages for fishing and trade of Richard Vines, his agent, steadily pursued for years in a ship purchased with his own money; the adventure of Dermer to the island of Monhegan, under his auspices in 1619; the aid he offered to Sir William Alexander, in 1621, to procure the patent of Nova Scotia; the grant obtained by John Mason and himself to the country between the Merrimac and Kennebec rivers, in 1622; and the subsequent grant in his own individual right, to the territory between the Piscataqua and Kennebec, . . . were all beneficial to New England, and hastened its settlement."[1] The policy of Gorges has been described as one of peace and good-will, and he himself has been styled the Father of New England Colonization.[2]

English traders and fishermen appeared on the coast of Maine with a strong force in 1611. There seem to have been scattered settlers in the region of Pemaquid, and doubtless the fur dealers had agents in this region who lived there the year round. Three years later the New England coast was visited by the father of the Virginia Colony, Capt. John Smith. He touched the coast near the Penobscot River, explored the bays and harbors, visited Cape Ann, Cape Cod, Massachusetts Bay, entered Saint John's Harbor and changed its name to Plymouth, and gave to this whole region the beloved name of New England.

Smith had printed in London, in 1616, "A Description of New England" which was "writ with his oune hand,"

[1] Sabine, p. 105.

[2] Winsor, III, p. 178.

and, together with maps of the region, was distributed throughout the western counties of England. In this account of his voyages Smith pays considerable attention to the condition of the fisheries, entering into elaborate details. He tells us that they were successful in taking 47,000 fish in the vicinity of Monhegan Island, although their intentions had been to "take whales, and make trials of a mine of gold and copper." The better quality of the fish of this catch were sold in England for "five pounds the hundredth, the rest by ill usage betwixt three pounds and fifty shillings." Another ship stayed on the coast to fit herself with dry fish, which later sold at a good price in Spain. Smith makes the prediction that the fisheries of New England would prove a greater treasure than the gold and silver mines of the king of Spain. He goes on to draw a contrast between the relative advantages afforded by Newfoundland and New England in the fisheries, always with the advantage in favor of the latter region. His references to the fisheries of Newfoundland would lead one to infer that annually 800 sail of vessels were freighted with fish from that place.[1]

The result of all these voyages and explorations, especially of those made since the opening of the seventeenth century, was directly influential in developing English fisheries in New England and in shaping and stimulating permanent colonization. Smith mentions six ships that came to these shores for fish in 1615. The next year eight ships from London and Plymouth made voyages to our coasts, carrying back their loads of fish and oil to Spain and Portugal. Writing in 1620, Smith speaks of twenty-six ships that have had success "within these six yeares" on this coast, while we learn from another writer that by 1624 the New England fishing was so profitable that forty

[1] Smith, A Description of New England.

or fifty vessels were employed there from England yearly.[1] Although there was, thus, a great stimulus to the successful prosecution of the New England fisheries there does not appear to have been a diminution in the number of English vessels visiting Newfoundland. Instead, there was an increase, for about this time the coast of Newfoundland was frequented by "two hundred and fifty sail of English vessels, estimated at 15,000 tons, employing 5,000 persons, and an annual profit of about 135,000 pounds sterling."

The work of preparation for extensive colonization in New England was complete by the time the Pilgrims settled at Plymouth. The coasts had been explored and carefully mapped out; many islands, headlands and bays had already received names; the island of Monhegan was famous across the ocean as a fishing station; one band of explorers had experienced the New England winter on the coast of Maine, and doubtless many agents of trading companies had passed successive winters on the coast. The people of England had been informed of the advantages that the new country possessed, they knew of the illimitable forest with its splendid material for ships and spars for the royal navy; they had heard of the climate, the soil, the kinds of vegetation; their traders had brought back packs of fur—the mink, the silver fox, and the beaver,—while sassafras-root had entered their harbors by the boat load. But more valuable to them than all others were the fisheries. They had learned of the marvelous wealth of the American seas; they had heard how fish were taken in abundance with line and with net; they had seen these reports confirmed many times when the returning ships anchored in their harbors deep-laden with cod; on many a fast day they had tasted with satisfaction the dishes pre-

[1] Weeden, Economic and Social History of New England, I, p. 10, note.

pared from fish that had been cured on the New England shore; and during the long winter months they had been fired with the zeal for this distant enterprise as the fishermen of London and Plymouth recounted their adventures or told of the wonders of the far-away land. Gradually the popular mind was prepared for emigration across the sea. The commercial and enterprising spirit of merchants was aroused to venture more capital, to risk life and fortune in the New World. When the days of religious persecution fell upon them, the people of England who left their native land for the distant New England shores had complete confidence that a comfortable living could be obtained in their new home.

NOTE.—Sabine's Report on the Principal Fisheries of the American Seas, 1853, has about ten pages devoted to the French and English fisheries in America previous to 1620, and is the only work of any importance on the subject.

CHAPTER III

THE STRUGGLE FOR EXISTENCE

There is little evidence that the pilgrims ever seriously contemplated engaging in the fisheries. The attractions of the lands across the sea and the profits arising from trade in fish and fur must have been well known to the English wayfarers in Leyden. The west counties of England already looked to the American fisheries for a large part of their support. The Virginia colony had passed the critical stage of existence, while every year the number of vessels sailing from English ports to fish at Newfoundland and on the coast of New England was on the increase.

When the agent of the Pilgrims went from Leyden in 1618 to secure the consent of King James to the proposed settlement in America the king asked, "What profit might arise?" The answer was the single word, "Fishing." But the answer is not conclusive proof that such were their intentions. No mention is made of the fisheries as one of the inducements to leave Leyden for a home in the New World. The profits of the New England fisheries were well known in England; but the Pilgrims had decided to settle in a region several hundred miles to the south of the region whence the fishermen of the west counties were making good catches annually. Only the chance of a treacherous sea, or perhaps the craftiness of the captain of the *Mayflower*, caused them to settle in the region of the fish rather than at the mouth of the Hudson River.

Taken at their best the Pilgrims were poor fishermen.

Upon their arrival at Plymouth they possessed neither boats, supplies, nor fishermen to engage in the fisheries. Eventually the salvation of the colonists rested upon the abundance of clams and other fish that were easily obtained; but their attempts to engage in the fisheries as a business undertaking always resulted disastrously.

This fact is a significant commentary on the business ability of the Pilgrims when it is remembered that at this time there were between forty and fifty vessels on the coast of Maine each year, making profitable trips in spite of having to sail thousands of miles to the fishing grounds, whereas the Pilgrims might fish within a few miles of their settlement. The lack of success in the fisheries at Plymouth is still more emphasized when contrasted with the success soon to be established by the Massachusetts Bay colony—a success that made the fisheries for one hundred and fifty years the "corner-stone of New England prosperity."

The Pilgrims were fortunate to settle at a place where both land and sea ministered to their wants. On the one hand were the abandoned corn fields of the Indians, which were easily cultivated by the new settlers; on the other hand was the sea with its abundance of clams to be obtained easily at low tide, the run of fish that came up into the brook in the month of April, and the supply of cod and bass to be had at all seasons and tides. Then, too, they were fortunate in having Squanto as a faithful Indian friend. He was a native of the place, had been taken to Spain as a slave, sold to a London merchant, spent some time as a fisherman at Newfoundland, and had come to the colonists with the opening of spring after the hardships of the first winter were over. Among other matters he instructed them how to catch the fish that came up into the brook and how to use them for dressing under the hills of corn.

When the summer of the first year had passed and the colonists began to gather the small harvest that rewarded their labors, part of their number "were engaged in fishing, aboute codd, and bass, and other fish, of which they tooke good store, of which every family had their portion." During the summer there was no want. As winter approached an abundance of wild fowl frequented the waters; wild turkeys, deer and other game were found in the forest. Edward Winslow, writing to a friend in December, 1621, says: "For fish and fowl we have a great abundance. Fresh cod in the summer is but poor meat with us. Our bay is full of lobsters all summer, and affordeth variety of other fish. In September we can take a hogshead of eels in a night, with small labor and can dig them out of their beds all winter. We have mussels and others at our doors. Oysters we have none near, but can have them brought by the Indians when we will." Apparently there was little suffering from lack of provisions during the second winter.

When the month of May came they were in hard straits again, being without means for supplying their needs. Fortunately for them, one of the fishing boats of Weston, who was engaged in fishing on the coast of Maine, came into the harbor bringing seven more colonists, but no provisions. The newcomers reported, however, that thirty English vessels were at Monhegan. Immediately Edward Winslow was despatched to secure means for relieving the famished colonists. He found that the fishermen had no provisions to sell; but, with a generosity characteristic of fishermen, they freely contributed such provisions as they could spare for the relief of their fellow countrymen.

Winslow relates that he found the colonists more reduced upon his return than when he left them. But, to free the people from the charge of negligence in not securing a good living when nature provided so bountifully with her products of the sea, he adds, "for though our bays and creeks are full of bass and other fish, yet, for want of fit and strong seines and other netting, they for the most part, brake through, and carried all away before them. And though the sea was full of cod, yet we had neither tackling nor hawsers for our shallops. And, indeed, had we not been in a place where divers shell fish are, that may be taken with the hand, we must have perished unless God had raised some unknown or extraordinary means for our preservation."[1]

Weston, the enterprising English merchant, engaged in the fisheries, and had attempted the settlement of a colony at Weymouth, a few miles north of Plymouth, but the attempt was a failure. Apparently his colonists lacked economy. When their store of goods was used up they sold their clothing to the Indians for food; some became servants of the Indians, others became avowed thieves. They deserted their dwellings, scattered up and down the coast in small groups, and were reduced to extreme want. Bradford records that "one in geathering shell fish was so weake as he stuck fast in ye mudd, and was found dead in ye place." Captain Standish with some men from the Plymouth colony was sent to the relief of Weston's men. He found them in wretched condition and offered to take them back to Plymouth. They preferred to attempt to reach England by going to the coast of Maine, there to hire out as fishermen on some vessel, and to return to their homes with the return of the vessel. This was the second attempt at colonization in Massachusetts that resulted in failure.

While the year 1623 was disastrous to the colony of Weston at Weymouth it was one of extreme need at Plymouth also. Their provisions became exhausted; their

[1] E. Winslow's Relation, in Mass. Hist. Soc. Coll., 1st series, VIII, pp. 246–247.

means of supply were so insufficient that it was the usual thing at night not to know where any supplies for the next day could be obtained. There was but one boat left them, and that was poorly fitted. This, with a net they had bought, was the principal means for securing a livelihood. The company was divided into gangs of six or seven each whose business it was to provide fish and other food for the colony for a stated time. The gangs took turns in using the boat and net, not returning from their trips until they had caught something even though they had to be absent five or six days at a time. They well knew that the members of the colony depended for their living on the results of these trips. When the boat returned the next crew set out immediately after the unloading of the fare. If they were away for a long time, or returned with a small fare, then the people at home resorted to the digging of clams for their food supply. In the spring when the fish came in abundance into the brook to spawn, the over-supply was used in the fields for fertilizing the corn and other crops. Even here there was need of nightly guards to keep their fields free from the depredations of the wolves.

In the middle of July the ship *Anne* came bringing supplies and an addition of sixty colonists. In about ten days there appeared a fine new pinnace of forty-four tons, built by the Company to be retained in the colony for the use of the people. Fishermen, also, had been sent, "with salt, etc." The boat was poorly manned, and the seamen refused to trade or to fish unless they were placed on wages instead of shares, the terms on which they had come. When these terms were conceded the boat was sent around the Cape to trade with the Narragansett Indians. The voyage was a failure, the boat being nearly wrecked on the return. As it was, her mainmast had to be cut away to prevent her from going ashore.

In March of the following year the boat, newly rigged, was sent eastward to fish. At Damarin's cove another disaster befell her, for she was driven upon the rocks in a storm and sank to the bottom. There the boat remained for a year when, at the instance of some fishing masters, she was raised by the use of many empty casks, which, being fastened to the boat at low water, buoyed her enough at high tide to enable the carpenters to take her to a convenient place to make the needed repairs. But she cost a great deal of money and "as she proved a chargable vessell to ye poor plantation" they sent her back to England.[1]

Other disasters in the fisheries were in store for the Plymouth people in 1624. They sent a ship to Cape Ann to fish, but the captain and crew were drunken and lazy, and "the loss was great." A ship-carpenter was sent to them this year from England, who built two strong shallops and a large lighter. In the heat of the summer he was stricken with a fever and died; his honesty and industry had endeared him to the people, who felt that his death was a "very great loss." In striking contrast to the ship-carpenter was the salt-maker who had been sent to them, "an ignorant, foolish, self-willed fellow" who, after great expense and bluster, made a failure of his salt works at Plymouth. The next year he was sent to Cape Ann to set up his salt pans there. Before the summer was over he burnt the works, most of the pans were spoiled, and "this was the end of that chargable business."

Although the fishing business at Plymouth met with little success its progress in other parts of New England was more encouraging. In 1620, Sir Ferdinando Gorges and other English noblemen received a charter—the patent of the Council for New England—which granted to them the territory lying between the parallels of 40° and 48°, with

[1] Bradford, History of Plymouth Plantation, pp. 188, 228–229.

a monopoly of the fishing industry in the waters adjacent to these lands.[1] This meant that both the land and the seas from Acadia to the Delaware were no longer free to English subjects but were placed under the exclusive control of the new company by the sanction of the king. By this patent, all persons without license first obtained from the Council were forbidden to visit the coast, under pain of the forfeiture of the vessel and its cargo. Fishermen were prevented from landing and procuring wood to construct stages for drying their fish.[2]

A spirited controversy arose immediately at the attempt of the Council to enforce such preposterous claims. The Council demanded that each fishing vessel should pay a sum equal to about eighty-three cents per ton, an amount estimated at more than one hundred dollars for every English fishing vessel that was accustomed to come to the New England shores.[3] Robert Gorges, son of Sir Ferdinando Gorges, was commissioned in 1623 to come to New England as lieutenant governor over all the country known by that name. In the same year the Council sent Francis West, as admiral of the seas, with power to restrain such ships as came without license either to fish or to trade on the coast. The matter of collecting tribute and of driving off unlicensed vessels was too great a task for the men to accomplish, especially as neither was an officer of the government. Captain West found the fishermen too numerous and too stubborn in maintaining their rights to be coerced by force; he left the region, and the right of freedom to fish was never again interrupted on the New England coast.

In England, the question of free fishing soon found its

1 McDonald, Select Charters and other Documents, 1606–1775, p. 28.

2 Winsor, III, p. 296.

3 Sabine p. 43.

way into Parliament, where many able statesmen took the side of the colonists, the principal advocate being Sir Edward Coke. The Commons were successful twice in passing a bill for the revocation of the original charter that granted the monopoly in the fisheries; but the bills failed of enactment. The efforts of the advocates of free fishing, however, were not without valuable results. The company found it impossible to withstand public sentiment on one side of the sea and the stubborn resistance of fishermen on the other; their plans were changed, and they continued to exist thereafter largely in the capacity of making grants of land in the New World.[1]

On the other hand the ill effects of the controversy over free fishing were felt on both sides of the Atlantic. Charles the First had scarcely ascended the throne before the Commons passed a bill concerning the maintenance and increase of shipping and navigation, and for the liberty of fishing on the coasts of Newfoundland, Virginia and New England. Although the bill was lost in the House of Lords, the Commons prosecuted the matter still further by presenting to the king a representation of grievances in which they insisted that the restraint of the subject in the matter of fishing was of national concern and required redress. The remonstrance was one of the causes that led to the dissolution of Parliament and the beginning of that personal rule of the Stuart monarch which led not only to the exodus of thousands of his subjects to the New World, but resulted ultimately in the untimely death of the king.

In America the ill effects of the dispute were seen in the great depression of the fishing industry itself, since "during the five years embraced in the struggle, the number of English fishing vessels on the whole extent of our coast diminished much more than one-half, or from four

[1] Sabine, p. 45.

hundred to one hundred and fifty; while it is certain that in the alarm which prevailed, the merchants who had purchased the island of Monhegan, and had provided there ample accommodations for the prosecution of their adventures, sold their property and retired from the business."[1]

While the decrease in the number of English vessels coming to our waters was due principally to the above named causes, other influences were at work to bring about such a change. Settlements were beginning to be made in New England outside of Plymouth, which were established for the purpose of carrying on the fisheries from fishing stations located on the shores adjacent to the fishing grounds. By 1620 it was an established fact that the best cod-fishing grounds in the world were on the New England coast. People who were acquainted with the conditions in each locality repeatedly asserted that the grounds of New England were superior to those of Newfoundland, and a credulous public listened eagerly to tales of successful fishing voyages made on the coast of Maine. Near the island of Monhegan in 1619, over an area of a few leagues only, a single ship got a fare that yielded 2,100 pounds in money. In the following year several ships made even more successful trips.[2] The successful establishment of the colony at Plymouth, also, had a direct bearing and influence on further settlement within the present limits of Massachusetts. When the fame of the Plymouth plantation was spread abroad in the west counties of England merchants were aroused to undertake another settlement on these shores.

Rev. John White, the Puritan rector of Trinity Church in Dorchester, England, was the chief promoter of the enterprise. At his instigation in 1623 merchants and other gentlemen about Dorchester formed a company for the

[1] Sabine, p. 45.

[2] Babson, History of Gloucester, p. 27.

double purpose of promoting colonization and engaging in the fisheries. They conceived the idea that, when the fishing season was over, the spare men above those actually required to navigate the vessel home again might be left behind in the region of the fishing grounds; when the ship should return they would be there to assist in fishing as in the previous year, and in the meantime be employed in constructing buildings and in planting corn for another year. With this purpose in mind the Dorchester Company sent out, in the year of its organization, a small ship of fifty tons. She arrived late on the coast of Maine and for that reason did not secure a full fare in those waters. The vessel later sailed to Massachusetts Bay where more fish were caught. The return voyage was made by way of Spain, where a bad market caused a loss of six hundred pounds for the venture.

Before setting sail for Europe the vessel left fourteen of her spare men behind at Cape Ann, in accordance with the instructions of the company. It is difficult to estimate accurately what influence this fact had on the colonization side of the company's project, for history fails to give further details of how the winter of 1623–24 was passed on Cape Ann by this plantation of fishermen. Doubtless some of them survived, if not all of them, and were found the following year when the vessel returned, else there would have been more difficulty encountered in finding men with courage to face a second attempt the next winter. The desolate experience of the fourteen men, shut off from the rest of the civilized world, can only be imagined. "The only other persons of the English race then in New England besides the people of Plymouth were a few at Nantasket; the remnant of Gorges' plantation at Weymouth; the settlers at Piscataqua River and Saco, who began these places the same year; a company at Monhegan, and perhaps one or two other residents on the coast of

Maine."[1] It was a task of exceptional difficulty, bravery and perseverance to establish English civilization on the New England shores.

In 1624 the Dorchester Company sent over two vessels but they met with little success. When the ships sailed away for home thirty-two men were left behind in the new country. In spite of the losses of the two previous years three vessels were sent forth in 1625. With the hope of better success a change was made in overseers. Roger Conant, who had withdrawn from the Plymouth colony because he did not sympathize with the Separatist views there, had come to Cape Ann on his own account. He was a man of good reputation, quiet and competent, and he was appointed overseer of the Cape Ann plantation by the Dorchester Company. But a change of management did not result successfully as far as the fisheries were concerned. To their own troubles at Cape Ann was added a quarrel with their neighbors. The Plymouth people had obtained a patent in 1623 to fish at Cape Ann. It has been shown already how disastrously their first venture at Cape Ann ended. Upon their return to the place in 1625, they found that the stage and other works that were built the summer before had been seized by the captain of an English vessel. He stoutly refused to give up the stage, whereupon Captain Standish was ordered by Governor Bradford to retake the works. The occupants were strongly entrenched behind a barricade of casks when Standish arrived, fully determined to carry out his orders. The affair was at the point of collision and bloodshed, when Conant and the master of another ship interposed their good offices with the result that the high-handed captain gave up his claim to the Plymouth works and used another stage.[2]

[1] Babson, p. 32.
[2] Bradford, p. 237.

By 1626, the adventurers at Cape Ann were so greatly discouraged that they dissolved the company on land, and sold their provisions and fishing apparatus. This "Fisher Plantation at Cape Ann" had proved a failure both to the Plymouth fishermen and to the Dorchester Company; to the former, partly because they made so poor a business of their fishing, partly because of the exorbitant rates charged by English merchants for forwarding their goods. To the Dorchester Company it proved a failure partly for the same reasons, but principally because the spot originally chosen was a poor one for the establishment of a new plantation.[1]

What was their loss, however, became another's gain. Cape Ann proved to be the stepping stone to Salem. When the discouraged settlers left the rocky cape with their cattle and tools in 1626, Roger Conant, still their leader, found for them a new and safer abode a few miles to the southwest at a place called Naumkeag. During all the years of struggle and discouragement, the Rev. John White had kept up a lively interest in the Dorchester adventurers. After they had settled at Naumkeag he wrote encouragingly to them and sent advice for them to remain there. In the meantime, in England, he set on foot a scheme for permanent colonization of a scale greater than any previously undertaken. A company of six persons obtained from the Plymouth Company, in 1628, a strip of land sixty miles in extent along the shore, and in September of the same year, John Endicott, one of the patentees, arrived at Naumkeag with a company of sixty persons. Endicott superseded Conant in the management of the colony, and the name of the place was changed to Salem.

After the coming of so eminent a personage as the new governor, Roger Conant became a less prominent figure

[1] H. Adams, Johns Hopkins University Studies, Vol. I, Art. IX-X, p. 4.

in the colony. He should be placed in the list of strong men who were instrumental in establishing the Commonwealth of Massachusetts. When the Fisher Plantation of Cape Ann was dissolved, it was Conant who still kept together enough of the colonists to form the nucleus of the settlement at Salem, which, maintained by him and strengthened by Endicott, not only became great itself, but led easily to the establishment of other settlements in Massachusetts Bay. This humble overseer of fisheries, "a pious, sober and prudent man," was a pioneer in establishing the fisheries in the New World. When it is remembered that the first founding of Massachusetts was for the establishment of fisheries the name of Roger Conant should find an assured place in early colonial history.

In 1623, the Laconia Company under the leadership of Mason and Gorges sent from England a company to plant a colony and to establish fisheries within the limits of the company's grant. This lay between the Kennebec and Merrimac Rivers. The settlers were divided into two companies, one of which took up their abode on the south bank of the Piscataqua River at Little Harbor. Immediately they erected salt works to furnish the salt needed for curing their fish. The men turned their attention so exclusively to the fisheries that agriculture was neglected, and as late as 1630 there had been only three or four dwelling houses built. Other settlements were made in the vicinity. Prominent among these was the settlement made on the Isle of Shoals, which later became a famous station for the prosecution of the fisheries.

When Laconia was divided, Mason obtained control of the part west of the Piscataqua, which he named New Hampshire in honor of his county in England. He was bred a merchant, was thoroughly familiar with the fishing business, and so intent on making a success of the colony at Portsmouth that he spent much of his time and fortune upon it. But the enterprise was so poorly managed on this side of the water that, after a decade of experience, it afforded the promoters no profit. For years the colony was in an unpromising condition, the growth of Portsmouth was slow, and during the remainder of the seventeenth century this region furnishes little of interest to the subject of the fisheries.[1]

The seat of the first permanent settlement in Maine was the island of Monhegan, a place famous in the Old World for its fisheries many years before the time of Plymouth and Massachusetts Bay. The first permanent settlement on the mainland, however, was at Pemaquid in 1625. Fishermen and hunters had settled at Cape Porpoise by 1630, and settlements of a similar character were made near Portland about the same time, the first house being built in old Falmouth in 1632. Other early settlements were made along the coast as far eastward as Penobscot Bay for the prosecution of the coast fisheries.[2]

In 1631, two merchants of Bristol, England, obtained a grant of land known as the "Pemaquid grant," which gave them the exclusive right to fish in their own waters. The grant included several thousand acres of mainland, the Damariscove islands, and all other islands within seven leagues of the shore, which included Monhegan. The grant lay wholly east of Gorges' claim, and, therefore, within the claims maintained by the French at that period. This region was already an old fishing resort for English vessels, but unfortunately there are only few records now remaining of the fisheries, and none of the doings of the fishermen as early colonizers. Thus it will be seen that by the time the settlement of Massachusetts Bay colony was well established, there was a chain of settlements stretching in a more or less interrupted line from Plymouth

[1] Sabine, pp. 287–289 passim.

[2] Winsor, III, p. 321.

on the south to the farthest bounds of territory claimed by the English on the coast of Maine.

It was the intention of the promoters of the Colony of Massachusetts Bay to profit by the disastrous experiences of many of their kinsmen on these shores. In a letter of instructions to Governor Winthrop, written from London early in 1629, it was ordered that a storehouse be built for shipwrights and their provisions, and that a responsible person be selected to have the matter in his special charge. Fishermen, salt and apparatus for fishing were sent over in 1629. Governor Winthrop took an active interest in the fisheries from the first. On the passage over he records the taking, "in less than two hours, with a few hooks, sixty-seven codfish, most of them very great fish, some of them a yard and a half long and a yard in compass." A few days afterwards he writes, "we took many mackerels, and met a shallop, which belonged to some English fishermen."[1]

The governor's interest in the sea was manifested in practical form when on the fourth of July, 1631, he launched a ship, built at Mystic, named the *Blessing of the Bay*. The little boat was of thirty tons only. She was used by the colonists in opening up commerce with the Dutch on the Hudson, and in maintaining intercourse with other parts of Massachusetts Bay. It is a matter of just pride, especially to those who go down to the sea in ships, that the first governor of Massachusetts thus dignified the calling of the sea, and by his influence helped to make Massachusetts become, in the fisheries, the empire state of the Union.

Activity in the fisheries and in shipbuilding continued in other places. In the fall of 1633, Isaac Allerton, a prominent citizen of Plymouth, set sail in the *White Angel* for Marblehead. There he helped to establish the first

[1] Winthrop, History of New England, I, p. 25.

fishing station in the town. It is reported that Mr. Allerton was fishing there with eight boats a short time after his arrival,[1] and that a ship's loading of fish was cured there in 1632 or 1633.[2] Mathew Cradock, a wealthy London merchant, though he never came to Massachusetts, established a station at Mystic, and built a house at Marblehead which was occupied by Allerton and fishermen in his employ.[3] A fishing station was set up at Scituate in 1633.

In addition to shallops and pinnaces, the kind of craft most generally in use by the early colonists, the fishermen and merchants now began to build more sea-worthy boats and barques. A vessel was built in Boston in 1633 called the *Trial;* another of one hundred and twenty tons was built by the people of Marblehead three years later. Salem followed with one of three hundred tons in 1640, and with another in 1642. The first vessel built at Plymouth seems to have been a barque of forty or fifty tons, in 1641, built at a cost of two hundred pounds.

The settlers at Salem and other nearby places had to pass through experiences of suffering and death not unlike that of the Plymouth colonists. About one hundred of the people of Salem died before the close of the year 1630. Among the number was the first minister of Salem, the Rev. Francis Higginson, who has left for us a glowing account of the treasures of the sea as found in Massachusetts Bay.

"The abundance of sea-fish," he writes, "are almost beyond believing, and sure I should scarce have believed it, except as I had seen it with my own eyes. I saw great store of whales and grampusses, and such abundance of mackerels that it would astonish one to behold, likewise

[1] Roads, History of Marblehead, p. 8.
[2] Weeden, Economic and Social History of New England, I, p. 133.
[3] Sabine, p. 124.

codfish in abundance on the coast, and in their season are plentifully taken. There is a fish called bass, a most sweet and wholesome fish as ever I did eat; it is altogether as good as our fresh salmon, and the season of their coming was begun when we first came to New England in June, and so continued about three months' space. Of this fish our fishers take many hundreds together, which I have seen lying on the shore, to my admiration: yea, their nets ordinarily take more than they are able to haul to land, and for want of boats and men they are constrained to let many go after they have taken them, and yet sometimes they fill two boats at a time with them. And besides bass, we take plenty of scate and thornbacks, and abundance of lobsters, and the least boy in the plantation may both catch and eat what he will of them. For my own part I was soon cloyed of them, they were so great and fat, and luscious. I have seen some myself that have weighed sixteen pounds; but others have had, divers times, so great lobsters as have weighed twenty-five pounds, as they assure me. Also here is abundance of herring, turbot, sturgeon, cusks, haddocks, mullets, eels, crabs, muscles and oysters. Besides, there is probability that the country is of an excellent temper for making salt; for since our coming our fishermen have brought home very good salt, which they found candied, by the standing of sea water and the heat of the sun, upon a rock by the sea shore; and in divers salt marshes that some have gone to, they have found some salt in some places crushing under their feet and cleaving to their shoes."[1]

Johnson, who came over in 1630, gives several instances of how the people appeased their suffering, in the absence of bread, by feasting on fish obtained from the seas; how women went daily to the flats for clams; with what pride they viewed their children "fat and lusty with feeding

[1] F. Higginson, New England's Plantation.

upon muscles, clams, and other fish, as they were in England with their fill of bread;" how some were providentially furnished with an abundance of fish in their nets or on their hooks; how others without hooks and nets caught the fish with their hands, and were thus provided for until other kinds of food came.[1]

The stories of hardship and suffering on land can be equalled, if not surpassed, by disasters that befell these seafaring people at sea during the period of the early years of struggle for existence. A remarkable accident befell a boat that was bound from Pemaquid to Boston with a barrel of powder aboard. As they were nearing Piscataqua on the way, one of the sailors wished to light his pipe. His companions remonstrated with him for taking so great a risk, but he replied that if the devil should take him away quickly he would take one pipe. In lighting the pipe the powder was set on fire and the boat blown to pieces, with a loss of two hundred pounds worth of commodities. None of the seamen were injured except the one whose foolhardiness had caused the accident; his body was found later with hands and feet blown off, "which was a very remarkable judgment of God upon him."[2]

Other accidents are recorded, due sometimes to carelessness, as when two persons from Roxbury left their boat unfastened as they went to gather oysters, the boat drifted away, and, since the men had no other means for getting to the mainland, they were drowned. Others were lost by their own rashness, as in the case of an old man who used to go to sea accompanied only by a dog which he had taught to steer. Once as he was putting out to sea he was warned that a storm was coming on, but he answered that he would go to sea though the evil one

[1] E. Johnson, Wonder Working Providence of Sion's Savior in New England.

[2] W. Hubbard, General History of New England, pp. 195–196.

himself were there. It is presumed that the latter was in evidence, for neither the man nor the dog was again heard from. More often disasters occurred to boats that were overtaken by storms at sea. The worst gale recorded in the early days of fishing occurred in the middle of August, in 1635. On the land houses were unroofed and turned over, crops were laid waste, thousands of great trees were broken off or twisted about like withes. On the sea the effects of the gale were no less terrible, a tidal wave, or exceptionally high tide, twenty feet "right up and down" causing many deaths. Several vessels that were caught out in the gale had miraculous escapes; many others were lost, among the number being the *Angel Gabriel* of Bristol.[1]

[1] Hubbard, p. 199.

CHAPTER IV

YEARS OF GROWTH AND EXPANSION

Down to the year 1635 the fishing industry of New England was carried on in a more or less desultory manner. The amount of home capital invested in the business was small, even in the aggregate. The returns from year to year were uncertain. The fishing stations were scattered along many miles of the coast. There was no central supply station where boats might be fitted out, nor was there a distinctive shipping port to which products of the sea could be brought before being shipped to other countries. However, men were beginning to see the need of greater capital invested in the industry, of central storehouses, and of more united action among the fishermen. Some began to foresee great possibilities both in development of the fishing industry, and in the consequent evolution of the scattered fishing stations into respectable, united settlements.

The Rev. Hugh Peter, a celebrated minister of Salem, was one of the first to advocate an increase of commerce and the fisheries through business enterprise. In 1635, he went about from place to place arousing men in public and in private to a more generous spirit for the general welfare. He was successful in raising a good sum of money "to set on foot the fishing business." He sent word to England asking that an equal amount be raised there for the purpose. He himself became actively interested in the business that his zeal had helped to create. It was his purpose "to set up a magazine of all provisions and other necessaries for fishing" in order that the fishermen might have a local supply of apparatus at a nominal cost.[1] As matters then stood, the merchants and others who brought goods over from England charged exorbitant rates for carrying freight between England and the colonies, sometimes the cost of the goods being doubled in value by the excessive charges of transportation. To the efforts of Mr. Peter is due the credit of establishing the commerce and fisheries of Salem on a substantial basis. For years this port of the colonies was without a rival in domestic and foreign commerce and in the fisheries.

The course of trade was now well fixed in its natural routes. The Dutch on the Hudson had been trading with the New Englanders for some time. The *Dove*, a pinnace of fifty tons, came from Maryland in 1634 with a cargo of corn to be exchanged for the fish and other commodities of the northern colonies.[2] The southern people were in need of the northern staples and gladly exchanged their tobacco and corn for New England fish. Settlers on the coast of Maine carried on a similar traffic with Massachusetts ports. A visitor to the coast of Maine about this time leaves a vivid account of the fisheries and coastwise trade. He describes the method of pursuing the fishing business thus:

"The fishermen take yearly on the coast many hundred quintals of cod, hake, haddock, pollock, etc., and dry them on their stages, making three voyages a year. They make merchantable and refuse fish, which they sell to Massachusetts merchants; the first for 32 ryals ($4) per quintal;[3] the refuse for 9 and 10 shillings ($2 and $2.25). The merchant sends the first to Lisbon, Bilboa, Marseilles, Bordeau, Toulon, and other cities of France; to Canaries,

[1] Sabine, p. 125.

[2] Weeden, I, p. 128.

[3] A quintal equals 112 pounds.

pipestaves and clapboards; the refuse fish to the West Indies for the negroes."[1] The fishermen used shallops for their fishing, there being four men in a crew. Often their share on a voyage was eight or nine barrels of fish per man. It was the practice of the merchants to buy of the planters their beef, pork, peas, wheat and Indian corn, which they sold to the fishermen or exchanged for cured fish.

Measures were taken at an early date to regulate the fisheries, and in several instances, to promote the industry by the passage of favorable legislation. In 1635, the General Court of Massachusetts appointed a commission consisting of Mr. Thomas Dudley and five others to have in charge the setting forth and management of the fish trade, all charges of the commission to be allowed out of the fishing stock. Three years later, in 1638, the Court allowed the commission the sum of £100 16s. 3d. to make up for the loss which followed from the governmental management of the fishing industry.[2]

The first measure for the protection of the infant industry in Massachusetts was passed by the General Court on May 22, 1639. "For the further encouragement of men to set upon fishinge," it was ordered that all vessels and other property employed in taking, curing, and transporting fish, according to the usual course of fishing voyages, should be exempt from all duties and public taxes for seven years; that neither cod nor bass should be used for manuring fish; and that all fishermen, during the season for their business, as well as all ship-carpenters, should be exempted from military training.[3]

Winthrop states that the order was not passed to encourage foreigners to engage in the fisheries among them,

[1] Folsom, Saco and Biddeford, pp. 36–37.
[2] Massachusetts Colonial Records, I, p. 230.
[3] Mass. Col. Records, I, pp. 158, 230.

for the gains then would be returned to the place where the promoters dwelt; but to encourage their own people to develop the industry. The same year a fishing trade was commenced at Cape Ann by Mr. Maurice Thompson, a London merchant. The people of Massachusetts hoped that Mr. Thompson would come to settle with them, but there is no record that he availed himself of the measure of encouragement passed by the General Court, or that he ever came in person to Cape Ann. Evidently his business was conducted by agents, for fishing was carried on at Cape Ann and some stages were built there in 1639.[1]

The fisheries now appear to be as prosperous and active as at any time in the history of their early development. Winthrop speaks of the great benefit derived by all the plantations from the abundance of large fat mackerel on the coast during the season of 1639. One boat with three men would take ten hogsheads of mackerel in a week, which found a market in Connecticut at £3 12s. per hogshead. In 1641 he states that the fishermen followed their calling so well that there were about 300,000 dry fish sent to market.[2]

Richmond's Island, on the coast of Maine, was becoming an important and noted place for the fisheries. Mr. John Winter, the superintendent, often employed as many as sixty men there, and several vessels were furnished with cargoes of cured fish annually. Upon the death of Mr. Winter, in 1645, a commission was appointed to examine into the affairs of his estate. The report of the commission was rendered in 1648. It is of interest to our subject in that it shows the extent of the fishing industry and the price of commodities at the time. During the six years between 1639 and 1645, Mr. Winter sent to his principal in England 3,056½ quintals of

[1] The Fisheries of Gloucester, p. 20.

[2] Winthrop, I, p. 308; II, p. 42.

merchantable and refuse fish, 38½ quintals of core-fish, eleven hogsheads of train-oil, and other products of the sea, valued at not less than £2,292. The inventory of the property belonging to the fishery showed that there were three boats, which, with their moorings and fittings, were worth £28; two old boats worth £2; the fishing stage and a quantity of casks worth £10 6s; six dozen hooks, at 16 shillings; five dozen lines, at £7; one seine and two old nets, £4 10s; about ninety hogsheads of salt, valued at £65 10s; and that there was due the concern the sum of £84 15s. 9d. for one hundred and thirty-three quintals of fish sold but not paid for.[1]

Public interest in the fisheries was further manifested in 1641. It was ordered by the General Court that fishermen should be served first at the weirs and have their bait at the same rate at which others secured it. By another act, a fishing station was established at Nantasket, and inducements were offered to the inhabitants of Hingham and other places to remove to the new station. Shore room for stages and flakes was to be furnished, while for every boat used in fishing, four acres of upland were allowed the owner, with a portion of meadow for the fishermen who owned cattle.[2] In the same year the Plymouth people allowed Mr. John Jenny certain privileges at Clarke's Island for the making of salt which was to be sold to the inhabitants at two shillings per bushel. In 1642, with the hope of further increasing the salt output, the colony granted the use of thirty acres at the island to five partners for the term of twenty-one years.[3]

The colonists now began their first attempts at fishing beyond the limits of the New England waters. It was the beginning of a movement which, at first directed to the

[1] Sabine, p. 107.
[2] Mass. Col. Rec., I, pp. 326–328.
[3] Mass. Hist. Soc. Coll., III, 2nd series, p. 183.

fishing grounds adjacent to Nova Scotia and Newfoundland, was destined to expand until the four corners of the earth were reached. The fishermen of New England pursued their industry at distances farther from home, and more intelligently, than other fishermen of their day or since. They advanced into the sea farther each voyage until the time came, during the colonial period, when they knew how to pursue their business on the coast of Labrador, or to follow the mightiest monarch of the deep amid the icebergs of the Arctic Circle, in the waters of the Southern Cross, and over the pathless wastes of the unexplored Pacific.

The inception of the movement of expansion was made in 1641, when some merchants of Boston sent twelve men to the Isle of Sable, off the coast of Nova Scotia. The men were left to carry on the fisheries as best they could, and to collect furs. Later in the year a vessel was sent to them which returned in three weeks with four hundred pairs of sea-horse teeth valued at $300. Twelve tons of oil and a quantity of skins were left behind on account of a storm that prevented the stay of the vessel at the island. The fishermen and their goods were taken off the island the following summer. The value of the oil, seal and horse hides, teeth and the fur of black foxes was about £1,500.[1]

In 1645, merchants of Boston and Charlestown sent several vessels to the Bay of Islands in Newfoundland. The venture proved disastrous, for the warring factions in England had partisan adherents in all waters. A captain by the name of Firnes was cruising in Newfoundland waters in the interests of the king when the New England vessels reached the island. Cargoes for the vessels had already been secured when the Englishman appeared and seized vessels, cargoes and crew. The men reached home

1 Winthrop, II, p. 67.

after considerable delay, but the enterprise was a failure.[1] It was some time before the New England fishery at Newfoundland became important.

As the fisheries developed, more legislation regarding the industry became necessary. The acts that were passed were generally for the encouragement of the men engaged in the enterprise; but sometimes repressive measures had to be enacted. Frequent depredations were committed by foreign fishermen who took wood and timber along the coast wherever it was most convenient for them. The inhabitants of Marblehead petitioned for redress. Accordingly, the General Court made it unlawful for any person, whether a fisherman or not, to enter upon land and to take wood and timber; but fishermen who were in that jurisdiction could use harbors and adjoining lands in the process of drying their fish, and also take what timber was necessary for firewood where it could be spared, upon making due payment to the owner.[2] The Marblehead petitioners were almost wholly sea-faring men, and their catch of fish for the season of 1647 was valued at £4,000.[3] It can readily be seen that their petition was instigated by reasons of justice rather than for their own convenience in following the practice of procuring wood.

This salutary protection to property owners was extended, in 1652, to cover similar conditions prevailing on the coast of Maine. At that time Pemaquid was the center of the fishing industry of Maine and the principal shelter for vessels passing to and from the French and English settlements that were scattered along the coast. Not only had the fishermen committed acts of depredation in the way of taking wood and timber, but they had also

1 Ibid, p. 236.

2 Mass. Col. Rec., II, p. 147.

3 Felt, Salem, II, p. 212.

fallen into inferior methods of curing and packing their fish. An act was passed providing for the appointment of officers to correct these abuses.[1]

Previous to the year 1650, the colony of Plymouth had permitted the inhabitants of Hull to seine for fish at Cape Cod. Some irregularities took place at the Cape about this time and the colony took the matter into their own hands. Mr. John Stone was forbidden to fish longer in that region. On the other hand, permission was given to three residents of Plymouth, and to others from Plymouth, Nauset, or Duxbury who should join them, to engage in the bass fishery at Cape Cod, and to use wood as they might need. The next year the privilege was given to the same persons and to William Bradford for three years' time. The regulations provided for the establishment of two companies only to engage in the fisheries there. The fisheries of Cape Cod were carried on by the Plymouth Colony, usually by annual leases, until the union of the colony with Massachusetts Bay in 1692. During the greater part of the period they were subject to continual annoyance from the Massachusetts fishermen. When the colonies became united, the inter-colonial troubles ceased.

During the last half of the seventeenth century the fishing interests of Massachusetts Bay grew into importance beyond all other industries. All business enterprises were affected by the fisheries. After the cessation of English immigration to New England with the overthrow of the Stuart monarchy, the people centered their attention upon fishing to secure cargoes for their outgoing vessels; and in shipbuilding, to provide ships to carry the cured products of the sea to the markets of Europe and the West Indies. Year after year the two industries developed side by side, for more than a century, until the time came when in each industry the New Englanders surpassed the mother country.

[1] Sabine, p. 127.

Even in the seventeenth century there were far-sighted statesmen who recognized the importance of the two industries not only as means for rivaling the commercial interests of England, but also as powerful agencies for bringing about the separation of the colonies from the mother country. Sir Josiah Child, writing of the people of New England in 1670, said that by means of their mackerel- and codfisheries these people were more proper for shipbuilding and producing seaman than the other colonies; and he added that nothing was more prejudicial to any mother country than the increase of shipping in her colonies.[1]

The colonists early recognized how closely the prosperity and very life of New England rested upon the fisheries. The first governors of Plymouth and of Massachusetts were interested not only in securing laws to promote the interests of the fisheries, but, in addition, each was actively interested in the business itself. The same kind of interest was shown by other men prominent in the colonies in the early days. Later, the Curwins of Salem and Boston took a prominent part in pursuing the fisheries from those cities and from other places in the colony.[2] Captain John Hull, the famous mint-master of Massachusetts, gives frequent reference in his diary to the success or failure of numerous ketches and ships engaged in the fisheries or in the carrying trade, in which he was interested either as principal or part owner.[3]

The legislation of the period shows with what jealous care the law makers regarded the fishing industry—the well-spring of their prosperity. In 1652, they provided for "the appointment of sworn fish-viewers at every fishing place within the jurisdiction, who were required to reject

[1] Anderson, Origin of Commerce, II, p. 513.

[2] Weeden, I, p. 247.

[3] American Antiquarian Society, III, pp. 141-317, passim.

all unmerchantable, all sun-burnt, salt-burnt and dry-fish that hath first been pickled, to be paid one-half by the dealer, the other half by the receiver."[1] Codfish, haddock, hake and pollock could not be killed during the months of December and January, because that was their spawning time. For the same reason, mackerel could not be taken before the month of July. Tortugas salt was condemned as being injurious to the best quality of cured fish.

By a noteworthy act the Plymouth Company, in 1670, granted all profits that should accrue annually from the fisheries at Cape Cod for mackerel, bass or herring to be employed "for and towards a free school in some town of this jurisdiction." The next year such a school was opened in the colony with John Morton as teacher. The school was maintained until 1677 when more liberal provisions were made for the support of other grammar schools in the colony.[2] For the seven years subsequent to 1677, the Cape Cod fisheries were rented at the rate of £30 per annum; the most of the income was granted to a school at Duxbury, to one at Plymouth, and to others elsewhere in the colony. Instances are found, also, where a part of the fund was expended in helping widows and orphans of men formerly engaged in the service of the colony.[3]

In 1673, Massachusetts found it necessary to pass laws regulating the packing of sturgeon for shipment. Heretofore, the fish had been packed in casks and kegs that proved unserviceable. By the new regulation, packers of the fish had to become licensed by the county courts; persons who examined and marked the casks as packed were bound by the same regulation. During the same year the

[1] Marvin, American Merchant Marine, p. 286.

[2] Plymouth Colonial Records, V, p. 108.

[3] Mass. Historical Society Collections, 2nd series, III, pp. 117, 220, 259.

law exempting fishermen from ordinary training in military service was modified to compel their attendance when at home for a period of time. To protect the masters and owners of vessels from loss occasioned by broken voyages a law was passed compelling all fishermen who shipped for a season's work to continue with the vessel to the end of the season; otherwise, they were liable for all damage resulting from the breaking-off of their voyage before the expiration of the season.[1]

A writer of the time describes Boston, in 1664, as having 14,300 souls, a great trade to the Barbadoes with fish and other provisions, and with a fleet of 1,300 boats that fished at Cape Sable. The towns of Salem, Ipswich, and Charlestown, also, were important places for trade in cod and mackerel, while the Isles of Shoals had fifteen hundred fishermen.[2] Even at New York a trade in fish had sprung up, stimulated by the discovery of several cod-fishing banks near Sandy Hook in 1669, and by the encouragement that the governor gave to the prosecution of the whale fishery at the east end of Long Island.[3]

The prosperous condition of the fisheries along the entire extent of the New England shores at this time is in striking and sad contrast to the manner of living and the low state of the morals of the fishermen, especially along the frontier coast of Maine. A vivid account of the state of affairs is given by Josselyn, who visited Maine in 1664, and remained in the colony for several years. Likewise, the commissioners of King Charles to New England in 1667, give an account of the morals of the persons connected with the fisheries on the Kennebec and Sheepscot rivers and at Pemaquid. "These people," they say, "for the most part, are fishermen, and never had any government

[1] Mass. Col. Rec., IV, part 2, pp. 252–53, 400, 450; V, p. 212.

[2] Calendar of State Papers, Col., 1661–68, pp. 532–33.

[3] Ibid, 1669–74, p. 20.

among them; most of them are such as have fled from other places to avoid justice. Some here are of the opinion that as many men may share a woman as they do a boat, and some have done so."[1]

A boat's crew usually consisted of four men, the master and two fishermen being on the boat, the fourth man, who stayed on the shore, cured the fish very much as fish are cured now, and did the cooking for the other members of the crew. Often the crew shared £8 or £9 to a man in a single voyage. The money, however easy or difficult it was in coming, was easily spent. Upon the arrival of a trading vessel laden with spirituous liquors, termed in that day a "walking tavern," the fishermen forsook their calling for a drunken carousal which often lasted several days. These are severe judgments passed by contemporary writers upon the moral condition of the fishermen of the day. It may be a safer judgment to regard the instances of gross laxity in moral conduct as the exception rather than the rule among the people of Maine—an exception that would be brought more sharply into view and remain longer in the memory in places where life was rough, hard and monotonous, with few if any relaxations in the struggle with elemental conditions.

In the boat fishing it was the prevailing custom for the fishermen to have one-half the catch.[2] In the winter season the boats went out in the morning to return at night, in the spring and summer they remained out until a load of fish was secured. The vessels used to make three trips to the banks in a season. The first fare of the spring produced the best quality of fish, they being large and thick. This kind was first salted and then dried; then it was kept alternately above and under ground until it became so mellow that it was called "dumb fish." When

[1] Sabine, p. 107.

[2] Bourne, History of Wells and Kennebunk, p. 182.

boiled it became red. This was eaten generally on Saturdays at the best tables of New England.[1]

Throughout New England the codfish were divided into three sorts—the merchantable, the middling and the refuse. The first of these was again sorted, the better fish being shipped to Bilboa, and, later, to Calais; the second, to Lisbon and Oporto. The middle sort was sent to the Canaries, the Madeiras, and to Jamaica. The Barbadoes and Leeward Islands received the shipments of refuse cod, it being the smallest in size, the thinnest and most broken. In addition to the codfish, some haddock, pollock and mackerel were sent to market. At the close of the seventeenth century there were exported from Massachusetts about 100,000 quintals of dried codfish, of the value of $400,000.[2] The merchants of Boston shipped, annually, about 50,000 quintals of codfish; about three-fourths of which went to the Biscayan port of Bilboa. In 1699, the fish sold, on the shore where it was taken, for 18 shillings a quintal, but the next year the price had fallen to 12 shillings. The cause for this decline in price was attributed to the glutting of European markets by the French fishers. The cod taken on the coast of New England brought two shillings more a quintal at Bilboa than the cod caught on the banks off Newfoundland. In New England the fishing was carried on all winter, the fish being of a better quality when cured in the colder season than in the warmer. When returns were made direct from Bilboa to Boston the merchants reckoned on a 50 per cent return of their money; if their money was returned by the way of London, and there invested for them in colonial supplies, the returns were 100 per cent.

About 1670, the New Englanders turned their attention

[1] Doc. Col. N. Y., IV, p. 116.
[2] Ibid, pp. 781–790, passim.

to the deep-sea fishing off the coasts of Newfoundland and Labrador. For several years previously they had visited the fishing grounds of Nova Scotia from Cape Sable to Canso. The development of their industry in this region brought them into contact with a class of maritime neighbors, who frequented the banks of Newfoundland as did their old-time rivals—the French fishermen of Acadia. Their relations with the English seem to have been friendly and needs but a passing glance. With the French it was a different matter, for the bitter rivalry between the French and the subjects of Great Britain continued with unabated fury until it became necessary, at last, for one nation to retire from the scene of action.

The New England fishermen arrived at Newfoundland in 1670. The condition of the English fisheries was low at that time, it having been on the decline for several years. In 1605, there were sent over from England to Newfoundland 250 vessels; in 1670, the number had diminished to eighty. The Massachusetts fishermen on the contrary made phenomenal progress in half a decade. To explain the failure of one class of fishermen during this period is to account largely for the success of the other.

The cause for the decline was attributed by a celebrated writer of the day, Sir Josiah Child, to the greater liberty in regard to eating fish in Lent and on other fish days; to the increase in the boat fishing on the coast by the Newfoundlanders; and to the great increase of the French fisheries at Placentia, Newfoundland. As a remedy it was suggested that the depopulation of the island was the best way to insure an increase in the number of vessels sent to the fishing grounds annually from England.[1] This view was finally adopted by the Lords of Plantations, who determined to break up and depopulate the colony. Sir John Barry was sent over with orders to drive out the

[1] Anderson, Origin of Commerce, II, p. 513.

fishermen and to burn their dwellings. The inhabitants of the island continued to suffer wrongs and indignities for six years, before this monstrous order was revoked.

In the meantime, the desired effect had been secured as far as increasing the number of English vessels sent to Newfoundland. But the hostility of the Newfoundlanders towards these vessels was intense. In "1676, there went thither from England, one hundred and two ships for the cod-fishery there; each ship having twenty guns and carrying eighteen boats, and for each boat five men; in all, nine thousand, one hundred and eighty men. Their convoy was two ships of war. And the total value of the fish and oil they made there, in that year, was computed at 385,400 pounds." Then the writer of the above adds, with no regard whatever for the outraged Englishmen living at Newfoundland, "how great a nursery, then, is this fishery for English sailors, and how beneficial for the employment of shipbuilding."[1]

The decline in the boat fishing from the shore and the extra cost of protecting the English vessels on the coast of Newfoundland could result only in increasing the price of fish to enormous figures. The New Englanders were attracted to the coast in the year of the edict of depopulation. They took no part in the quarrel between the other two classes of English fishermen, although it does appear that many of the residents of the island migrated to New England each fall after the fishing season was over. Massachusetts profited most in the economy, the rapid expansion and the excellent returns of her fisheries during the period of high prices. Five years after the first vessel went from New England to Newfoundland, they had in the employment of the Newfoundland fishery 665 vessels, which measured 25,660 tons, and were navigated by 4,405 seamen; at that time they caught between 350,000 and 400,000 quintals of

[1] Ibid, p. 535.

fish annually, in striking contrast to the boasted returns that England was securing from the same fisheries.[1] The New England people employed less than half the number of men in the service than the English, while the returns from their fish must have been equal to, if they did not surpass, those of the English vessels.

The relations between the fishermen of New England and their French neighbors were critical and hostile. A brief explanation of the political conditions at the time will readily account for this animosity. Acadia was the scene of strife, partly because of the excellent fisheries in its waters that each party wished to monopolize, partly because the land had been held alternately by each country, either through seizure or concession. When Acadia passed into the possession of France in 1667 the bounds of the province were not clearly defined; the English claimed that the western boundary was the Penobscot River, the French laid claim to the land as far west as the Kennebec River. Consequently, the region was again open to strife and quarrels between the two classes of fishermen. The French took pains to embitter the Indians against the English; and in several instances used the Indians as their allies in making attacks upon the settlements, especially in the frontier region between the Merrimac River and Pemaquid. This stretch of the coast of Maine was the scene of atrocious warfare during the forty years that followed.

De Bourg, the French governor of Acadia, in 1675, not only prohibited his people from intercourse with the Protestant fishermen, but also levied a tax of four hundred codfish on every English colonial vessel found fishing on the coasts of Acadia. The French officers seized all who did not pay the fine, and took from them whatever fish they had aboard, together with the provisions and outfits. Two years later, about twenty fishermen from Salem were

[1] Seybert, Statistical Annals, IV, p. 333.

captured on the coast of Maine by the Indians. In the fight that ensued, several men were killed and a score were wounded. When news of the engagement reached Salem, a large vessel was fitted out and sent to recapture the boats and men; but the natives, plundering and abandoning the ketches, left before their pursuers could do them harm.[1] "In the lapse of a few years," says Sabine, "the fishermen at Cape Porpoise were either slaughtered or driven off, and the settlement there laid desolate; a fishing smack was intercepted near Portland, three of her crew killed, and the remainder carried into captivity; eight fishing vessels were captured at the Fox Islands; the coast for more than a hundred miles was abandoned, and the wretched men who depended upon the sea for support, without shelter, and too scattered for concert and resistance, were compelled to suspend their employments."[2]

It had now become evident to the thoughtful people of New England that the only way to secure the peaceful prosecution of their fisheries along the coasts of Maine and Nova Scotia was by the expulsion of the French. The accession of William and Mary to the throne of England brought on war between that country and France. The struggle was taken up by the colonists on this side. Sir William Phipps, a native of Maine, led an expedition against Nova Scotia in 1690, which he succeeded in capturing from the French. The victory was fruitless, however, as the province was re-taken by the French the next year.

The French, realizing the importance of the American fisheries, renewed their efforts to retain hold upon their colonies. Under the leadership of Baron de Castin, a trading fort was established at the mouth of the Penobscot, now the town of Castine. The French officials and Jesuits

[1] Felt, Annals of Salem, p. 258.

[2] Sabine, p. 109.

enlisted the sympathy and services of the Indians, who assisted them in carrying on their depredations against the English colonists. The fort at last was taken by Iberville and Castine in 1696, thus extending the frontier of the French dominion into the heart of Maine, and putting to an end, for a time, the fisheries in that region.[1] This condition of affairs emphasized rather than weakened the conviction in the minds of the New Englanders that the French should be driven from the North Atlantic coast of America.

By the treaty of peace at Ryswick, in 1697, the French retained possession of the whole coast, islands and fishing grounds, from the Penobscot River to beyond Labrador, with the exception of the eastern half of Newfoundland, which was held by the English. The country lying between the Penobscot and Kennebec Rivers was left in dispute, but it was claimed by the French in spite of the fact that no settlements had been made there to substantiate their claim.

The evil effects of the treaty were soon manifest. Scarcely a year had elapsed before the French laid claim to the sole ownership of all the fisheries. The king of France sent over an order for the seizure of all vessels other than French found in Nova Scotia waters. A rigid enforcement of this order was attempted. Bonaventure, of the ship-of-war *Envieux,* boarded and sent home every English colonial vessel found trespassing within his jurisdiction. From Villabon, governor of Nova Scotia, came an official despatch to Massachusetts that he had received instructions from his royal master to seize every colonial fisherman found east of the Kennebec River.[2]

At the end of the century the strife between the French and English "for the monopoly and the mastery" of af-

[1] Bancroft, History of the United States, II, p. 183.

[2] Sabine, p. 11.

fairs in America had no immediate prospect of settlement. Of the two, the French had the advantage. They laid claim to all the fishing grounds between Maine and Labrador. The only fisheries that were open and free to the English subjects were those of the eastern half of Newfoundland and that part of New England between the Kennebec River and Long Island Sound. The supremacy of the French in the fisheries was due no more to their zeal and daring than to the stupid policy of the English government towards her fishermen in American waters. The edict of 1670, "to burn and destroy," had a most disastrous effect upon the fisheries of Newfoundland, even if it increased the number of English vessels coming to the Grand Bank. At the close of the century the government allowed about one thousand persons to reside permanently in Newfoundland, since that number could be of use in constructing boats and fishing stages.

In 1698, the British Parliament enacted a law providing that the master of the first vessel from England arriving on the coast should be admiral of the harbor in which he cast anchor, with authority to administer justice and to regulate the general concerns of the fisheries and the colony.[1] The resident population was ignored entirely. Each autumn, at the close of the fishing season, the admiral, with the fleet, disappeared. With them went every semblance of law and order, while the residents were left to pass the winter as best they could. "The triumph of the English merchants over their fellow-subjects, in this lone and desolate isle, was as complete as that of the warrior who storms a city. In fine, the 'admirals' selected the best fishing stations, displaced at will the resident fishermen who occupied them, drove the inhabitants from their own houses, took hush-money and presents of fish in adjusting cases brought before them for adjudication, and,

[1] Sabine, pp. 50–51.

in their general course, were as arbitrary as the leaders of the banditti. There were exceptions, it may be admitted; but the accounts are uniform that, as a class, the 'admirals' were both knaves and tyrants. Yet the law which authorized these iniquities bore the title of 'An act to encourage the trade of Newfoundland.' "

The attitude of the English government towards the fisheries of New England was less severe, but this, probably, was occasioned more from necessity than from choice. The statement of Sir Josiah Child, referring to the northeastern fisheries and shipbuilding, that nothing was more prejudicial to any mother country than the increase of shipping in its colonies, was only one instance of the growing jealousy with which English merchants viewed the prosperous condition of maritime affairs in New England. Our fishermen were too energetic, too willful, and too far removed from the source of authority to allow all, or a large part, of the profits of the fisheries to go over the seas.

They built their own smacks and vessels, and master and crew were natives of the colonies. As early as 1676 there hailed from Boston and other nearby ports thirty ships of between fifty and one hundred tons, and five hundred smaller vessels. Twenty-four years later Boston is credited with 194 sea-going ships. Richard, Earl Bellamont, who was colonial governor in Massachusetts at the close of the century and who, an enemy of the Stuarts, was in sympathy with the fishing interests of the people, wrote in admiration of the spirit of the New Englanders, "I believe I may venture to say there are more good vessels belonging to the town of Boston than to all Scotland and Ireland, unless one should reckon the small craft, such as herring boats."

Within seventy-five years of the settlement of the country the energy of the New Englanders had built up an industry that was to be, if it had not already become, the backbone of colonial life and trade. Their fishing vessels ventured from the confines of Cape Cod and Cape Ann to the coast of Maine, from there to Sable Island and the adjacent coast of Nova Scotia, and finally swarmed by hundreds upon the banks of Newfoundland. Every year they sent out from New England harbors more than ten million pounds of cured fish. They outdid the mother country in the quantity and quality of the catch. Bilboa and Calais bought their best fish from the holds of American vessels; Lisbon and Oporto bartered the products of their land on board the craft of New England masters; the Madeiras, Jamaica, and the Barbadoes willingly exchanged their wines, sugar and molasses for the third-grade product of the northern seas.

Their prosperity, due to the expansion of the fishing industry, had led the northeastern fishery into contact with England and France. Towards the former they had already demonstrated that an English admiral was quite helpless in attempting to enforce unpopular decrees upon a seafaring people. Furthermore, in disregard of the navigation act of England, they maintained a large contraband commerce with Europe. France, on more than one occasion, had learned to dread her New England rivals, who already had furnished men and vessels for the conquest of Nova Scotia. Embittered against the French for their policy of exclusion from all fishing grounds, and indignant at the home government for the disgraceful consequences of the treaty of Ryswick, the fishermen of New England at the close of the seventeenth century were waiting impatiently an opportunity to use their men and supplies in helping the mother country drive the common foe from the fishing grounds of North America.

CHAPTER V

THE SURVIVAL OF THE FITTEST

During the course of their development in the eighteenth century the fisheries of New England attained a commanding position in commerce, in legislation, and in international affairs. In extent and value, the fisheries reached their culmination at the outbreak of the Revolutionary War when over a thousand vessels gave employment to upwards of ten thousand men. For three score years they played an active and aggressive part in the struggle between England and France for mastery in North America, indirectly by keeping England awake to the necessity of securing the provincial fishing grounds for her own subjects; directly, by the capture of Nova Scotia and the taking of Louisburg, the stronghold of France in America. When the common foe had been driven from the field the colonies and the mother country turned upon each other. The colonies openly violated measures that were passed prejudicial to their interests. England looked upon her colonial subjects as means for developing her industries and influence at home, she attempted to enforce her laws upon the people of America, and, failing in that, passed a law that was designed to cripple the commerce of New England and to drive her fishermen from the sea.

The beginning of the reign of Queen Anne found France and England involved in a new war. One of the causes of the struggle was the claim of France to a part of Maine and to the whole of the fishing grounds from the Kennebec River to Labrador. The French seemed determined to defend all of their claims. They had between 400 and 500 vessels engaged in the fishery; they were well armed; they outdid their competitors in the quantity of fish that they caught, claiming, at the opening of the century, that their catch of codfish was equal to the supply of all continental or Catholic Europe; their fishermen were first in European markets, and their fish sold at a larger profit than that of their rivals.[1] The outlook for England to retain what few fishing privileges she possessed in America was nearly as dubious as was the prospect that she could regain in war the advantages that had been lost through shameful treaty.

The people of New England, feeling keenly the loss of a province that had been won very largely by their valor, and stirred to action by the loss of the privilege to fish upon the Acadian fishing grounds, needed little urging from the mother country to enter heartily into the contest. They employed armed vessels of their own; they swept the coast of Nova Scotia; they equipped a fleet at Boston and twice attempted the conquest of that province; and they furnished four battalions of fifteen hundred men and thirty transport vessels to Nicholson when, in 1710, he captured Port Royal, thus finally winning Nova Scotia as a province under the crown of Great Britain.

Peace was concluded by the treaty of Utrecht in 1713. By this treaty the British statesmen attained, or supposed that they had attained, what had been their ambition for many years,—the supremacy in the fisheries of the American seas. All Nova Scotia, or ancient Acadia with its boundaries, was made over to the Queen of England and her successors. The French were excluded from fishing on the coast of that province or within thirty leagues of it, from Sable Island to the southwest. Newfoundland with the adjacent islands came wholly into the right of Great Britain, as did Hudson Bay with its borders. On

[1] Isham, The Fishery Question, pp. 15–16.

the other hand, the French were given possession of, and the right to fortify, the Island of Cape Breton and all other islands both within the mouth of the Gulf of Saint Lawrence and in the Gulf itself. At Newfoundland they were allowed to catch fish and to dry them on land from Cape Bonavista to the northern point of the island, and thence down the western coast as far as Point Riche. They were prohibited from settling or fortifying any place on the island of Newfoundland.[1]

When all things are considered one must conclude that the treaty was greatly to the advantage of Great Britain. If adhered to strictly, it meant that the fishermen of New England would secure a monopoly of the fisheries of the coast of Maine, the Bay of Fundy, and the shore and bank fisheries of Nova Scotia, with equal chance to compete with French fishermen in all other fisheries. The narrow limits of the French possessions would result in a serious curtailment of their fisheries. Bitter opposition, however, was aroused in England because France had been given any privileges whatever in American waters. Lord Oxford was impeached because he dared to advise his sovereign that "the subjects of France should have the liberty of fishing and drying fish in Newfoundland." This was a new sentiment in diplomacy for an English subject to express; but since his day the great principle that "the seas of British America are not to be held by the British subjects as a monopoly, and to the exclusion of all other people," has never been entirely disregarded by British statesmen.[2] We shall see shortly how the treaty resulted neither in excluding the French from the coasts of Maine and Nova Scotia, nor in diminishing the extent of their fisheries; but, on the contrary, it gave them exceptional advantages

[1] McDonald, pp. 229–233.
[2] Sabine, p. 14.

in the new fishing grounds granted to them on the coast of Newfoundland.

The period between the treaty of Ryswick, 1698, and that of Utrecht, 1713, was one of general depression for the New England fisheries. The accounts of the time are few and discouraging. On the coasts of Maine and New Hampshire the fishing was practically suspended; and on the Nova Scotia coast it was reduced greatly in value. In 1699, the governor of Massachusetts gave passes for fourteen sloops and ketches of 25 to 36 tons and carrying five or six men each. Marblehead had not ventured extensively into this industry in which, before the century was over, she became recognized on both sides of the Atlantic as leader. The town of Salem saw the number of ketches that she had engaged in the fisheries dwindle from 60 or 70 to a mere half dozen.[1] The coast from Cape Cod to Cape Sable was rendered extremely dangerous to our fishermen by the depredations of the Indians and French.

At the opening of the century Gloucester had a population of about seven hundred. This plantation, just entering upon its second half century, had as yet gained no importance in maritime affairs. There is no evidence to show that before 1700 Gloucester had a single vessel engaged in the fisheries as far east as Cape Sable; a Gloucester sloop fished there in 1711. Activity in the ship-building industry began with the century. Ships and brigantines were built for Boston merchants, and sloops for the townspeople. The sloops were used at first for carrying wood and timber to Boston; later they were used in a wider coasting trade and in the fisheries, which began to develop rapidly after 1720.[2]

It was at Gloucester that the first schooner was con-

[1] Weeden, I, p. 373.

[2] The Fisheries of Gloucester, pp. 9–22, passim.

structed. In 1713, Captain Andrew Robinson, the father of the schooner, contrived a new rig for sailing vessels. Up to that time the square rig was the characteristic sail of all vessels with more than one mast. Even with the fishing ketches, square-rigged sails were the most in use. Captain Robinson built a two-masted vessel and rigged her with fore-and-aft sails on booms, with a jib forward. At the launching of the craft, a bystander exclaimed as she left the stocks, "Oh, how she scoons!" To which Captain Robinson replied, "A schooner let her be."[1] The schooner was destined to become the most efficient vessel of its size for the practical mastery of the seas. A popular variation from the Robinson model for many years was the topsail schooner, which carried on the foremast a lower, a topsail and sometimes a topgallant yard; otherwise, it was rigged in the regular fore-and-aft style. Captain Robinson possessed at least two of the qualities that have distinguished New England fishermen,—ingenuity in contriving apparatus to better their condition, and activity while aboard ship. He is said to have been so industrious on the banks when fish were plentiful that he would not leave his place on deck even to eat. When hungry, he had ship-biscuit brought to him which he contrived to eat by working it around in his mouth with his lips and teeth, while he attended to hook and line with his hands.

With the return of peace, in 1713, a new wave of settlement advanced northeasterly. The old towns in Maine that had been abandoned were rebuilt, new ones were founded, and the frontier pushed farther to the east and deeper into the forest. From Cape Porpoise to the Kennebec River these settlements went on. Brunswick, Topham, Georgetown and Augusta were begun. The fisheries were set forward by the ingenuous Dr. Oliver Noyes, of Boston. This generous man also built a stone garrison for soldiers

[1] Sabine, p. 130; Babson, p. 250.

at Augusta and maintained it for several years at his own expense.[1] On the Kennebec River a sturgeon fishery was set up and carried on with such success that many thousand kegs of the cured sturgeon were made in a season, esteemed as good as any that came from Hamburg or Norway. A trade with Boston and foreign ports was established, with high hopes for continued success.[2] This success did not continue very long. The merchant, originally from Boston, who established the business, contracted to furnish some London fishmongers five thousand kegs of fish annually. Fish twelve feet long, weighing four hundred pounds, were caught. In 1721, the merchant shipped to England 1,500 kegs weighing forty to fifty pounds each. The fish, although good when caught, had been oversalted in curing them. They were unmerchantable, and the business in Maine was discontinued.

The fisheries of New Hampshire increased after the treaty with France. In 1717 the Council of the Province laid an embargo on all outward bound vessels, except those engaged in fishing voyages.[3] No less than twenty vessels were engaged in foreign commerce in 1721. There were about one hundred fishing vessels belonging in the province, with about four hundred seafaring men.[4]

When the French missionaries saw the revival of colonization along the coast and noticed the hopeful growth of the struggling frontier towns, they were stirred with envy and jealousy toward their English rivals. They soon aroused the Indians to action against the English subjects by insinuating that the land belonged to the Indians and that the English had invaded their property. The frontier once more smoked with burning cabins, while the settlers

1 Niles' History of Indian and French Wars.
2 Mass. Hist. Soc. Coll., 4th series, V, p. 338.
3 Prov. Papers, N. H., II, p. 701.
4 Doc. N. Y., V, p. 595.

fled from their new homes. The garrison at Augusta was neglected, the inhabitants withdrew, and the Indians advanced unmolested to the sea-coast. Towns were attacked, fishermen dispersed, and the fisheries had to be abandoned. Sixteen fishing vessels on the coast of Maine were captured by the enemy. When the news reached the colonists of Massachusetts two sloops were fitted out in less than half a day, manned with crews under the command of Captain Eliot and Captain Robinson. They succeeded in taking ten of the captured vessels and at least twenty-four prisoners.[1]

The English and colonial vessels hardly dared to venture on the coast of Nova Scotia to cure their fish. France had practically a monopoly of the fisheries, their fleet being reported to be larger than ever before.[2] The privileges she had received on the Newfoundland coast were proving very remunerative, she was firmly intrenching herself at Cape Breton Island, and every day breaking the terms of the treaty that she had made in good faith less than a decade before. It was no wonder that numerous complaints were heard from New England that the mother country, by the treaty of Utrecht, had granted the best fisheries to their bitterest foes.

Better days, however, were in store for the colonists. In 1725, several eminent sachems from Maine arrived at Boston to negotiate a treaty with the Massachusetts government, after which there was a revival of the fisheries. From a journal written in 1726 it is learned that forty large fishing vessels put into Portland Harbor for refuge from a storm, indicating that there was a quick recovery from the ravages of the Indians as soon as hostilities ceased.

During the twenty years that succeeded there was activity in all phases of the fishing industry. The town of Marblehead now came to the front as a fishing port, a

[1] Mass. Hist. Soc. Coll., V, p. 342.

[2] Doc. N. Y., V, p. 593–94.

position that it maintained until after the Revolution. To the Rev. John Barnard credit should be given for awakening an interest in the fisheries among the people of Marblehead. When he began his ministerial labors among them in 1714 he found not only a lack of first class workmen among them, but also a people who were content to do menial labor and leave "the merchants of Boston, Salem and Europe to carry away the gains." The town was in debt, the inhabitants a rude, swearing, drunken class.[1] Mr. Barnard associated with the masters of English vessels to learn their methods of business. Then he attempted to arouse his fellowmen to engage in the fishing industry. After considerable trouble and discouragement he finally persuaded a young man named Joseph Swett to put his ideas into practice. The business was a success from the start, and in a short time the young merchant was engaged in a prosperous carrying trade even to European markets. Others followed his example, engaging in fishing or in the carrying trade. The town prospered, the good Mr. Barnard lived among them fifty years, long enough to see the morals of the place changed for the better, and the infant industry that he started reach its high-water mark in this seaport town of New England. Before Mr. Barnard's death, Marblehead had the reputation of shipping off more dried codfish than all the rest of New England together. In 1732, which was a good year for the fisheries, Marblehead had about 120 schooners of about fifty tons burden engaged in the fisheries, with about 1,000 men employed from the town. This number did not include seamen who were upon vessels that carried the fish to market.[2]

At the Isles of Shoals about 1730, and afterwards, the fisheries increased to such an extent that three or four ships used to load annually with cargoes of winter and spring

[1] Sabine, pp. 129–130.

[2] Douglass, Summary View, pp. 300–304.

merchantable fish for Bilboa and other places that demanded fish of the first quality. Their fish of an inferior quality were carried to Portsmouth for shipment to the West Indies.[1] Mackerel and herring, also, were sent to the West Indies, but their principal use at this period was for bait. "Choice Hook Mackrill for Bait" were advertised in a Boston paper of 1728 as early as February 19th.[2] How the merchant obtained mackerel for bait at this season of the year is difficult to understand. Salmon were being caught and salted on the Connecticut River in the early part of the century. In New Hampshire petitions for salmon weirs were issued as early as 1729.[3]

The fisheries of New England employed between five thousand and six thousand men in 1731. They were reckoned to produce annually 230,000 quintals of dried fish which yielded in the European markets about $700,000. As their salt, rum, molasses, also their provisions and utensils, were purchased from the refuse fish unfit for European markets and from fish oil, "the said rum may be said to be all gained out of the sea." There were about 1,300 tons of shipping engaged in the whale fishery this year in addition to the codfishery tonnage.[4]

With the return of more peaceful relations with the French and Indians after 1725 the colonists pursued their calling with increasing activity on the fishing grounds of Nova Scotia, Cape Breton Island and Newfoundland. An English fishing station had been erected at Canso, at the eastern extremity of Nova Scotia. This place, commanding the entrance to the Strait of Canso between Nova Scotia and Cape Breton Island, soon became an important trade center. In 1721, twenty thousand quintals of codfish were

[1] Mass. Hist. Soc. Coll., 1st series, VII, p. 242.

[2] New England Journal, Feb. 19 and 26; Mar. 4, 1728.

[3] Weeden, II, p. 596.

[4] Anderson, Origin of Commerce, II, p. 172.

cured there.[1] New England merchants sent their vessels to Canso to be loaded with cargoes of dried codfish for European and other markets. Among the most famous of these merchants was Peter Fanueil of Boston. So important a base of supplies did the Strait of Canso become for New England that Fanueil soon sent an agent to reside at Canso. At times he kept another assistant stationed at Louisburg. These agents were kept informed of the price of fish in the markets of Massachusetts; they studied closely the condition of the English and French supply of fish on the coast, and acted as Fanueil's advertising agents. Other Massachusetts fishermen, notably those from Marblehead, became jealous of the importance and strength which the Canso trade acquired within a few years.

Fanueil was a great merchant of the day, shrewd, active, and ready to engage in many kinds of business that gave promise of good returns. He was a descendant of Huguenots who had been driven from France. The business of his uncle fell to him and he pursued it in original ways. It was he who settled a vexatious dispute among Boston tradesmen concerning markets and the best way to conduct them by building and giving to the city that famous structure known as Fanueil Hall. The lower floor contained stalls for a market, the upper story contained the hall afterwards called the Cradle of American Liberty.

The trade interests of Fanueil were wide and varied. Now he gave directions for a cargo of rum to be returned in sole-leather; again it was a cargo of fish sent to the West Indies for "a straight negro lad, fifteen years old, having small-pox if possible"; or it might be that another load of fish was sent to London to be returned in pepper. Whether rum, leather, negroes, or pepper, the important fact to be noticed is that fish not only entered into every phase of life and form of industry, but also that fish was the

[1] Douglass, Summary View, pp. 300–304.

basis of these industries, the ever-present article of trade with the West Indies, Africa, Southern Europe and London,—the corner-stone of New England prosperity. Perhaps no better opportunity is afforded for learning of the cosmopolitan character of the New England cod and its influence in shaping trade and commerce at home and abroad than is to be found in the writings, correspondence and business activities of Peter Fanueil.[1]

The schooner type of vessel which had its advent at Gloucester in 1713 rapidly superseded the old-time shallop as a fishing vessel. At Gloucester a few schooners had been added to the fleet by 1720, many of them being of a burden of fifty tons and suitable for trips to the Grand Bank of Newfoundland and other distant offshore banks. They were two-masted, with short high-pointed bowsprit, a low bow, straight keel and high sterns. About seventy schooners belonged at Gloucester in 1741, being engaged principally in the Grand Bank fishery. The crews of vessels were then accustomed to "go on their own hook," which meant that an account was kept daily of the number of fish caught by each member of the crew. At the end of the voyage each man was paid according to the number of fish he had caught.[2]

At this time Marblehead owned 160 vessels of an average of fifty tons. The whole number of fishing vessels in Massachusetts was not less than four hundred, with an equal number of ketches, shallops and undecked boats.[3] The outbreak of another war with France, in 1744, checked the colonial fisheries for a brief time and turned the attention to the more exciting experiences of foreign conquest and military campaigns.

After the treaty of Utrecht, in 1713, the French lost no

1 Weeden, II, pp. 614–615.
2 Fisheries of Gloucester, pp. 22–24.
3 Sabine, p. 131.

time in again setting forth their fisheries and in strengthening their defenses in the New World. French fugitives from Nova Scotia and Placentia settled on Cape Breton Island and resumed fishing operations there. The government of France began the construction of a stronghold at Louisburg, sparing neither money nor material to make it impregnable. After working on the defenses for twenty-five years at an expense of more than five million dollars the military authorities looked upon Louisburg as the strongest fortress on the west shores of the Atlantic. With the outbreak of the war in 1744 the French had no hesitation in sending a company of men from the fort to take Canso from the English. The place was easily captured, the buildings set on fire, and the enemy moved to the siege of Annapolis. The timely arrival of two sloops from Boston broke up the siege, the French retired to Louisburg, and Nova Scotia was saved to the English.

The state of affairs filled the colonists with alarm. They feared that in subsequent attacks the French would be successful in again securing a foothold in Acadia. Our fishermen regarded the seizure of Canso and other hostile acts of the French as matters of small importance compared with the significance of the prosperous condition of their fisheries. They had beheld with astonishment the ease and rapidity with which their rivals had recovered after being driven from Acadia. For a quarter of a century the New Englanders had been helpless witnesses to the construction of a fortress in the very heart of the fishing industry of America. They were keenly aware of the growing importance to the French of the coast fisheries of Newfoundland where cod were so plentiful and so near the shore that their fishermen "caught them with a kind of grappling," without the charge or trouble of bait and line.[1] They felt humiliated and disappointed not only because

[1] Douglass, I, p. 6.

their rivals were forging ahead of them so fast and far in this their boasted occupation, but also because their mother country had shown none of that maternal instinct to foster and protect colonial industries, which had been expressed so often in such significant ways by the government of France toward her colonies in America. A brief comparison of the condition of these rival industries at the outbreak of the fresh hostilities in 1744 will serve to explain more fully why New England wished to drive France from the North American fisheries.

Accounts of the extent of the fisheries vary considerably, and doubtless are exaggerated. Thus, one writer states that the French at Cape Breton Island "carried on an unbounded fishery, annually employing at least one thousand sail, from two-hundred to four-hundred tons, and twenty thousand men."[1] From another account it is learned that they had "five hundred and sixty-four vessels in all, and twenty-seven thousand five hundred yearly employed from France on the banks of Newfoundland and the adjacent shores;" the catch was no less than 1,149,000 quintals of cod, with nearly four million other fish, exceeding in value, $4,500,000 annually.[2]

The returns of the English fisheries at Newfoundland for 1701 were 216,320 quintals of fish; in 1716, there were exported to Spain, Portugal and Italy 106,952 quintals of fish, and in 1724, 111,000 quintals.[3] The annual value as reported in 1731 was placed at $600,000, derived from 200,000 quintals of fish.[4] The total number of men employed in the Cape Breton fishery in 1745 was 5,260, while the yield was 186,000 quintals.[5] Another account places

[1] Auchmuty, Importance of Cape Breton, pp. 3–4.

[2] Bollan, Ancient Right of the English Nation to the American Fishery, p. 53.

[3] Holmes, American Annals, II, pp. 55, 92, 114.

[4] Anderson, Origin of Commerce, III, p. 172.

[5] MacPherson, Annals of Commerce, II, year 1745.

the number of persons employed in the English Newfoundland fishery at this time as ten thousand, and the annual profit of the industry as $2,000,000. These computations include all persons employed in fishing, curing, packing, etc., except those belonging to the colonies.[1] The annual catch of the New England fisheries at this time, as already indicated, was about 230,000 quintals of dried fish.

An important agency in arousing the people of New England to the gravity of the situation was a pamphlet entitled "the importance of Cape Breton to the British nation and a plan for taking the place." Robert Auchmuty, an eminent lawyer of Boston and judge of admiralty, was the author of the paper. The ideas set forth in the paper were, doubtless, nothing more than the public opinion of New England on the important subject expressed in attractive form. It was instrumental in arousing other men to devise plans for the capture of Cape Breton.

Among the foremost of these was William Vaughn, of Damariscotta, Maine. He was a graduate of Harvard College, had established a fishing station at Matinicus, and was a prosperous dealer in fish and lumber. His acquaintance with Louisburg and his hatred for the French helped him to devise a plan of campaign for the capture of the city. This "mad scheme" was laid before Governor Shirley, of Massachusetts. After some hesitation on the part of the authorities the scheme was set on foot and all New England joined in the contemplated expedition. Enthusiasm was unbounded. In Massachusetts men enlisted as in a crusade. Merchants hoped to increase their fortunes by making better markets possible in Europe. Military officers were ambitious for distinction. Fishermen looked for revenge on the French for past insults and injuries. For commander of the expedition the governor

[1] Bollan, pp. 54–55.

selected William Pepperell of Kittery, Maine, a merchant and the son of a fisherman. At first Pepperell was reluctant to assume command, realizing his lack of fitness for the position; but Shirley's choice probably was the best that could have been made, "for Pepperell joined to an unbounded popularity as little military incompetency as anybody else who could be had."[1]

The story of the capture of Louisburg is one of enthusiasm, of hardihood and exposure, of splendid achievement. The credit belongs wholly to New England and to the people of those colonies who lived on the seaboard of New England. A thousand men, mostly fishermen, came from Maine to join the expedition. Connecticut sent five hundred or more, New Hampshire four hundred and fifty. A government sloop was provided from Rhode Island and ten cannon were loaned by New York. Early in April, 1745, this expedition of New England farmers and fishermen assembled at Canso before entering upon the active work of besieging Louisburg.

The daring and hardihood of the men who composed this expedition seemed incredible. There was lack of sufficient tents for shelter; in their stead the men used old sails and when these failed huts were built of sod, with roofs of spruce boughs. Many days were spent in landing cannon and supplies from flatboats. To do this the men waded through ice-cold water to their waists, then, after the day's labor was over, went ashore to sleep with no change of clothing, exposed to the chill of foggy nights, and yet "cheerfully underwent all of these difficulties for the sake of executing a project they had voluntarily undertaken." Teams consisting of two hundred men, with straps over their shoulders, dragged cannon and sledges over marshes where cannon on wheels had actually sunk from sight in the mud. This kind of work, too, was done at night or

[1] Parkman, Half-century of Conflict, II, p. 99.

in a thick fog so as to avoid exposure to fire from the enemy's cannon. Thirteen years afterwards when Amherst besieged Louisburg a second time he made use of eleven thousand British regulars supplied with all possible equipments, to accomplish results similar to what Pepperell secured with his four thousand New England fishermen and volunteer soldiery, supplied only with what scanty apparatus their own ingenuity could devise.

For forty-nine days Louisburg was besieged closely by land and by sea. At one time, fever and exposure placed fifteen hundred of the besiegers on the sick list, yet they toiled on with indomitable pluck and cheerfulness, "doing the work oxen could not do, with no comfort but their daily dram of New England rum." Nine thousand cannon balls and six hundred bombs were fired into the French entrenchments. Pepperell wrote that never was a place more mauled with cannon and shells, nor does history give an account of troops behaving with greater courage.[1] When the victorious New Englanders finally entered the fortifications they first fully realized the stupendous task they had undertaken and were astonished at their own success.

Thirty years afterwards a member stated in the House of Commons that the colonists "took Louisburg from the French single-handed, without any European assistance—as mettled an enterprise as any in our history—an everlasting memorial to the zeal, courage and troops of New England." The historian Smollett regarded the capture of Louisburg "the most important achievement of the war of 1744," since the outcome of this New England expedition "proved the equivalent of all the success of the French on the continent."[2]

The peace of Aix la Chapelle, in 1748, was dishonorable

[1] Parkman, II, p. 131.

[2] Smollett, History of England, III, p. 155.

to England and most unjust to New England. The chief conquest of England—Louisburg with Cape Breton Island—was made possible through the valor of the farmers and fishermen of New England.[1] Yet the king of England yielded this important province in exchange for Madras, in India, greatly to the chagrin of the people who captured it. This was but another instance when the people of New England were prevented from reaping the fruits of a victory over their rival and foe by the discreditable statesmanship of the home government,—one more stroke that helped to alienate the affections of these people from Great Britain.

It was not until 1763 that England became practically mistress of the North American fisheries. In the meantime she and her colonies had to pass through a life and death struggle with her old foe, carried on in three continents. In the end she was triumphant, but victory was gained too often at the sacrifice of colonial trade and development. Foreign wars break up the industries of the sea more quickly and more completely than those on land. There was a noticeable decline in the fisheries of New England during the twenty years between 1743 and the close of the French and Indian War. The causes of the decline were emigration from Gloucester, Marblehead and Cape Cod to the coast of Maine where new towns were founded, the call for seaman to man privateers and to enter the naval service of England, and chiefly the two wars in which England and consequently the American colonies were engaged. We have just seen how, in the first of these wars, the fishermen left their lines and flakes to take up guns and to engage in the expedition against Louisburg. The activities

[1] It is interesting to note that a monument was erected on the site of the old fortress at Louisburg by the Society of Colonial Wars in 1895, "to Commemorate the Capture of Louisburg, A. D. 1745." No mention is made of the taking of Louisburg in 1758.

of the Seven Years' War had the same detracting results. The carrying trade between New England and the West Indies actually increased during part of this time; but this was due not so much to the quantity of fish and lumber to be exchanged as to the enhanced value of these commodities through the fortunes of war. When in 1758 Louisburg was again besieged by an army, under Lord Amherst, the victory that attended the British arms was in a large measure due to the courage and energy of the men of the New World. Nearly one-third of the effective men of Massachusetts were engaged in this second expedition. In the House of Commons it was stated at this time that of the seamen employed in the British navy ten thousand were natives of America.[1]

It will be well to consider the economic condition of the fisheries during the score of years previous to 1763. The condition of the British colonial fisheries at the time of the treaty of Aix la Chapelle in 1748 is best set forth by William Douglass, who gives us an insight from personal investigations into the state of the larger fisheries of cod and whale and of smaller fish as well. The New England winter dry cod were of prime quality and would bear watering; they were best for Bilboa market, since they retained their mellowness and would stand land-carriage to Madrid. In Great Britain and Ireland, fish were not reckoned marketable if under eighteen inches from the first fin to the setting on of the tail. August and September were the best months for selling a cargo in the Roman Catholic countries, since their Lent supply was exhausted by that time.

Salt for the New England fishery came from Salt-Tortugas, Cape de Verde Islands, Turks Island, Lisbon and Bay of Biscay. The salt from Tortugas generally reached New England about the middle of April. More salt-burnt

[1] Elliott, The United States and the Northeastern Fishery, p. 19.

dried cod came from New England than from Newfoundland because the Tortugas salt used at the former place was more fiery than the milder salt from Lisbon and Bay of Biscay that was in use at Newfoundland. The New England fishermen generally cured or dried their fish on hurdles or brush, the apparatus being called flakes. All codfish caught from the beginning of June to the beginning of October were called summer fish; the others, spring or fall fish. The staple food for fishermen was salt pork, biscuit and rum.

The fisheries of Marblehead, at this time the most extensive of any in the colonies, employed about 120 schooners in 1747, and only fifty-five the following year. There were twenty at Cape Ann, eight at Salem and six at Ipswich. They used to make five fares yearly; the first to Sable Island, made in March; the second to Brown's Bank and other banks near Cape Sable, for spring fish; their third and fourth fares were to George's Bank for summer fish; the last fare, to Sable Island again for winter cod. The schooners employed in this deep-sea fishery were usually of fifty tons burden, manned with a crew of seven and, one year with another, they made six hundred quintals of fish per schooner.[1]

As it was the custom for the men to "go on their own hook" the returns for the season's work depended fully as much on individual exertions as on fisherman's luck. To prevent the fishermen from deserting before the end of the season for which they had been engaged the General Court of Massachusetts, in 1755, enacted a law providing that no man should receive any share unless he continued for the full term for which he had shipped.[2] At Boston, in 1753, a sum of money was raised by subscription for the encouragement of the codfishery. Sixty dollars was to be

1 Douglass, Summary, pp. 537–538.

2 Weeden, II, p. 650.

given to the crew of the vessel returning the most fish for the season in proportion to the size of the crew, fifty dollars to the crew second highest, forty to the third highest and the remainder to be divided among the rest of the fleet.[1]

In 1747 there were cured in all places of British North America about 300,000 quintals of dry merchantable cod. This quality was carried to the markets of Spain, Portugal and Italy; the refuse cod were shipped off "for the West India Islands to feed the negro slaves."[2] For the year ending at October, 1748, one hundred and thirty-one vessels cleared at the Salem custom house for foreign voyages. In them were shipped to Europe 32,000 quintals of dry codfish; to the West Indies, 3,070 hogsheads of refuse codfish, at six and seven quintals per hogshead.[3] In addition to the cod, which was the staple of Massachusetts, many smaller kinds of fish were being used for food and becoming of economic importance. Haddock, hake and pollock, called scale fish, were caught and cured similarly to codfish. They were used in the trade with the West Indies. Herring were caught in seines, or mesh nets; they, too, were pickled and barreled for negroes, but as they proved unmerchantable in that market the business had to be suspended.

Mackerel, which with herring were first used exclusively for bait, now began to be saved for trade with the sugar islands of the West Indies. The fish came upon the coast much as they do now, one school of them putting in appearance about the middle of May, lean and wild, only to vanish after two or three weeks; another school appeared in the summer and fall, of better size and quality. They were caught with hooks,—although some sort of seine was used

1 Boston Evening Post, Feb. 18, 1754.

2 Douglass, pp. 300–304.

3 Ibid, pp. 537–538.

and mesh nets,—and, after being split and salted, were barreled for the negro trade.

In 1750, two hundred vessels were employed in the mackerel and other small fisheries for the trade with the West Indies; in the codfishery, four hundred vessels; in the whale fishery in the North Atlantic, especially in the Gulf of Saint Lawrence, one hundred vessels.[1] Half of the catch of codfish was the refuse fish suitable only for the West Indies. It was estimated in 1755 that the Barbadoes took merchandise from New England amounting annually to 100,000 pounds sterling.

The growing dependence of Great Britain upon the American colonies for articles of trade and commerce, and the important place that the fisheries and allied industries held in this trade is clearly indicated by a writer of the day, who says, "It is from American Colonies our Royal Navy is supplied in a great Measure with Masts of all Sizes and our Naval Stores, as well as our Merchant Ships, it is from them we have our Vast Fleets of Merchant Ships, and consequently an increase of Seamen; it is from them our Men of War in the American World are on any occasion man'd, and our Troops there augmented and recruited; it is from them we have our Silver and Gold either by their trade with foreigners in America, or by the way of Spain, Portugal, and Italy, in payment for their immense Quantities of Fish, Rice, &, it is from them we have all our tobacco, Rice, Rum, and most of our sugars. Dyeing and other valuable Woods, Cotton-Wool, Ginger, Indico, Whale, and Liver Oil and Whale-bone, Beaver and other Furs, Deer Skins, and innumerable other articles."[2]

After the fall of Louisburg, in 1758, and the reduction of Quebec the following year had prevented the French from making further monopoly of the fisheries of the Gulf

[1] Weeden, II, pp. 641, 644.

[2] Huske, The Present State of North America, pp. 56–57.

of Saint Lawrence the British fishermen eagerly and boldly entered those fishing grounds from which they had been excluded. By the time peace was concluded between the two countries, in 1763, they were carrying on the business of capturing whales, sea-cows, and seals to a greater extent than it had been conducted by the French.[1] A revival of all kinds of fisheries began in New England as soon as the fall of the French strongholds made the colonists safe in pursuing their vocation on the distant fishing banks. Thirteen thousand men had been raised in 1759 for the siege of Quebec. With the fall of that city the men returned to engage in their former occupations. How zealously they plied their trade is to be seen in the gains made immediately upon the revival of the fishing business. The revival of the fisheries at the close of the French and Indian War was the beginning of a period of great prosperity on the sea which continued uninterrupted until the opening of the Revolutionary War, a dozen years later.

Salem claimed thirty fishing vessels in 1762 which brought in 11,177 quintals of merchantable fish and 17,498 quintals of Jamaica fish. The year that peace was concluded, 1763, there were three hundred vessels in the cod-fishery which brought in 240,000 quintals of fish valued at $600,000; ninety mackerel vessels caught 18,000 barrels worth above $80,000; one hundred and eighty vessels engaged in the whale fishery had returns valued above $392,000; other fish and fish by-products yielded enough more to make the total value of the fisheries for that year above $1,200,000.[2]

When the time came for determining conditions of peace in 1763 there was a strong party in England opposed to granting to the subjects of France any privileges in America. The causes for so determined an attitude are

[1] MacPherson, III, p. 371.

[2] Mass Hist. Soc. Coll., VIII, pp. 202–203.

not difficult to be seen. Time and time again France had been the favored party in negotiations referring to rights and privileges in North America. Beginning with the treaty of Saint Germain, in 1632, concessions were made favorable to the subjects of France. These were continued by the terms of the treaty of Breda, in 1667, confirmed by the treaty of London twenty-one years later, increased by that of Ryswick, in 1697, and so changed by the treaty of Utrecht that French fisheries, instead of being seriously curtailed, actually increased under its provisions during the half-century preceding the close of the Seven Years' War.

The prosperity of their rivals was viewed by the fishermen of New England and Great Britain with envy and with dread. The record of their own military triumphs in Nova Scotia, Cape Breton and Quebec made the New Englander feel that the French were privileged to enjoy what by right of conquest should have belonged to them fifty or more years previous. Their recent success in the Gulf of Saint Lawrence had demonstrated clearly the true value of those fishing grounds and served to increase the feeling of resentment. But beyond the question of justice and economy was the more significant one of permanent menace. The people of New England never could feel safe in pursuing this industry of the sea nor could they be assured of its continuance as long as the French remained in permanent possession of the land and much of the sea in regions adjoining their own, on the north and east. The colonists had hoped for more than a half-century that the idea of excluding the French from North America might prevail in English politics. England at heart hoped for the same, but heretofore her continental interests had led her always to sacrifice the welfare of her colonies in America when settling accounts with France. From the Stuarts down to the American Revolution it was the policy of England not only to neglect but even to oppress and restrain the commerce and navigation of New England. The continued neglect of the colonists' interests and the strong protests of the people served to arouse and maintain a sentiment in England for the exclusion of the French from America.

The terms of the treaty were extremely harsh towards the French. By a stroke of the pen France transferred her empire in America to the crown of Great Britain. She lost forever the control of the vast region extending from the mouth of the Mississippi to the Rockies, northward to the Hudson Bay region, and eastward to the Alleghenies, through the valley of the Saint Lawrence, and out to the shores of the Atlantic. The only territory retained by France was the small islands of St. Pierre and Miquelon, situated on the Newfoundland coast. The labors of Champlain, Hennepin and La Salle in establishing a mighty French power in America were set at naught in a single day. Added to the losses of power and territory was the serious curtailment of her fishing privileges. Henceforth France was practically excluded from shore fisheries at Nova Scotia, Cape Breton Island, the islands of the Gulf of Saint Lawrence and the southern coast of Newfoundland. The three league limit applied to the Gulf of Saint Lawrence would exclude the subjects of France from its richest shore fisheries, leaving them only the deeper waters of that inland sea.

The dreams of France for supremacy in America were over, her hopes for a monopoly of the fishing industries were shattered. The ravages of war, the loss of territory, and the restraint placed upon her fisheries were too powerful agents for her merchants and fishermen to combat successfully. France never again regained her prestige in American waters, although she had 259 vessels employed at Newfoundland in 1768.

CHAPTER VI

A HARBINGER OF THE REVOLUTION

At the close of the Seven Years' War the colonists of New England found themselves in a position to pursue their great industry unmolested by the attacks of the French and undisturbed with thoughts of another war. With renewed energy and vigor they resumed the fisheries. The dreams of Auchmuty were about to be realized, and the lapse of time in their fulfillment was to be compensated for amply in the greater returns to be derived from a larger field of action. Apparently there was nothing to hinder the people of New England from becoming in a very short period of time the greatest fishermen in the world. War clouds no longer hung over their borders; hostile privateers no longer infested their waters; a friendly sea invited their ships to quieter and wider scenes of activity; unexampled prosperity lay before them, when, like a bolt from a clear sky, the voice of the ministry in England thundered forth its decree that was heard with dread and consternation on the shore and aboard the ship,—*the navigation laws must be enforced!*

By the navigation laws were meant the so-called Molasses Act of 1733 and the proposed measure of the ministry, called the Sugar Act of 1764. The northern English colonies in America from the very first had found their most important and lucrative trade in the exchange of their fish, lumber and agricultural products for the sugar, rum and molasses of the West Indies, and it was mainly by this avenue of trade that money was obtained for the purchase of English manufactured goods. In 1717, France had adopted a more liberal commercial policy which enabled the sugar of the French West Indies to displace the British product in the European trade; further, by passing a law prohibiting the importation of rum into France her colonies were compelled to seek new markets for their molasses in the northern English colonies. It is said that when the Americans first went after their molasses they found it so plentiful and cheap that the planters had been accustomed to throw much of it away.[1] The prosperity of the French planters interfered greatly with the English planters of the sugar islands, who sent numerous protests to England asking for protection. The result of the protests of the planters was the passage of the Molasses Act of 1733.

This act imposed heavy import duties on all rum, molasses and sugar—the product of colonies other than English—imported into any of the American colonies. The duties were so large[2] that they were practically prohibitive. A storm of protest arose from the northern colonies, who feared for the safety of their fisheries and carrying trade if these exchange products were excluded from the country. The provisions of the Act were systematically disregarded by the merchants and remained largely a dead letter on the statute books. The Molasses Act was to continue in force for five years; it was renewed five times, before being made perpetual by the passage of the Sugar Act. During the thirty years that it remained on the statute books inoperative it probably produced little injury to the codfishery of New England.

At the close of the Seven Years' War it was estimated that 300,000 pounds sterling would be needed by Great

[1] Sabine, p. 135.

[2] The duty on molasses was three pence per gallon, on foreign white sugar one pound two shillings per hundred weight; the importation of rum was forbidden under heavy penalties.

Britain to man the forts in America that had been vacated by the French, and to maintain regiments to hold the Indians in check. The ministry proposed to raise one-third this amount—100,000 pounds—by laying duties on certain colonial imports and exports, to modify the provisions of the Molasses Act by reducing the duty on sugar and molasses one-third, and to enforce vigorously the new act. Rumors of the intention of the ministry to enforce the provisions of the Sugar Act, as it was called, had led the Massachusetts assembly early in 1764 to instruct their agent in London that in the execution of the act "the consequences would be ruinous to the trade of the province, hurtful to all the colonies, and greatly prejudicial to the mother country." [1]

The instructions of the Civil Court of Massachusetts to their agent, Mr. Manduit, show what the enforcement of the Molasses Act meant to the people of that colony. The instructions stated that the business of the fishery, which it was alleged would be broken up by this act, was estimated in Massachusetts at 164,000 pounds sterling annually; the vessels employed in it, which would be nearly useless, at 100,000 pounds; the provisions used in it, the casks for packing fish, and other articles, at 22,700 pounds and upwards; to all of which there was to be added the loss of the advantage of sending lumber, horses, provisions, and other commodities to the foreign plantations as cargoes, the vessels employed in carrying fish to Spain and Portugal, the dismissing of five thousand seamen from their employment, the effect of the annihilation of the fishery upon the trade of the colony and of the mother country in general, and its accumulative evils by increasing the rival fisheries of France. "This was forcibly urged as it respected the means of remittance to England for goods imported into the province, which had been made in specie to the amount

[1] Winsor, VI, p. 26.

of 150,000 pounds sterling, besides 90,000 pounds in the treasurer's bills for the reimbursement money, within the last eighteen months. The sources for obtaining this money were through foreign countries by means of the fishery, and would be cut off with the trade to their plantation.''[1]

Keen interest in the proposed measure was felt in a sister colony as well. The month of January had not passed before a strong remonstrance was issued by the people of Rhode Island setting forth the state of the colony and the effect that the enforcement of the navigation laws would have upon their trade. British manufactures to the value of 120,000 pounds were imported annually into Rhode Island for consumption. There were 184 vessels used for foreign voyages, 352 used in colonial carrying trade, which were navigated by 2,200 seamen. The 150 vessels engaged in the trade with the West Indies imported 14,000 hogsheads of molasses, of which not more than 2,500 hogsheads came from the British islands; in fact, the total molasses product of all the British West Indies would not equal more than two-thirds the amount of molasses annually imported into Rhode Island. There were more than thirty distilleries in the colony that had been erected at great expense. ''This distillery is the main hinge upon which the trade of the colony turns and many hundred persons depend immediately upon it for subsistence,'' they stated in their report. The molasses was distilled into rum, which was sent to Africa where it supplanted French brandies. For thirty years past Rhode Island had sent thither every year eighteen vessels loaded with 1,800 hogsheads of rum and provisions to be sold for slaves, gold dust and other articles. By this means the annual remittances from Rhode Island to England amounted to about 40,000 pounds.

The cargoes sent to the French West Indies consisted of

[1] Minot, History of the Province of Massachusetts Bay, II, p. 147.

horses, lumber and fish. The last named were not fit for European markets. If a duty were placed on molasses, the protest continued, it would amount to a prohibition. Distilleries would have to be closed to the ruin of many families, the rum trade of Africa would cease, two-thirds of their vessels would be rendered useless and perish on their hands, the nursery of seamen would be destroyed, and the mechanics who depend upon the merchants would be compelled to seek employment elsewhere. The French West Indies would get their lumber from the Mississippi region, erect their own distilleries, and export rum to the Indians and to Africa. The British West Indies would not consume near the New England produce; therefore, if it could not be sold, a great part of the produce would be lost. Jamaica, the only English island that supplied Rhode Island with molasses, frequently failed to do so for want of sufficient supply. The planters of the British West Indies had no grounds for complaint if the people of New England purchased of others what British merchants could not furnish.[1]

Massachusetts imported, in 1763, fifteen thousand hogsheads of molasses, of which only five hundred hogsheads were from British islands.[2] Of the 29,000 hogsheads of molasses used in Massachusetts and Rhode Island in that year only 3,000 hogsheads, or about eleven per cent, were the produce of the English sugar islands. Thus nearly ninety per cent of the trade with the West Indies, the basis of which was the fisheries, was imperiled to the point of annihilation by the enforcement of the navigation laws. At that time the West India trade in fish was equal to about sixty-four per cent of the total value of the New England fisheries, not including the whale fishery which was little affected by the Act.

[1] R. I. Colonial Records, VI, pp. 378–383.

[2] Bernard, Letters on Trade, pp. 2–7.

The answer of the Council and House of Representatives of Massachusetts to a speech of Governor Bernard in November, 1764, gives a concise but vigorous statement of the effect of increasing the duty on molasses; the paper states, "We shall instance in the fishery and lumber trade only. In respect to the first, it is greatly diminished by means of the duty on foreign molasses. Our pickled fish wholly, and a great part of our cod fish, are fit only for the West Indies market. The British islands cannot take off one-third of the quantity caught; the other two-thirds must be lost, or sent to the foreign plantations, where molasses is given in exchange. The duty on this article will greatly diminish its exportation hither; and being the only article allowed to be given in exchange for our fish, a less quantity of the latter will of course be exported. The obvious effect of which must be the diminution of the fish trade, not only to the West Indies, but to Europe; fish suitable for both these markets being the produce of the same voyage. If, therefore, one of these markets be shut, the other cannot be supplied. The loss of one is the loss of both, as the fishery must fail with the loss of either."[1]

From the foregoing facts can be seen the state of the New England colonies, the condition of the fisheries, the importance that the industry occupied in the life of the time and as an agency for promoting other industries of the northern colonies, as well as the influence that the discussion of this measure had in arousing the colonists to united, and at times, to unlawful acts in resisting proposed legislation of Parliament. It has been the popular idea that the Revolution began with the Stamp Act. The temper of the people had been tried before that time. Patrick Henry's famous declaration[2] in the "Parson's Case" had raised

[1] Mass. State Papers, 1765–1775, p. 19.

[2] Henry stated that the King, by disallowing acts of a salutary nature, forfeited all right to his subjects' obedience.

the question of the supremacy of the colonial assemblies in judging matters of internal taxation, while James Otis, in arguing upon the writs of assistance, had assumed the natural right of the colonists to independence from external control. Other issues arose that provoked discussion at the time, but, says one famous historian, "the promulgation of none of these theories of abstract rights accounts for the general outbreak in 1765. Its most potent influence was the enforcement of the navigation acts in the great commercial centres, and the ruin threatening New England through the breaking up of trade with the French West Indies and the Spanish Main by the modification of the Sugar Act in 1764."[1]

The interest in the measure was not confined to the northern colonies. When the news of the passage of the act reached America "the strongest apprehension arose from the publications of the orders for the strict execution of the Molasses Act, which is said to have caused a greater alarm in the country than the taking of Fort William Henry did in the year 1757."[2] From the pen of John Dickinson, of Pennsylvania, came a strong protest to this and other acts. In his pamphlet Dickinson's object was to show that, as all the profits from the monopoly that would confine American trade to English islands went to British merchants, it was the height of folly for those islands to give up the trade. America had submitted quietly to all extortion because the British trade connection was valuable to them.[3]

When the time came for the Act to be enforced the custom house officers were called to their posts, other officers were appointed, and the strictest orders were given that they should pay close attention to their duty in enforcing

1 Winsor, VI, pp. 24-25.

2 Minot, II, p. 140.

3 Works of John Dickinson, I, Political Writings. (Ford.)

the provisions of the Act in the strongest manner. All commanders and other officers of British ships of war in American waters received authority from the crown to act in the capacity of customs officers. The naval officers were in most cases unfamiliar with the laws that they were called upon to administer; the result of their ignorance, coupled with their zeal to execute the laws against smuggling, led to many serious blunders.[1] Boston, Salem, Gloucester and Portland, especially, were the scenes of collision between the novices of the customs service and the indignant masters of trading vessels. The British naval officers "seized indiscriminately, and confiscated all ships, whether American or foreign," that were engaged in the carrying trade in which they had been unmolested for years. Before the close of the year seizures and confiscations had been made to the extent of over $15,000, one-third of which was appropriated by Governor Bernard, with the effect of heightening the animosities already raised against him.[2]

This determined effort on the part of the ministry and customs officials to exterminate illicit trade apparently stimulated rather than checked its continuance. The native shrewdness of Yankee skippers was usually a match for the bustling zeal of British officials. One scheme that was resorted to for avoiding the payment of revenue was the use of what were called "Anguilla clearances." As usual the cargoes of fish would be taken to the French West Indies and exchanged for molasses. Then the vessel would be taken to the little English island of Anguilla. This island, with an area of thirty-four square miles, was too small to furnish more than a single cargo of molasses yearly,—a fact well known to the collectors of New England. The governor of the island was also collector, and

[1] Grahame, History of the United States, IV, pp. 175–176.
[2] Minot, II, p. 168.

from him the New England captain would purchase a clearance stating that the cargo had been secured at the island. For a considerable time "Anguilla clearances" were winked at by the New England officials until the government, learning of the state of affairs, put an end to the practice.[1]

The effect of the enforcement of the navigation laws was felt distinctly by commercial interests. Dickinson, writing in 1765, complains of the scarcity of money for use in trade, due to this unwise policy of the ministry. Other instances of increased sluggishness in trade arising from lack of money can be noted within the next three or four years, because of the duty imposed on goods imported into America.[2] In spite of all, however, the fisheries were not ruined, neither was the carrying trade of New England seriously affected by the passage of the Sugar Act. The effective enforcement of the act would undoubtedly have brought about an alarming state of affairs. Neither the duties laid in 1764 nor the collection of the taxes anticipated from the Stamp Act of 1765 would have produced a tithe of the evil to the colonies that would have followed its strict enforcement.[3] Such troubles were averted partly by the continuance of smuggling; but mostly, as Hutchinson states, "by connivance or indulgence in the officers," who, when the act was carried into execution, seldom exacted a duty of more than one-half the prescribed amount on molasses.[4] The merchants, regarding this as a fair tax, were usually willing to pay the duty without complaint.

The decade between 1765 and 1775, following the attempt to enforce the Sugar Act, was one of general activity

[1] Sabine, pp. 137–138.
[2] Weeden, II, pp. 759–760.
[3] Winsor, VI, pp. 25–26.
[4] Hutchinson, History of Massachusetts Bay, III, p. 109.

in commercial lines throughout New England. The fisheries in particular felt the beneficial influences of peaceful conditions. While the annals of the industry are brief for the period, they show that at the opening of the Revolution the fisheries of New England were more extensively carried on and of greater economic importance than at any previous period of colonial history, a condition of prosperity that was not equaled again until after the establishment of the Federal Government.

During this decade there were twenty towns in Massachusetts that were engaged in the codfishery. It is estimated that for the period as a whole, there were 665 vessels employed in New England in the fishery, of a total of 25,630 tons, and carrying 4,405 men. The fisheries were carried on most extensively in Massachusetts. In Maine, the number of vessels employed was 60, of a tonnage of 1,000 for all, carrying 230 men; the balance of the fishery belonged to Massachusetts. Marblehead was still the leading town, although Gloucester ranked a close second. The other principal fishing towns were Plymouth, Salem, Chatham, Ipswich and Manchester. The state of the codfishery in New England, from 1765 to 1775, is shown in the table on the following page.[1]

From this table it can be seen that each fishing vessel averaged about forty tons burden and carried a crew of seven men. One-half the number of men were required on shore to cure the fish. It required 350 vessels to carry the fish to market, each vessel carrying eight hands. There were, then, between 9,000 and 10,000 men engaged in the codfishery, or in business relating to it, at the outbreak of the Revolution. The value of the catch, or its extent, cannot be ascertained accurately. Based upon the state of the New England fisheries in 1763, the value of the codfishery in 1775 may be estimated at about $1,300,000, and

[1] Sabine, p. 174.

TOWN.	Vessels Annually Employed.	Tonnage.	No. of Men.
Marblehead	150	7,500	1,200
Gloucester	146	5,530	888
Plymouth	60	2,400	420
Salem	30	1,500	240
Chatham	30	900	240
Ipswich	50	900	190
Manchester	25	1,500	200
Yarmouth	30	900	180
Beverly	15	750	120
Other towns	69	2,750	497
In Maine	60	1,000	230
Total	665	25,630	4,405

of other kinds of small fish enough more to make the total value of the fisheries not far from $1,800,000.

This decade marks, also, the critical period of colonial administration of the northern colonies. Oppressive acts on the part of the mother country were met with protests, more or less vigorous, on the part of colonial legislatures, or by open acts of violence against the English attempts to execute unpopular laws. Acts of Parliament compelling submission and obedience to the mandates of that body were offset, on the part of the colonists, by the formation of associations whose members agreed to use such goods only as were of American growth and manufacture. The English government found it was necessary that radical measures be employed not only for asserting the supreme legislative authority and executive power of Great Britain, but also to compel the Americans to cease boycotting goods of British growth and manufacture. Accordingly, in February, 1775, Lorth North moved for leave to bring in a bill "to restrain the trade and commerce of the province of Massachusetts Bay and New Hampshire, and colonies of Connecticut and Rhode Island and Providence Plantation, in North America to Great Britain, Ireland, and the British islands in the West Indies; and to prohibit such provinces and colonies from carrying on any fishery on the banks of Newfoundland, or other places therein mentioned under certain conditions and limitations."[1]

The bill further provided that after July 1, 1775, no goods, wares, or merchandise, the produce of Massachusetts, New Hampshire, Rhode Island or Connecticut, were to be exported to any other British province except to Great Britain, Ireland, and some of the islands of the West Indies. Vessels sailing with goods were under bonds not to land any part of their cargoes except as required above; all goods going to these four colonies were to be laden and shipped in vessels in Great Britain, except that goods, the growth and produce of the British West Indies, could be imported to the colonies.

Lord North recognized the great importance of the fisheries to the New England people; and he made it plain in his bill that he could conceive no quicker or surer method of securing obedience to British laws than to strike directly at the corner-stone of New England industries. The conditions of the bill that referred to the fisheries were very explicit. All vessels and ships which were the property of persons living in New England were prohibited from carrying on the fisheries upon the banks of Newfoundland, on the coast of Labrador, within the Gulf of Saint Lawrence, or upon the coasts of Cape Breton, or of Nova Scotia, or any other part of the coast of North America. The penalty for disobedience was forfeiture of the vessel, her cargo, apparatus and supplies. British war ships were encouraged to visit, examine and search all vessels suspected of carrying on these fisheries contrary to the provi-

[1] American Archives, 4th series, I, p. 1691.

sions of the bill. The prohibitions and restraints imposed by the act upon trade, commerce, and the fisheries of these colonies were to cease when it was made clear that the trade and commerce of his Majesty's subjects could be carried on without interruption.[1]

Lord North supported his measure by declaring "that, as the Americans had refused to trade with Great Britain, it was but just that they should be deprived of the right to trade with any other nation. In particular, he said that the fisheries upon the Banks of Newfoundland, and the other Banks in America, were their (Great Britain's) undoubted right, and that, therefore, such disposition might be made of them as the government pleased. The two houses, he continued, had declared that a rebellion existed in Massachusetts, and that it was just to deprive that province of its fisheries; that though a government still existed in New Hampshire, the royal authority was weak; that a quantity of powder had been taken out of a fort there by an armed mob; and that, besides, the vicinity of that province of Massachusetts Bay was such, that if it were not included, the purpose of the act would be defeated. Rhode Island, he stated, was not in much better situation than Massachusetts; that several pieces of cannon had been taken and carried into the country, and that the people were arming to aid any colony that should be attacked. With regard to Connecticut, he observed that a large body of her men had marched into Massachusetts on a report that the soldiers had killed some inhabitants of Boston, and that that colony was in a state of great disorder and confusion. To this he added, that the river Connecticut afforded the people of that colony an opportunity of carrying on the fishery, and that the same might be said of Rhode Island. . . . But he was willing, he said, to admit of such alleviations of the measure as would

[1] McDonald, pp. 369–374.

not prove destructive to its great object, and would therefore move it as only temporary, and would permit particular persons to be excepted, on certificates from the governor of their good behaviour, or upon their taking a test of acknowledgment of the rights of Parliament."[1]

The motion of Lord North and his defense of the measure called forth a spirited discussion. The defenders of the bill who came forward to the support of the ministry based their arguments on the fact that rebellion already existed in Massachusetts, on the spirit that continued to prevail in the colonies, on the evident purpose of the colonists to ruin the British merchants and manufacturers and to starve all the West Indies, and because the colonies had prohibited trade with the mother country.

For the opponents, Dunning was of opinion that the Americans had a right to the banks of Newfoundland, that there was no rebellion in Massachusetts, but even if so, he failed to see why New Hampshire, Rhode Island and Connecticut should be punished for acts of another province. He declared that "the ministry were the best authors for a receipt to *make* rebellion."

Governor Johnstone said that the measure was absurd and cruel, that the project to starve a whole people, except such as the governor should think proper to favor, was inhuman; the God of nature had given these fisheries to *New* England and not to *Old* England; his long experience in the British navy had taught him that it was a constant practice of the service for the British cruisers to spare the fishing craft of an enemy's coast, thinking it savage and barbarous to deprive the miserable inhabitants of the seacoast of their daily food and the means of procuring it.

Sir George Saville ridiculed the bill as a measure that would deprive one province of its subsistence because,

[1] Sabine, p. 140.

possibly an ill-defined rebellion lurked in it; of punishing another because it was a near neighbor of that rebellion; a third, because nothing would be accomplished if a third were allowed to escape; and a fourth, because the ministry needed that to square their plan.

Burke made a severe attack on the ministry in his opposition to their bill. He thought that the results of this bill would fall especially hard on British merchants and manufacturers who had property in New England, as these colonies could pay their debts only by means of the fisheries and trade depending upon them; consequently the bill would beggar the British manufacturers and merchants.

When put to a vote, Lord North's motion was agreed to by a ratio of about three to one. When the bill was taken up on the 28th of February, several persons who were acquainted with the fisheries were examined as to the value of the fisheries and the probable effect of their suspension on the people of New England. This evidence is of particular interest as showing the state of the British and New England fisheries during the decade preceding the Revolution.

Probably the most efficient witness was Mr. Brook Watson. He had been called before the bar of the House of Commons in 1765 and 1766 to give information concerning the North American fisheries; since that time he had studied the subject and in 1766 went to America to corroborate his testimony by personal investigation. From his testimony and that of other persons who had been governors of Newfoundland or merchants at that place or in New England it is learned that the fisheries of New England increased greatly between 1766 and 1775, although fish were cheaper in the latter year. There were about 700 vessels engaged in the codfishery, five hundred of which were of a tonnage of from forty to seventy tons, the others from fifteen to forty tons. These vessels carried each a crew of seven men, on an average. One-half that number were needed on shore for curing the fish. Three hundred and fifty vessels were employed to carry the fish to market, of a tonnage ranging from 70 or 80 to 170 or 180 each, carrying eight hands. There were, then, between nine and ten thousand men in New England directly engaged either in the codfishery or in business depending immediately upon such fisheries.

The number of whale ships was 309, of which 122 were from Nantucket. There was an extensive shore fishery carried on at New England, the shad, herring and mackerel being important. Very little of the salt cod and salt mackerel was consumed by the people there; therefore they were not necessary to the support of the people as far as preventing starvation was concerned. About one-fourth the inhabitants of the seaport towns of Massachusetts lived on fresh fish. Shad and alewives were necessary in some parts. The New England fishermen received higher wages than those of Newfoundland; their fish were of a better quality, and usually sold better than the product of the British fishermen.

The effect of the bill, the testimony continues, would be to destroy the deep-sea fisheries of New England. The people of Nantucket would be ruined. The trade with the West Indies would be greatly interfered with, if not wholly stopped. The merchants of New England would be unable to pay the bills that they were owing in Great Britain. But the people would not migrate to Halifax or other parts of Canada for the purpose of carrying on their industry. Neither would the measure necessarily "starve them into submission," as there was agriculture to which many of the people could easily turn their hand for support.

On the other hand it was shown that Newfoundland could not carry on the fisheries as New England was doing. The Newfoundland fisheries were carried on cheapest from

New England. England would lose annually between £200,000 and £300,000 in returns from the codfishery, as all apparatus, fittings and ship chandlery used in New England came from the mother country, being paid for in cod and whale products. In addition, there would be the loss of £1,000,000 that was due the British merchants.[1]

When the bill was finally taken up for consideration on the 6th of March, the discussions were resumed with vigor. Among those to take an active part in the opposition were Fox, Burke, Townsend, Lord John Cavendish and Lord Camden. The last named spoke with great feeling, characterizing the measure as a "bill of pains, penalties, and coercion, not of commercial regulations." And further he stated, "The true character of the bill is violent and hostile. It is a bill of war; it draws the sword, and in its necessary consequences plunges the Empire into civil and unnatural war." Lord North, however, was not without a good following and the bill was passed by the House of Lords on the 21st of March by a considerable majority.

Twenty-one peers who were in the minority entered protest which embodied the principal arguments they had used during the course of the debates. The opening paragraph of their document is well worth repeating for its earnest ness, its eloquence, and the spirit of humanity that it expresses. The passage is as follows: "We dissent because the attempt to coerce, by famine, the whole body of the inhabitants of great and populous provinces, is without example in the history of this or, perhaps, of any civilized nation, and is one of those unhappy inventions to which Parliament is driven by the difficulties which daily multiply upon us from an obstinate adherence to an unwise plan of government. We do not know exactly the extent of the combination against our commerce in New England and the other colonies; but we do know the extent of

[1] American Archives, 4th series, I, pp. 1639–1675, passim.

the punishment we inflict upon it, which is universal, and includes all the inhabitants; among these, many are admitted to be innocent, and several are alleged by ministers to be, in their sense, even meritorious. That government which attempts to preserve its authority by destroying the trade of its subjects, and by involving the innocent and guilty in a common ruin, if it acts from a choice of such means, confesses itself unworthy; if from inability to find any other, admits itself wholly incompetent to the ends of its institution.''

Lord North was not content with the destruction of the American fisheries. His next move was the introduction of a bill for the encouragement of the fisheries of Great Britain and Ireland. The bill granted bounties to vessels engaged in the cod and whale fisheries, and in other ways favored the fishermen of Great Britain. There can be little doubt that the measure was intended to stimulate British merchants and fishermen to supply the domestic and foreign markets with the products of the sea. The colonists retaliated by agreeing to refuse to supply British ships in American waters with any outfits or provisions that would be of use in their fishing.

Within less than a month of the passage of the act to restrain the commerce of the New England colonies and to destroy their fisheries the conflict at Concord Bridge marked the beginning of active hostilities between the two people. Open war, and not the provisions of the bill, brought about the cessation of the fisheries, which remained inactive during the period of the Revolution. It is very clear that the ministry were resolved to ''starve the people into submission'' by destroying their fisheries, which were regarded in Great Britain as the basis of the industrial system of the New England colonies. Had there been a fair chance for the provisions of the bill to be enforced under conditions of peace, it is altogether probable that the effect upon the industry would have been quite as disastrous as that produced by the ravages of war; but, on the other hand, the ulterior aim of the ministry would not have been secured. Not only did the colonists of New England manage to get a living after their industry of the sea was destroyed, but they were largely instrumental in bringing the war for independence to a successful close. Under peaceful conditions, their living could have been secured with greater ease than under the burden of war times.

CHAPTER VII

WAR AND ITS RESULTS

With the opening of the Revolutionary War the colonial era of the fisheries came to a close. For the first time since its beginning in the early part of the seventeenth century, this ancient industry of the sea was wholly suspended. For a decade the annals of the fisheries give place to records of war, to feats of daring on land and sea. What could be accomplished neither by raiding Indians, nor by hostile French rivals, nor by restrictions of Parliament, nor by two centuries of battling with the storms of the ocean, was accomplished in a single season by the war. The doughty schooners fled for refuge to their native harbors; lines, tubs and sounding-lead were laid away in storehouses; cargoes of fish and salt were unloaded upon the wharves; captain and crew threw off their oiled barvels, and the seas were undisturbed by the white sails of fishing craft that had dotted their surface for more than five generations.

But a wonderful transformation was at hand! Almost in a night the change took place, for, on another day, the largest vessels in the fleet were speeding out of harbor once more to scour the seas in search of a new prey. Lines and tubs had given place to cutlasses and swivels; out of sounding-leads bullets had been melted; the hold of the vessel, once filled with salt and fish, furnished commodious quarters for a score or two of fighting seamen; barvels had been exchanged for American uniforms for men who were as eager now to train their guns upon

British men-of-war as they had been but a few weeks before to cast their lines on the Grand Bank of Newfoundland. In this way hundreds of fishing schooners were fitted out as privateers, manned by fishermen in numbers reaching into the thousands. Many fishermen enlisted in the land service, too, and gave good account of themselves at Bunker Hill, Trenton, and other places. But the sea held by far the greater number of enlisted fishermen. This was but natural, as the sea was their chosen home, and men who had been driven from the scene of their daily toil in securing a living for themselves and families found no better place for redressing their wrongs than at the place where the wrong was committed. How readily and successfully the fisherman of New England transformed his vessel into a speedy privateer and himself into an effective fighting seaman speak volumes of praise in favor of his character as a patriotic citizen.

While the burdensome effects of the war fell most severely upon the larger port-towns of the coast, particularly Marblehead, Gloucester and Salem, there was no fishing village of New England, however humble, that did not realize from bitter experience the hard lot that befalls maritime industries in time of war. At the Isles of Shoals, the Whigs were called upon by their enemies for articles of sustenance and for naval recruits; having been commanded to abandon their homes, the majority of the people moved to the mainland, never to return to their native isles. Newburyport, which set up a rival claim to Portsmouth for the honor of sending out the first privateer, despatched a fleet of twenty-two ships from her harbor, manned with over one thousand men, which never again were heard from.[1] War put a stop to the fisheries of Gloucester. Some of her schooners were fitted out as privateers, others rotted at the wharves. Two com-

[1] Goode, Sec. II, p. 682.

panies of Gloucester fishermen shared in the glories of Bunker Hill; on the ocean they were found in larger numbers. It is reported that thirty married men perished in the wreck of a single privateer. When peace was declared it was found that the number of dead and missing seamen from Gloucester was three hundred—one-third the able-bodied men of the town.[1]

In 1774, there were twenty-seven vessels engaged in the codfishery from Chatham. In 1783, four or five vessels only were left in the harbor, but "the town was filled with widows mourning the loss of their husbands and sons."[2] Between April and September of the first year of the war, the town of Salem lost $70 000 in the depreciation of vessel property and apparatus, and in the loss of a season's fishing. The people of Beverly joined with the people of Salem in their zeal in pursuing the war upon the sea, and from the opening to the close of the contest they were extensively engaged in fitting out and manning privateers. In a single season they despatched to sea, to plunder upon British commerce, fifty-two vessels chiefly owned in Salem and Beverly, which mounted about seven hundred and fifty guns and carried crews of nearly four thousand men.[3]

For several years previous to the opening of the war, Marblehead held the honor of being the largest fishing port in the New World in respect to the number of vessels engaged in the business and the extent of their enterprise in this direction. The services of these people are entitled to particular consideration. They had generously proffered their wharves and storehouses to the people of Boston when that port was closed in 1774. One of their merchants, Elbridge Gerry, had framed a measure that

[1] Marvin, Am. Merchant Marine, p. 288.

[2] Mass. Hist. Soc. Coll., 1st series, VIII.

[3] Sabine, p. 200.

was passed by the Provincial Congress of Massachusetts in November, 1775, authorizing captures upon the seas, the first avowal of active hostility against England. The honor of hoisting the first American flag, and of causing the first British ship to strike her colors, are both credited to a citizen of Marblehead. For service in the field, Marblehead furnished one entire regiment, which won for itself great credit and lasting fame on the eve of the battle of Trenton by transporting Washington's army safely across the Delaware, then filled with floating cakes of ice. At the close of the struggle for national independence it was found that the historic fishing town of Marblehead had paid noble sacrifice at the altar of freedom. The nature of the sacrifices are summarized by Sabine, who says:

"To remark, now, that, in 1772, the tonnage of Marblehead was upwards of twelve thousand, and the number of polls was twelve hundred and three; that in 1780 the polls were but five hundred and forty-four; and that the tonnage of the place was only fifteen hundred and nine; to state that nearly every able-bodied citizen was abroad, engaged in the public service, either 'upon land or water'; to show from a document presented to the general court of Massachusetts, that, at the close of the contest, there were within the borders of this single town four hundred and fifty-eight widows, and nine hundred and sixty-six fatherless children—is to sum up its sufferings in the cause of freedom, and to prove that, as has been averred, 'it was a mere wreck and ruin,' when we emerged from the war. No other town in the United States, of the same population and property, lost so large a proportion of both, probably, as Marblehead."

The effects of the War of the Revolution upon the fishing industry cannot be ascertained accurately. It will never be known how many fishermen enlisted at different times in the Continental service, either on land or on sea.

Neither can it be learned how many lost their lives in the war. The pecuniary losses resulting from the decay of idle vessel property, wharves and fishing apparatus must have amounted to several hundred thousand dollars. The only measure for estimating the loss of the products of the annual catch of fish is the value of the catch at the opening of the war multiplied by the number of years that active hostilities continued. Even then the estimate may be low, as the fisheries could not recover immediately upon the declaration of peace. Had the New England fisherman been unmolested in the pursuit of his calling during the period of the war, he would have secured fifteen million dollars worth of fish from the sea.

"The fisheries and the Mississippi" were the two important questions which entered into the peace negotiations of 1783. The statesmen of the Continental Congress expected that England, when negotiating for a peace, would exclude the United States from the fisheries of the banks of Newfoundland and the Gulf of Saint Lawrence. Their grounds for such a belief rested on the exclusive right of fishing which England held as successor to the claim exercised by France previous to the treaty of 1763, and to the policy that she had followed under the terms of that treaty of excluding the French from approaching nearer than three leagues to the shores of the Gulf of Saint Lawrence and fifteen leagues on the coast of Cape Breton Island. Important as the fisheries were to the colonists, the members of the Congress did not make the right to the fisheries an ultimatum in their instructions to John Adams, August 14, 1779. His instructions on that question were as follows:—

"Although it is of the utmost importance to the peace and commerce of the United States that Canada and Nova Scotia should be ceded, and more particularly that their equal common right to the fisheries should be guaranteed to them, yet a desire of terminating the war has induced us not to make the acquisition of these objects an ultimatum on the present occasion."

But in the instructions given to Mr. Adams the same day for negotiating a treaty of commerce with Great Britain it was expressly stated that he was to consent to no treaty of commerce unless the United States should be unmolested and undisturbed in taking fish on the banks of Newfoundland and other fisheries in the American seas, excepting within three leagues of the shores of the seas, or of the coast of the Gulf of Saint Lawrence, unless a nearer distance could be obtained through negotiation.[1]

Adams, who at first was appointed sole commissioner to negotiate with Great Britain, was joined subsequently by Franklin, Jay and Laurens as associate commissioners. The principal labor of securing treaty rights concerning the fisheries rested with Adams. This was but natural, perhaps, as he was from Massachusetts, and was better acquainted with their importance than were his associates. He was deeply concerned that the fisheries should be recognized in the treaty of peace, and in the course of the negotiations he proposed to the conference that an article be inserted in the treaty setting forth the right of the Americans to enjoy the fisheries.

The paper was subjected to critical examination, in the course of which Strachey, one of the British commissioners, proposed that the word "right," in its connection with the entire fishery, should be changed into "liberty." Fitzherbert, another British commissioner, said that the word "right" was an obnoxious expression. Upon this, Adams arose, and, with the concentrated power that he possessed when excited, said, "Gentlemen, is there or can there be a clearer right? In former treaties—that of Utrecht and

[1] Snow, American Diplomacy, p. 56.

of Paris—France and England have claimed the right and used the word. When God Almighty made the banks of Newfoundland, at three hundred leagues distant from the people of America, and at six hundred leagues distant from those of France and England, did He not give us as good a right as to the latter? If Heaven, at creation, gave a right, it is ours at least as much as yours. If occupation, use and possession give a right, we have it as clearly as you. If war and blood and treasure give a right, ours is as good as yours. We have been continuously fighting in Canada, Cape Breton, and Nova Scotia for the defense of this fishery, and have expended beyond all proportion more than you. If then, the right cannot be denied, why should it not be acknowledged and put out of dispute? Why should we leave room for illiterate fishermen to wrangle and chicane?"[1]

The argument presented by Adams was unanswerable. Eventually the whole of his proposal was embodied as an article in the treaty of peace that was signed September, 3, 1783. Article III of that treaty follows:—

"It is agreed that the people of the United States shall continue to enjoy unmolested the right to take fish of every kind on the Grand Bank, and on the other banks of Newfoundland; also in the Gulph of Saint Lawrence, and at all other places in the sea where the inhabitants of both countries used at any time heretofore to fish. And also that the inhabitants of the United States shall have liberty to take fish of every kind on such part of the coast of Newfoundland as British fishermen shall use (but not to dry or cure the same on that island) and also on the coasts, bays and creeks of all other of His Britannic Majesty's dominions in America; and that the American fishermen shall have liberty to dry and cure fish in any of the unsettled bays, harbours and creeks of Nova Scotia, Magdalen Islands, and Labrador, so long as the same shall remain unsettled; but so

[1] Works of J. Adams, III, p. 334.

soon as the same or either of them shall be settled, it shall not be lawful for the said fishermen to dry or cure fish at such settlements, without a previous agreement for that purpose with the inhabitants, proprietors or possessors of the ground."[1]

The rights thus secured for the American fishermen were generous beyond expectation. It was little in the nature of a concession for Great Britain to grant the "right" of fishing on the different banks to the inhabitants of the United States. The possession of those rights can never be monopolized, any more than the navigation of the ocean can be. But the granting of the liberty to take fish on all the coasts of the British provinces in America equally with British fishermen was more than our statesmen had hoped to secure; with this liberty of taking all kinds of fish went the privilege of curing the same in certain parts of the different provinces—another concession that the American statesmen had little grounds for demanding. When it is considered how the French were excluded from the privileges of the shore fisheries of these colonies by the treaty of 1763, it is a matter of considerable wonder that the Americans were granted, by the treaty of 1783, privileges as an independent government almost equal to what they had enjoyed as colonies of Great Britain. Great credit should be given to our negotiators, and to Adams in particular, for the granting of such satisfactory terms by the British government. The dictatorial attitude that Adams assumed during part of the negotiations, his evident transgression of his instructions as an accredited commissioner to make terms for a treaty of peace, and his bold stand for "the fisheries or no peace," undoubtedly were the means for securing the insertion of the third article in the treaty of 1783.

[1] Snow, p. 65.

CHAPTER VIII

THE RENAISSANCE OF THE FISHERIES

The history of the fisheries subsequent to the Revolutionary War falls naturally into three divisions: the first embraces the period of our Federal Government down to the treaty of 1818; the second extends from 1818 to the close of the Civil War; and the last reaches from that conflict to the present time. The first period is characterized by a war with Great Britain which destroyed the fisheries during the continuance of active hostilities, by a steady increase in the extent and importance of the industry during the years of peace, but by no radical changes either in methods of catching fish or in establishing new industries. The second period marks the expansion of the deep-sea fisheries, the rise and development of the mackerel fishery, and important beginnings in the oyster, herring and menhaden industries. The last period witnesses the introduction into general use of revolutionary methods for taking ground-fish, mackerel and menhaden, the origin and growth of the canning industry, the evolution of processes of fish-culture chiefly by the establishment of a national fish commission, and the introduction of new methods for preserving and transporting the products of the sea.

The recovery of the fisheries after peace had been declared was slow. In the earlier years there had been rapid recoveries after colonial wars in which there had been more or less interference with the progress of the business. Such a recovery, however, was not possible after the Revolutionary War. The fishing industry had been shaken to its foundation by a decade of inactivity and suspension. There had been rapid and disastrous depreciation of the property used for the furtherance of fishing interests. Wharves had fallen into decay, mainly through lack of trade to keep them in repair. Many vessels, too, had become valueless for the same reason; others had been employed in the privateering service, never to return as fishing vessels. Flakes and other shore apparatus used in curing fish had long since disappeared. Men, too, had lost the habit of their old vocation in following varying fortunes of service in the army or navy. The younger generation of boys had received little training in the shore fisheries such as their fathers had, and none of them had acquired practical experience in deep-sea fishing by a trip to the Grand Bank as "cut-tail" aboard a New England schooner.

Added to the discouraging domestic conditions was an order, proclaimed by the British council the very year of the treaty, prohibiting American fish from being carried to the British West Indies. The order worked injury to our fishermen in two ways; first, in distinctly narrowing the markets for the larger part of our fish; secondly, in encouraging the rival fisheries of Nova Scotia and other British possessions in the vicinity of the Gulf of Saint Lawrence. The Congress of the Confederation declared that retaliatory measures were necessary to prevent our commerce from falling into the hands of foreigners, and asked the States for power to provide suitable remedies. But no such power was granted to the Congress.[1]

The condition of the codfishery of Massachusetts for the

[1] Sabine, p. 155.

period between 1786 and 1790 is shown in the table below.[1] During that time there were 539 vessels employed in the fishery, of a total tonnage of 19,185 tons, and carrying 3,292 men. The average size of the vessels was 35 tons, and of the crews six men. Marblehead, with over one-fourth of the total tonnage employed, still held the position of the first in the number of vessels employed. The other principal fishing towns were Plymouth, Salem, and Beverly, the three together doing about one-fifth the fishing business of New England at this time.

The exports of the products of the New England cod-fishery for the five years 1786 to 1790 inclusive show that there were annually sold in European markets 108,600 quintals of fish valued at $325,800, and in the markets of the West Indies 142,050 quintals valued at $284,100.[2] Allowing, on a liberal estimate, that one-half the value of the sales went as wages to the fishermen, the average amount

[1] The state of the codfishery of Massachusetts between 1786 and 1790:

Town.	Vessels An. Employed.	Tonnage.	No. of Men.
Marblehead	90	5,400	720
Gloucester	160	3,600	680
Plymouth	36	1,440	252
Salem	20	1,300	160
Beverly	19	1,235	157
Chatham	30	900	240
Yarmouth	30	900	180
Manchester	15	900	120
Ipswich	56	860	248
Other towns	53	2,350	415
Maine	30	300	120
Total	539	19,185	3,292

[2] Pitkin, Statistical View, p. 84.

received per year for five consecutive years would be $93 a man,—and there must have been many others on shore employed in curing the fish whose wages would not be much higher, if as high. In 1789, the sale of our fish in foreign markets was very low. Both England and France made efforts to induce our fishermen to move to their ports, fearing that American competition under conditions of peace would interfere with their fisheries. In addition to excluding the fish of other nations from their ports both England and France at that time gave bounties to their fishermen. For a number of years France had paid a bounty to her fishermen in the codfishery at a rate per quintal larger than the average price at which American codfish had been sold.[1] Under circumstances such as these there is little wonder why the codfishery of New England shows a slow development after the War of the Revolution.

The state of the fishery and the wretched condition of the people who secured their living from it appealed to the sympathies of patriotic Americans. Men in New England who knew the state of affairs felt that the need of assistance for the relief of fishermen and the encouragement of their ancient industry was urgent. In other parts of the country the need was clearly recognized and maintained. Charles C. Pinckney of South Carolina declared in the convention of his state that met to ratify the Constitution that "the eastern states had lost everything but their country and freedom; . . . it was notorious that some ports at the eastward which used to fit out one hundred and fifty sail of vessels do not now fit out thirty; that their trade of shipbuilding, which used to be considerable, was now annihilated; that their fisheries were trifling, and their mariners in want of bread;" and that the people of the

1 Hunt's Merchant's Magazine, V, p. 22.

South were called upon by "every tie of justice, friendship and humanity, to relieve their distress." [1]

The first measure of relief for the fisheries came, July 4, 1789, through the second act passed by Congress after the establishment of the Federal Government. The subject of the fisheries came up in the course of the debate on the bill to levy duty on imports. The deplorable condition of the industry and the necessity of immediate assistance were set forth at length by the representatives from Massachusetts: Ames, Gerry and Goodhue. Fisher Ames took a leading part in the discussions, during which he said: "We exchange for molasses those fish that it is impossible to dispose of anywhere else; we have no market within our reach but the islands from whence we get molasses in return, which again we manufacture into rum. It is scarcely possible to maintain our fisheries with advantage, if the commerce for summer fish is injured, which I conceive it would be very materially, if a high duty is imposed upon this article; nay, it would carry devastation throughout all the New England states; it would ultimately affect all throughout the Union. When gentlemen contemplate the fishery, they admit its importance, and the necessity we are under of encouraging and protecting it, especially if they consider its declining situation; that it is excluded from these advantages which it formerly obtained in British ports, and participates but in a small degree of the benefits arising from our European allies, whose markets are visited under severe restrictions; yet, with all these discouragements it maintains an extent which entitles it to the fostering care of government. . . . In short, unless some extraordinary measures are taken to support our fisheries, I do not see what is to prevent their inevitable ruin. It is a fact that near one-third of our fishermen are taken

[1] Sabine, pp. 155–156.

from their profession—not for want of skill and abilities in the art, for here they take the rank of every nation on earth—but from the local, chilling policy of foreign nations, who shut us out from the avenues to market. If, instead of protection from government, we extend to them oppression, I shudder for the consequences. . . . I contend they are poor; they are in a sinking state; they carry on their business in despair. But gentlemen will ask us, 'Why, then, do not they quit the profession?' I answer, in the words that are often used in the eastern country respecting the inhabitants of Cape Cod—they are too poor to live there, and are too poor to remove.''[1]

The act that was passed in 1789, in lieu of a drawback of the duties imposed on the importation of salt employed and expended in the curing of fish, allowed a bounty of five cents on every quintal of dried fish, and the same sum on every barrel of pickled fish, the product of the fisheries of the United States, exported to any country.[2] This act inaugurated the system of drawbacks or bounties which continued, with a single interruption, down to the reciprocity treaty of 1854. In August, 1790, the allowances were increased to ten cents a barrel on pickled fish, and the same increase on dry fish per quintal. These acts continued in force, with modifications contained in the acts of the 18th of February and 8th of July, 1792, second of March, 1799, 12th of April, 1800, and were finally repealed by the abolition of the salt duty, March 3, 1807. There were no bounties or allowances on vessels engaged in fishing from 1807 to July 13, 1813, when allowances to fishing vessels were restored; these rates were increased by the act of March 3, 1819, and were in force for thirty-five years thereafter.[3]

However, the assistance that was given by the Congress

1 Annals of Congress, I, pp. 291–335, passim.

2 Tariff Acts, p. 11.

3 Andrews' Report on Lake Trade, p. 633.

of 1789 was inadequate for the purpose for which it was designed, that is, relief to the men engaged in the fishery. Since the average price of fish per quintal was about two and one-half dollars, the additional allowance of five cents was only about one-fiftieth the value of the fish. To be sure it did afford extra income,—perhaps a total of $12,000 for the year 1790; but that meant assistance to the fisherman to the extent of less than three dollars per man. Further demands for assistance came in April, 1790, when a representation of the government of the Commonwealth of Massachusetts on the subject of the cod and whale fisheries was laid before Congress. That body promptly referred the matter to the Secretary of State, Thomas Jefferson. After a ten months' consideration of the subject the secretary made a report, February 4, 1791, in which he enumerated the advantages and disadvantages of the American fisheries. Mr. Jefferson summed up the advantages thus:

"1. The neighborhood of the great fisheries; which permits our fishermen to bring home their fish to be salted by their wives and children.

"2. The shore fisheries so near at hand as to enable the vessels to run into port in storm, and to lessen the risk, for which distant nations must pay insurance.

"3. The winter fisheries, which, like household manufactures, employ portions of time which would otherwise be useless.

"4. The smallness of the vessels, which with the shortness of the voyage, enables us to employ, and which, consequently requires but a small capital.

"5. The cheapness of our vessels, which do not cost above the half of the Baltic fir vessels, computing price and duration.

"6. Their excellence as sea boats, which decreases the risk and quickens the returns.

"7. The superiority of our mariners in skill, activity, enterprise, sobriety, and order.

"8. The cheapness of provisions.

"9. The cheapness of casks, which of itself, was said to be equal to an extra profit of fifteen per cent."

Of the disadvantages encountered by the American fishermen those that depended on the Americans themselves were:

"Tonnage and naval duties on the vessels employed in the fishery.

"Import duties on salt.

"Import duties on tea, rum, sugar, molasses, hooks, lines, and leads, duck, cordage, and cables, hemp, and twine, used in the fishery; coarse woollens, worn by the fishermen, and the poll tax levied by the State on their persons. . . .

"To these disadvantages, add ineffectual duties on the importation of foreign fish."[1]

The additional relief that was desired came through an act passed February 9th, 1792, "for the more immediate encouragement of the said fisheries." The bounty on dried fish exported was abolished; and, in place thereof, a specific allowance was to be paid annually to vessels engaged in the codfishery, graduated according to the tonnage of the vessel. Vessels of a tonnage below twenty tons received one dollar per ton; those between twenty and thirty tons, one dollar and fifty cents per ton; if above thirty tons, two dollars and fifty cents per ton. In order to receive the allowance vessels must have been engaged in the codfishery at least four months of the year. Three-eighths of the allowance was to accrue to the owner of the vessel, and the other five-eighths to be divided among the vessel's crew proportionally to the individual catch of fish. But the bounty allowed on any one vessel for a season was not

[1] Jefferson's Works, VII, pp. 542–544.

to exceed one hundred and seventy dollars.[1] These rates of allowance were further increased one-fifth by the passage of an act May 2nd of the same year which provided that "from and after the first day of January next, there shall be an addition of twenty per centum to the allowances, respectively granted to ships or vessels employed in the bank or other codfisheries," and in the terms of the above act.[2]

An act of a different nature was passed in 1793 for the further encouragement of the fisheries. Collectors of customs were empowered to grant vessels duly licensed permits "to touch and trade at any foreign port or place," to procure salt and other necessary outfits, without being subject to the payment of duties. This act continued to be of great convenience to the fishermen for many decades although its privileges were admittedly liable to abuse.

In 1797, the allowance on pickled fish exported was changed to twelve cents a barrel, and the yearly allowance to vessels engaged in the codfishery was increased one-third over the former rates; by this act, the maximum allowance was two hundred and seventy-two dollars for a single vessel.[3] These allowances were to continue as long as duties on salt were paid. The bounty act was repealed by the abolition of the salt duty, March 3, 1807.[4] From 1807 to an act of July 29th, 1813, there were no bounties or allowances to fishing vessels.[5] By this last act, a duty of twenty cents a bushel was placed on foreign salt imported; after December 31, 1814, the annual allowances (four dollars per ton for vessels above thirty tons) according to tonnage were restored to vessels engaged in the bank and other cod-

[1] Annals of Congress, 1791–1793, pp. 362–363.
[2] Tariff Acts, pp. 30–31.
[3] Sabine, p. 160.
[4] Benton, Thirty Years' View, II, pp. 194–198.
[5] Andrews' Report, pp. 632–633.

fishery, and a bounty of twenty cents per barrel was granted on pickled fish exported, if it had been cured with foreign salt. These regulations were to continue one year after peace with Great Britain. By a subsequent act they were continued in force without limitation.[1] After September 30, 1817, the above allowances were granted only when the officers and three-fourths of the crews of the vessels were proved to be American citizens.[2]

The duty on salt, which formed the basis for the above-mentioned bounties, was in the several tariffs down to 1818, as follows: Act of August 10, 1790, twelve cents per bushel; Act of July 8, 1797, twenty cents per bushel; Act of July 1, 1812, forty cents per bushel; Acts of July 29, 1813 and April 27, 1816, twenty cents per bushel.[3] The quantity of salt imported between the years 1791 and 1818 inclusive was 73,928,614 bushels, and the amount of duty accruing on this was $12,928,528.[4]

A glance at the statistics of tonnage engaged in the cod-fishery during this period shows that there was a gradual increase from the 19,185 tons of 1789 to 50,163 tons in 1793. The next year the tonnage fell to 28,671, only to rise again until it reached 42,746 tons in 1798. For the last two years of that century the tonnage kept between 29,000 and 30,000; but from 1800 to 1807 there was a gradual increase annually until the high-water mark for this period was recorded as 69,306 tons, in 1807. The passage of the Embargo Act in 1807 proved disastrous for the fishing industry, as it did in other maritime pursuits. Within two years the tonnage fell off to 34,486 tons, a loss of over fifty per cent for those years. The fisheries had just begun to recover from the effects of the Embargo Act

1 Seybert, U. S. Statistical Annals, p. 338.
2 Ibid, p. 389.
3 Tariff Acts of the U. S.
4 Niles' Reg. (1819), p. 53. Sabine, p. 176.

when they were subject again to a more serious decline occasioned by the opening of the second war with Great Britain. By 1814 the tonnage had fallen to the lowest point recorded since the establishment of the Federal Government, 17,855 tons only being engaged in the codfishery. The recovery of the industry after the return of peace in 1815 was very rapid, the growth being so great that in the year 1818 there was a tonnage of 69,107. The tonnage thus employed in the fisheries included that engaged in the mackerel and codfishery, although the tonnage is given under the latter industry.[1] The average yearly allowance paid fishing vessels during this period of thirty years was $83,075, and the total amount of allowance paid fishing vessels from 1793 to 1818 inclusive was $2,166,894.33.[2]

For the eight years ending with 1798 the average annual amount of dried fish exported was 394,198 quintals; for the ten years ending 1808 the average exports were 438,453 quintals; and for the decade ending with 1818 the annual average of exports was 200,437 quintals. The total exports of dried fish from 1791 to 1818 were 9,532,466 quintals. The total value of dried fish thus exported between 1803 and 1818 inclusive was $17,983,000, and the value per quintal during these years was a little above five dollars. At the same rate per quintal, the total value of exported dried fish from 1791 to 1818 inclusive exceeded $49,000,000.

The statistics of pickled fish, which consisted in part of

[1] The average tonnage employed in these fisheries for the decade 1789–1798 was 34,024; for 1799–1808, it was 48,208; for 1809–1818, it was 40,071. The average for the period was 40,768 tons. The average yearly allowance paid on vessels engaged in the codfishery for the six years, 1793–1798, was $80,840; for the decade, 1799 to 1808, the average was $127,787; and for the six years of the decade ending with 1818 (there were no allowances paid in 1811, 1812, 1813, and 1814), the average was $67,659, or $40,595 for the decade.

[2] Niles' Reg., X, pp. 226–227. Sabine, p. 176.

mackerel, show that there was exported from 1791 to 1800 inclusive an average of 57,916 barrels annually; during the following decade 61,538 barrels were exported annually; and for the eight years ending with 1818 the exports were 33,078 barrels annually. In addition, there were exported from 1796 to 1818 inclusive 191,158 kegs of pickled fish. The value of pickled fish exported in barrels and kegs for the sixteen years ending with 1818 was $4,473,000. Reckoned on the same basis and in the same ratio of values, the value of pickled fish exported previous to 1803 exceeded $4,000,000. The total value of pickled fish exported during the period of these thirty years, as far as records show, was between $8,000,000 and $9,000,000. The total value of all kinds of fish, dried and pickled, exported between 1789 and 1818 inclusive was, thus, in excess of $57,000,000.

The bounty on pickled fish and salted provisions exported, from the commencement of the government to September 30, 1819, amounted to $486,930.73;[1] and the allowances to vessels employed in the fisheries from their inception in 1792 to the close of the year 1818 were $2,166,894.33. The amount of the duty accruing on salt imported from 1789 to 1818 inclusive was $12,928,528. Thus the excess of the salt duty over the amounts paid to the fishing industry in allowances and bounties for the period of thirty years was more than $10,000,000.

The direction which exports of fish took subsequent to the Revolution was along the same trade routes as had been established for many years previous to the war. The West Indies took the larger part of the exports, perhaps sixty per cent on an average, but the fish exported to Europe found better markets and better prices. The total exports of dried and smoked fish for the year 1800 were 392,726 quintals. Sixty-two per cent of this amount, or 244,353

1 Niles' Reg. (1819), p. 53.

quintals, was sent to markets in the West Indies, and 144,-493 quintals were sent to continental Europe. Spain took seventy-six per cent of the latter amount. France took none.

In 1807, the total exports were 473,924 quintals. Fifty-six per cent of this, or 268,332 quintals, went to the West Indies; 192,981 quintals were sent to Europe. France and Spain took eighty-eight per cent of this amount, the exports to France exceeding those to Spain by less than 4,000 quintals. In 1816, the total exports, now on the increase again after the depressions caused by the Embargo Act of 1807 and the second war with Great Britain, were 219,991 quintals, the West Indies again taking fifty-six per cent. France consumed forty-five per cent of the 89,192 quintals sent to Europe. The average price per quintal for exported fish during these years ranged from $3.25 to $4.80. The pickled fish that were exported were usually sent to the West Indies for the negroes.

An account of the fisheries of Marblehead during the early part of the century gives an excellent description of the course of exports at that time. "The fisheries of Marblehead were most profitable during the first six or eight years of the century. An immense quantity of codfish was then exported to France, Spain, and the West Indies, if not to other countries, and those exportations were made mostly in the winter in vessels that had been fishing in the summer, some going direct from the banks to foreign ports. To Spain the fish were carried as often wet as dry, but that sent to the West Indies was always well dried and packed in casks or 'drums.' As a general thing, no return cargoes were brought from Spain. The fish sold at Bilboa and other Spanish ports were paid for in doubloons, and our vessels would often proceed to the Cape de Verde Islands and there purchase cargoes of salt. From France we receive our pay in silks, wines, olive oil, and other articles, all of which found a ready sale. Sugar, molasses, coffee, rum, pine apples, oranges, lemons, and other tropical products were brought from the West Indies, and disposed of without delay. There were times in the Spanish and French harbors when fish commanded an extremely high price. I was in conversation not long since with an old fisherman, who informed me that he once went to Bilboa as a mate of a fish laden schooner, and that the cargo was sold at the rate of twenty dollars a quintal. 'We got,' he said, 'about one dollar for every fish we carried out.' He added that he had known the article to bring a still higher price, but this was soon after the termination of the war of 1812."[1]

The mackerel fishery of New England did not occupy a position of economic importance previous to the year 1815. The total number of barrels of pickled mackerel inspected in Maine from 1804 to 1818 inclusive was 6,553, and in Massachusetts for the same period 231,085 barrels, the total amount for New England being 237,638 barrels. Nearly sixty per cent of this quantity was caught during the last four years of the period. In Maine, the business was carried on principally from Portland and Eastport. In Massachusetts, it centered in Boston; other principal towns engaged in the industry were Hingham, Cohasset, Scituate and Newburyport.[2]

The average price of fresh mackerel in Boston markets from 1804 to 1822 was six to eight cents apiece, sometimes ten cents; they were always sold by the count. The pickled mackerel were exported to the West Indies principally, but the quantity exported could not have been great, as the total catch of pickled mackerel was only thirty-seven per cent of the total exports of pickled fish. A beginning of the southern mackerel fishery was made in 1817 by the

[1] Goode, Sec. II, p. 706, quoting Marblehead Ledger of 1860.

[2] Report, U. S. Fish Commission, 1881, pp. 280, 292.

schooner *Defiance*, of Rockport, Massachusetts, which went as far south as Cape May and took sixty barrels of mackerel; but this fishery did not become general until many years after.[1]

There now remains a consideration of the growth and expansion of the fishery industry in the towns and hamlets along the New England sea-coast during this period. Local records of the importance of the industry for a period of years are exceedingly rare; it is only here and there that materials for history may be found, barely enough to weave into a connected narrative, yet in the light of subsequent history the tendency of the times becomes evident. That tendency was for fishermen to cling to their lines and vessels, to leave to others the development of lands beyond the Alleghenies, and to extend their interests farther and farther along the eastern coast of Maine or into the deep recesses of the shores of the Gulf of Saint Lawrence and the unfrequented stretches of the Labrador coast.

The system of granting allowances to vessels engaged in the codfisheries, inaugurated by the government in 1791, acted as a stimulus to the industry. Evident proof of this fact can be seen from a comparison of the tonnage engaged annually in this fishery from the passage of the first act until the discontinuance of allowances in 1807. The expansion of the industry was not confined to a few of the leading fishing ports of New England, it was widespread throughout the coast towns of the country from New Bedford to Eastport. All along the coast of Maine new settlements were being made—mere fishing hamlets in the earlier part of their history. The easternmost part of the state was settled in 1780, when a trading post was established at Moose Island in Passamaquoddy Bay, later incorporated (1798) as the town of Eastport. From its first settlement, this place became an important post for English and Ameri-

[1] Ibid, pp. 308, 310.

can traders, including as it did many of the nearby villages in what is now called the Passamaquoddy district. Lubec, settled at the same time, remained a part of Eastport until 1811.

The fisheries of Maine, carried on with some degree of success before the settlements at Jamestown and Plymouth, were our country's first enterprise. During the colonial period Maine never became a permanent base for the pursuit of the fisheries, neither did it establish its real importance in the industry. There were famous fishing stations, as at Pemaquid and Monhegan, but these served more as rendezvous for outside fleets than as centers for a fish trade that was the product of the natives of the coast. The fisheries of the state, instead of being developed locally and in turn helping in the establishment of other forms of industry in the region, were for too long a period the plunder-ground for fishermen of other places who came only to carry off the treasure of the sea and left the fishing coast, like a plundered province, poorer because of its natural wealth. With the establishment of settlements along the whole water front of the state the fisheries were developed locally more and more until, by the opening of the nineteenth century, the people began to realize that their commercial and political independence from Massachusetts lay in securing for themselves the natural wealth of the adjacent waters.

The settlement of Eastport formed the last link in a chain of fishing hamlets that stretched from the Piscataqua to the Saint Croix. Fishing and lumbering, the two industries that made possible the settlement and civilization of Maine, were carried on hand in hand from the first. When the timber line retreated from the shore, the coastwise farmers joined hands with the fishing interests so intimately that for nearly a century it was difficult to tell, in the case of hundreds of the inhabitants of the Maine coast, whether they were farmers or fishermen. This attenuated line of fishing hamlets, even in the early part of the nineteenth century, had begun to have economic importance. It is safe to say that in the half-century between 1830 and 1880 there was not a village bordering on the sea from Kittery to Calais which did not have important fishing interests.

The condition of the fisheries during the period now under consideration shows that for several years before Maine was admitted as a state the industry was pursued generally by the coast towns. At Machias in 1793 between seventy and eighty tons only were employed in the fisheries, and not above five hundred quintals were exported.[1] About 1810, persons went to Gouldsboro to engage in the whale fishery which was carried on in small boats from the shore whenever whales were to be seen.[2] Deer Isle sent a vessel to engage in the codfisheries in the Gulf of Saint Lawrence as early as 1815; but in 1818 there were only two vessels from that place with a tonnage greater than forty tons burden, and twelve or fifteen Chebacco boats from the island. Many of the residents were employed on fishing vessels from Newburyport. Vinal Haven had about the same number of Chebacco boats in 1817, ranging from fifteen to thirty tons, the smaller ones fishing along the shore while the larger ones went to Seal Island grounds and to Brown's Bank. As early as 1800 the island of Matinicus, always famous for its fisheries, was engaged in the herring fishery in the Bay of Fundy. At the same time Booth Bay sent a fleet of small vessels to the Bay of Fundy and to Sable Island for cod. Shore fisheries also were carried on. In 1817, large schooners were built to engage in the Labrador trade, the *Ruby*, a schooner from North Booth Bay being the first one sent out. Enough is

[1] Mass. Hist. Soc. Coll., VIII, p. 146.
[2] Goode, Sec. ii, p. 30.

gleaned from these records to show that the recovery of the fishing interests after the Revolution was very slow; however, the industry was generally carried on along the whole water front at the opening of the next decade. The war of 1812 was disastrous to the industry but the recovery from this war was rapid. By 1818, ships of large tonnage were being built to engage in fisheries in more distant regions.

The fishery at Piscataqua and its neighborhood, according to Belknap, for the year 1791, not including the fisheries of the Isles of Shoals, employed in the cod and scale fishery twenty-seven schooners and twenty boats, aggregating six hundred and thirty tons, and carrying two hundred and fifty men. The products of the New Hampshire fisheries for the year 1791, including those of the Isles of Shoals, were 5,170 quintals of merchantable fish, 14,217 quintals of Jamaica fish, and 6,463 quintals of scale fish; making the total 25,850 quintals.[1] It was estimated that the total number of seamen belonging to New Hampshire for the year 1791 was 500 employed in foreign trade, fifty in the coast trade, and 250 in the fisheries. Some of the seamen who in summer were employed in the fisheries were, in winter, engaged in the coast and foreign trade.

While there was a general revival of the fisheries along the whole water front of Maine and in New Hampshire after the establishment of the Federal Government, the greatest development was to be found in Massachusetts—the mother-state of this industry. The deep-sea industries of Massachusetts since the Revolution have continuously outrivaled those of all other states to the present time, as they did for a century and a half previous to that war. From Newburyport to New Bedford the revival of this industry went on, in places recovering slowly and

[1] Belknap, History of New Hampshire, III, pp. 157–160.

with great difficulty from the effects of the war; in others developing beyond previous volumes of business and forming the nucleus of a new industry for the majority of the inhabitants of some towns.

The first vessel sent out from Newburyport to engage in the Labrador codfishery was in 1794, and for eighty-five years scarcely a fishing season was allowed to pass without one or more vessels being sent. In 1806, this fleet numbered 45 sail; in 1817, it numbered 65, including sixty schooners, one brig, and four sloops.[1] A mercantile Company was formed in 1816–17 for prosecuting the bank fishery, and paid good dividends the first year of its establishment.

Essex had forty vessels engaged in the bank and shore fisheries in 1814, chiefly in the latter. The best account of the condition of the fisheries of Gloucester immediately after the Revolution is given by Babson, who says: "No means exist for ascertaining how many vessels engaged in the bank fishery immediately upon the return of peace. One statement says that sixty were employed in 1788 and fifty in 1789. Another, in giving an account of fish caught by vessels from the town in the fall of the last-named year, shows that forty-four vessels took 426,700 fish. . . . Concerning the revival of the fishery, it may be further stated that the average earnings of the fishermen were so small that they were kept in a condition of poverty. It is not surprising, therefore, that the number of vessels engaged in it decreased from year to year till 1804, when we find that only eight of more than thirty tons burden were engaged in the Gloucester fisheries. This small number had probably dwindled to less in 1819, when an effort was made to put new vigor into the business by the establishment of a corporation to carry it on. . . .

"The shore fishery of Gloucester had risen to some impor-

[1] Goode, II, p. 135.

tance before the Revolution, and upon the return of peace the enterprise of the people was again directed to this pursuit, to which some encouragement was given by early acts of the General Government. In 1792, one hundred and thirty-three Chebacco boats, measuring in the aggregate 1,549 tons, were engaged in it. These boats resorted to the ledges and shoal grounds near the coast, where they found, at different seasons, cod, hake, and pollock, and pursued their fishery with such success that in twelve years from the last-named date the number of boats engaged in it had increased to about two hundred, while the tonnage had nearly doubled. At this time the boat fishing was chiefly carried on at Sandy Bay and other coves on the outside of the cape; but the advantage of a good harbor for their large boats drew a few people from these localities to settle on Eastern Point soon after 1800. The business, however, was not profitable enough even with additional encouragement from the General Government, to attract many new adventurers, or even to stimulate much of the enterprise of the old ones, and it had a slow growth for the quarter of a century, the annual average increase of tonnage during that time having been only about one hundred and twenty-five tons."[1]

A brief statement of the condition of the fisheries at Salem shows that between 1786 and 1799 the annual average of bank fishing vessels was only twenty, of a tonnage of 1,300 tons burden, and employing one hundred and sixty men.[2] No good account exists of the Marblehead fisheries previous to 1815, although they are reported as being at their best during the few years preceding the Embargo Act of 1807, when large cargoes of fish were exported annually to European and other markets. About fifty schooners went to the banks

1 Babson, pp. 383–385.

2 Goode, Sec. II, p. 701.

in 1815, and did well. Sixteen schooners were added to the fleet during the next two years, all built in Essex. Marblehead continued to hold the distinction of being the principal fishing town of the New World for half a century after the Revolution, when Gloucester took the lead, a position that it has claimed ever since.

At the time of the treaty of 1783, the fisheries of Plymouth were of little account. By the opening of the nineteenth century they had expanded to include the cod-fishery, which was carried on to the best advantage, and the mackerel and herring fisheries. During the proper season the shores for two miles above and below the town were devoted to this industry. The products of the catch were sold to Spain or to Portugal, or to islands belonging to those countries. In 1790, the inhabitants of Provincetown employed about twenty vessels in the codfishery. This number had increased to thirty-three in 1802, with an aggregate tonnage of 1,722 tons. These vessels went to the banks of Newfoundland, the coast of Labrador, and the Bay of Chaleur. The annual catch was about 33,000 quintals of fish worth $100,000. About three hundred men and boys were engaged in this business. One-half of the fish were cured in Provincetown. The provisions and stores were purchased in Boston, where the greater part of the catch was sold. Eight thousand hogsheads of salt were used in the fisheries. The taking of mackerel, herring, and bass was a local business, and was done by means of seines, each worth about one hundred dollars; there were about fifty seines owned in the town.

Wellfleet had twenty-five vessels in 1802 engaged in cod and mackerel fishing and in the oyster trade. The inhabitants took special pride in the quality of oysters to be obtained from Wellfleet, a writer of the time stating, "No part of the world has better oysters than the harbor of Wellfleet." At the opening of the century, Duxbury had a few vessels employed in the codfishery; and a profitable business was carried on in shipbuilding. Yarmouth had only ten boats in the fisheries, but Chatham did a considerable business in cod fishing with twenty-five schooners. Whale fishing was carried on from Truro and New Bedford.

The cod and mackerel fisheries were carried on in Rhode Island and Connecticut to a small extent only. Their combined tonnage in the year of greatest prosperity, 1807, was 6,000 tons, which was about 10 per cent of the tonnage of the other New England states for that year. Outside of New England these fisheries were carried on to an extent hardly worth notice. New York led the list, but the tonnage from that state during this period, 1784 to 1818, never reached nine hundred tons. The deep-sea fisheries of the country during the period were practically limited to New England, and of these states Massachusetts usually possessed more than ninety-five per cent of the total tonnage.

The conditions surrounding the life of the fisherman at the opening of the last century were far better and more attractive than at any previous period of the country's history. The disastrous results of the Revolution had been either overcome or outgrown. Seventeen years of peace following the Revolution were not without beneficial results upon all industries, most of all upon the deep-sea fisheries. The liberality of the General Government in granting bounties and allowances was encouraging and stimulating to this industry. The migration of hundreds of New Englanders into the new fields of central New York, Ohio, and Kentucky had no attraction for the fisherman. Never were his prospects brighter for success in his own calling than in the half-dozen years at the opening of the new century. His business increased in volume; his vessels grew in size as additions were made to the fleet; the area of his field of activity enlarged, for he over-ran the islands and bays of the Gulf of Saint Lawrence, he pushed through the Straits of Belle Isle to the eastern shores of Labrador, and extended his fisheries to Cumberland Island and the entrance of Hudson's Bay.

Many influences combined to hold the fisherman to his calling when men of other occupations were leaving their eastern homes for the uncertainties of life in the new West. He was held fast-bound to the life on the sea by the chains of habit. The traditions of the family and the education of the children were confined chiefly to nautical affairs. "Long before a lad could nib a quill, or make a pot-hook, or read half the precepts his primer contained, he knew the name of every brace and stay, every sail and part of a Grand Banker and a Chebacco boat, all the nautical terms. . . . By the time he had seen his tenth birthday he was old enough not to be seasick, not to cry during a storm at sea, and to be of some use about a ship, and went on his first trip to the Banks. The skipper and the crew called him 'cut-tail,' for he received no money save from the fish he caught, and each one he caught was marked by snipping a piece from the tail. After an apprenticeship of three or four years the 'cut-tail' became a 'header,' stood upon the same footing as the 'sharesmen,' and learned all the duties which a 'splitter' and 'salter' must perform." Entering upon his period of apprenticeship at an early age and pursuing the fishing industry continuously throughout his youth, the young New Englander found himself, at the age of twenty, fully equipped in experience, in strength, and in knowledge to engage in the fisheries with the hope of earning a place for himself above the rank of one of the crew. He aspired to the position of skipper of a schooner, possibly as owner of the craft that he was to command, and it was this ambition that led him, summer and winter, to face the storms of the Atlantic in preference to the fields and forests of the Middle West.

The fishing opportunities of the Gulf of Saint Lawrence were gradually making themselves known to the enterprise and vigilance of the New England fishermen, and for a few seasons prior to 1808 these fishing grounds had become an object of attention for all fishing towns between the Thames, at New London, and the Schoodic, in Maine. The boats, both small and large, went there by the hundreds, and if we can believe some accounts, by the thousands. On Sundays, the American fishermen "swarmed like flies upon the shore," for at that time the Sabbath was observed as a day of cessation from all fishing among the fisherfolk. In the year 1807 or 1808, probably the former, jealous British merchants stationed a watchman in some favorable station near the Strait of Canso to count the number of American fishing vessels that passed those straits. He counted nine hundred and thirty-eight that passed through, and doubtless many others were missed in the fog and at night.[1] The products of the American catch in Canadian waters and off Labrador for several years previous to 1808 were computed to furnish three-fourths of all the dried fish exported from the United States. Merchants were beginning to send large ships to the coast of Labrador to load from the small ones there and sail direct to European markets. To the jealous British subjects the activity of the Americans was of exaggerated importance. They looked upon this industry as "one of the greatest resources of the Eastern States," in which two thousand schooners were annually employed,

[1] Adams, Duplicate Letters, pp. 210–213, quoting a letter from J. Lloyd.

and men obtained employment to the number of from fifteen thousand to twenty thousand.[1]

A more reliable statement of the condition of New England fisheries during the period from 1790 to 1810, and of their relation to the rising discontent between American and British subjects is obtained from a letter written by a Boston merchant to Adams in 1815. The gentleman was well acquainted with the business and took pains to make his statement correct. The letter is as follows:—

"My calculation is that there were employed in the Bank, Labrador, and Bay fisheries, in the years above mentioned (1790 to 1810), one thousand, two hundred and thirty-two vessels yearly, namely, 584 to the Banks and 648 to the Bay of Chaleur and Labrador. I think that the 584 bankers may be put down at 36,540 tons, navigated by 4,627 men and boys (each vessel carrying one boy). They take and cure 510,700 quintals of fish, and average about three fares a year, and consume annually 81,170 hogsheads of salt. The average cost of these vessels is about $2,000 each; the average price of these fish at foreign markets is six dollars per quintal. These vessels also make from their fish annually 17,520 barrels of oil, which commands about $10 per barrel. Their equipment costs about $900 each, annually, exclusive of salt.

"The 648 vessels that fish at Labrador and the Bay, I put down at 41,600 tons, navigated by 5,832 men and boys. They take and cure annually 648,000 quintals of fish. They go but one fare a year, and consume annually 97,200 hogsheads of salt. The average cost of the vessels is $1,600, and their equipment, provisions, etc., $150 each. This description of vessels is not so valuable as the bankers, more particularly that class which goes from Maine, Connecticut, and Rhode Island, as they are mostly sloops and of no great value. Most of the vessels cure a part of their

[1] Coloquhoun, British Empire in America, p. 296.

fish near the place where they catch them, on the beach, rocks, etc., and the rest after they return home. Several cargoes of dry fish are shipped yearly from Labrador directly for Europe. The usual markets for these fish are in the Mediterranean, say Alicante, Leghorn, Naples, Marseilles, etc., as small fish are preferred at these markets, and the greater part of the fish caught in the Bay and at Labrador are very small. The average of these fish is $5 per quintal. These vessels also make from their fish about 20,000 barrels of oil, which always meets a ready sale at a handsome price, say from $8 to $12 per barrel. Most of it is consumed in the United States. . . .

"There are many persons who assert that in one year there were at Labrador and up the Bay more than 1,700 vessels, besides the bankers, but I am very confident that they are much mistaken."[1] This statement of the number of the vessels may be true when it is considered that each trip was reckoned as a single vessel by some, although many vessels made two trips to the Bay and even to Labrador each season.

The war of 1812 produced an entirely different effect upon the fisheries of New England than was felt from the Revolutionary War. In the first struggle our fisheries were so completely checked that it was years before they recovered their former status, but treaty rights were secured that were liberal in the extreme. In the war of 1812 the deep-sea fisheries were stopped for a couple of years only, and their recovery was rapid after hostilities ceased; but the former liberal treaty rights were lost, the new ones were designed to restrict the movements of American fishermen in British waters, and the interpretations that have since been placed upon them have greatly curtailed the progress of our fisheries in the provincial territories of Great

[1] Adams, Duplicate Letters, quoting a letter written to him by a Boston merchant, May 20th, 1815.

Britain. During the progress of the war it became evident that the British government would never again ratify a treaty of peace without a very strong protest against granting liberties to American fishermen in the waters of the British dominion. They were determined to wrest from the Americans the rights and privileges granted to them under the treaty of 1783.

The peace commissioners from the United States, J. Q. Adams, Clay, Bayard, Gallatin and Russell were expected to conclude a treaty of peace; and yet, rather than to allow the subject of surrending the fisheries to come up for discussion, they were instructed to break off peace negotiations and to return home. The peace commissioners of Great Britain were determined to keep Moose Island—the site of Eastport—and the fisheries. The assumption was made by them that the effects of the war were to terminate any and all privileges that the subjects of the United States held under the treaty of 1783. The views of the American commissioners were set forth in a proposition offered by Clay that we held our rights in the fisheries by the same tenure as we did our independence; that the treaty of 1783 was an instrument recognizing the rights and liberties enjoyed by the people of the United States as an independent nation, and containing the terms and conditions on which the two parts of one empire mutually agreed thenceforth to constitute two distinct and separate nations.[1]

The British commissioners found these grounds of maintaining the fisheries by the Americans to be impregnable. To offset this condition, they made a demand for the free navigation of the Mississippi River. No justifiable reasons for such a demand were presented, but it caused a certain amount of consternation and confusion among the American commissioners, for these subjects—the fisheries and the navigation of the Mississippi—were matters that would al-

[1] Adams, Duplicate Letters, p. 55.

low no compromise. Clay, who represented the West, was as steadfast in upholding the interests of his people as Adams was in insisting upon full privileges in the fisheries. The result of the deliberations was that the treaty of Ghent was agreed to by the commissioners, and afterwards ratified by the respective governments, and the two important issues—the fisheries and the Mississippi—were omitted altogether.

With the return of peaceful relations the American fishermen again pursued their industry with their old-time vigor on the banks, in the Gulf of Saint Lawrence, and on the Labrador coast. The tonnage employed in these fisheries rose rapidly from 17,855 tons in 1814 to 64,807 tons in 1817. The principle that was maintained by the American commissioners at Ghent, namely, that no privileges of the treaty of 1783 were abrogated by the war of 1812, was adhered to by the fishermen of the United States. The bays and harbors of the British dominion were again resorted to by them as places for shelter from storms and for the curing of their fish. It was assumed by the British government and by her colonists, however, that such a procedure was in violation of public law. In a letter issued by British authority in June, 1815, the right of the Americans to fish upon the high seas was recognized as being permanent; the privileges which they held under the treaty of 1783, however, were declared to be forfeited by them, and it was further declared that "the subjects of the United States can have no pretense to any right to fish within the British jurisdiction." Every indulgence was to be given to American fishermen who already had begun to cure fish upon shores previously occupied by them so as not to cause them too severe loss by being compelled to remove apparatus. This leniency, however, was to apply only to the summer of 1815.[1]

[1] Niles' Weekly Reg., X, p. 58.

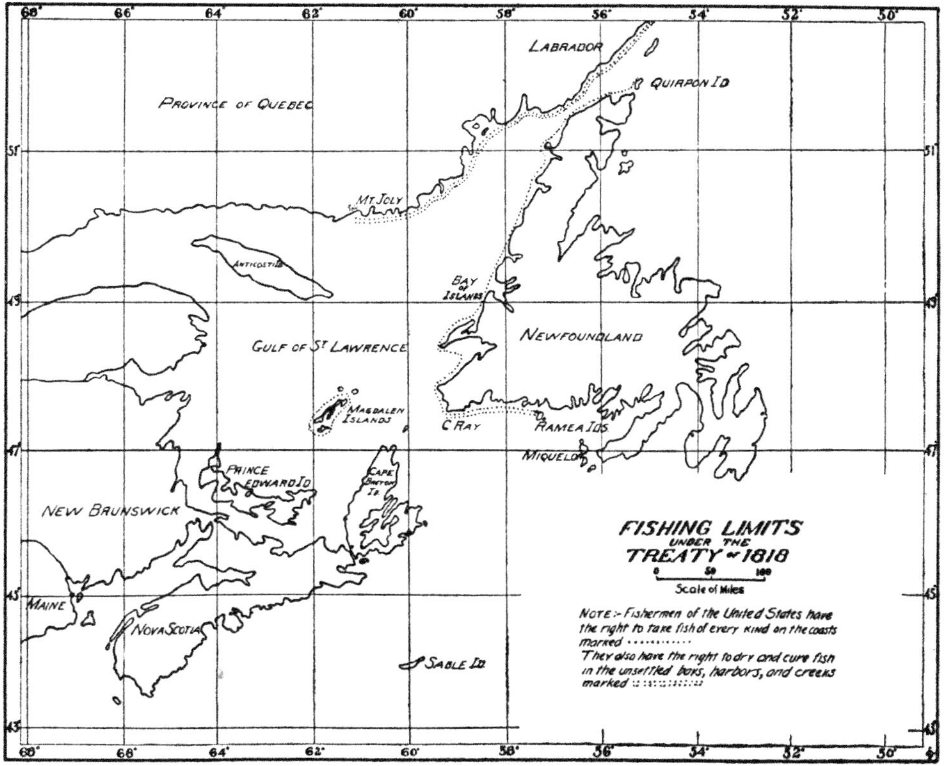
FISHING LIMITS
UNDER THE
TREATY of 1818
0 50 100
Scale of Miles
NOTE:- Fishermen of the United States have the right to take fish of every kind on the coasts marked ....
They also have the right to dry and cure fish in the unsettled bays, harbors, and creeks marked
LABRADOR
QUIRPON ID
PROVINCE OF QUEBEC
MT JOLY
ANTICOSTI
BAY OF ISLANDS
NEWFOUNDLAND
GULF OF ST LAWRENCE
MAGDALEN ISLANDS
C RAY
RAMEA IDS
MIQUELON
PRINCE EDWARD ID
CAPE BRETON ID
NEW BRUNSWICK
MAINE
NOVA SCOTIA
SABLE ID

Before the order had been received in Nova Scotia, zealous British officers already had begun to exclude the fishermen of New England from their territory and waters. On the 28th of June, 1815, eight American vessels were brought into Halifax by a British cruiser. After a short detention the vessels were released, having their papers indorsed forbidding them to fish on the western shores of that province. Upon protest by President Monroe, the act was promptly disavowed by the British government, but the outrage was the means of leading the two governments to enter into negotiations which resulted in the Convention of 1818.

Before the convention took place other seizures were made by the British. In 1816, several vessels in the Bay of Fundy were seized and sent to Newfoundland.[1] The next year one British sloop of war alone sent twenty sail of American fishermen to Halifax for trespassing in the waters of the coast.[2] More seizures followed in 1818, even at the time that the treaty negotiations were going on.[3] Great indignation was felt in the United States over these seizures, especially those of 1818. It was evident to the public mind that Great Britain was determined, if possible, to exclude American fishing vessels from waters adjacent to British territory in America.

By the terms of the Convention of 1818 relating to the fisheries our fishermen received fewer rights and privileges than they had been enjoying under the former treaty with Great Britain. The first article of this convention defines the extent of the right of fishing as follows:—

"Whereas differences have arisen respecting the liberty claimed by the United States for the inhabitants thereof, to take, dry, and cure fish on certain coasts, bays, harbours, and creeks of His

[1] Ibid, X, p. 372.
[2] Ibid, XII, p. 299.
[3] Ibid, XIV, pp. 344, 360.

Britannic Majesty's dominions in America, it is agreed between the high contracting parties, that the inhabitants of the said United States shall have forever, in common with the subjects of His Britannic Majesty, the liberty to take fish of every kind on that part of the southern coast of Newfoundland which extends from Cape Ray to the Rameau Islands, on the western and northern coasts of Newfoundland, from the said Cape Ray to the Quirpon Islands, on the shores of the Magdalen Islands, and also on the coasts, bays, harbours, and creeks from Mount Joly on the southern coast of Labrador, to and through the Streights of Belle Isle, and thence northwardly indefinitely along the coast, without prejudice, however, to any of the exclusive rights of the Hudson Bay Company: And that the American fishermen shall also have liberty forever, to dry and cure fish in any of the unsettled bays, harbours, and creeks of the southern part of the coast of Newfoundland hereabove described, and of the coast of Labrador; but so soon as the same, or any portion thereof, shall be settled, it shall not be lawful for the said fishermen to dry or cure fish at such portion so settled, without previous agreement for such purpose with the inhabitants, proprietors, or possessors of the ground. And the United States hereby renounce forever, any liberty heretofore enjoyed or claimed by the inhabitants thereof, to take, dry, or cure fish on, or within three marine miles of any of the coasts, bays, creeks, or harbours of His Britannic Majesty's dominions in America not included within the above-mentioned limits; Provided. however, that the American fishermen shall be admitted to enter such bays or harbours for the purpose of shelter and of repairing damages therein, of purchasing wood, and of obtaining water, and for no other purpose whatever. But they shall be under such restrictions as may be necessary to prevent their taking, drying or curing fish therein, or in any other manner whatever abusing the privileges hereby reserved to them."[1]

The important differences between this treaty and that of 1783 are, first, that the Americans gave up the inshore fishing along certain parts of the coast, and secondly, that

[1] Snow, American Diplomacy, pp. 78–80.

facilities for drying and curing fish were enlarged in favor of American fishermen. The importance of the concession of inshore privileges was not realized by the American commissioners at the time. The mackerel fishery in the Gulf of Saint Lawrence did not become of economic value to our fishermen for more than a decade later. The treaty, while it settled the main question at issue—that of defining the right of American fishermen in British waters—did not enter into the details of the controversy sufficiently to make clear matters that appeared to be unimportant at the time. The free navigation of the Strait of Canso, the purchase of bait and other supplies, the landing and transshipment of fish, and other privileges were soon to be demanded by the Americans, only to be met with refusal on the part of Canadian officials. The fishery question, instead of being settled had really begun the course of its unenviable career.

CHAPTER IX

A HALF-CENTURY OF GROWTH

The fifty years from the close of the second war with Great Britain to the abrogation of the Reciprocity Treaty in 1866 are filled with many important events in the political history of the country as well as being the years during which the internal improvements of the country were fostered and developed. It is the period of slavery agitation, of the war with Mexico and the Civil War, of increased transportation facilities by the construction of highways, canals, and railroads, of the growth of the Middle West and of the beginnings of American civilization upon the Pacific coast.

With the exception of taking an active part in the naval history of the Civil War, the fisherman of New England appears to have played an unimportant rôle in the national movements of this half-century. Neither the rich valley of the Mississippi nor the gold of California attracted him from his nets and boats. His was the oldest industry of the land—one not to be exchanged easily for any of the alluring vocations that were opening daily in the new West. But if he was unwilling to participate actively in the new movements going on about him there was no escape, and no desire for escape, from the influences that enhanced his own calling. With increased facilities for transportation came the expansion of the Baltimore oyster industry beyond the Alleghenies. The opening of the Erie canal was the signal for the entrance into the Mississippi Valley of the best quality of Gloucester codfish, which formerly had been exported to Bilboa. The settlement of the shores of the Great Lakes led to the development of an inland fishery that was worth a quarter of a million dollars per annum within a decade of its establishment.

Within the sphere of his own industry the fisherman never was more active than during this period. The ancient calling of codfishing was pursued with greater zeal and wider area. The mackerel fishery rose within a few years from an insignificant calling to become a formidable rival of the codfishery. The oyster industry increased with each year until it became an active occupation for thousands of people along the gulfs and bays of the Atlantic. Menhaden and shad were taken by the million from the Carolinas to Eastport, while the new employment of taking and curing the herring of the Magdalen Islands and of the Passamaquoddy region laid the foundation for scores of fortunes.

On the other hand, there were years of losses as well as years of prosperity in the fisheries,—that is one of the characteristics of the fishery industry, especially of the cod and mackerel fisheries which necessitate a considerable outlay of capital for establishing and maintaining them. An unfavorable season for catching fish might be partially offset another season by securing large gains, either from a large catch or from higher prices. But the loss of men and vessels at sea never could be remedied. International trouble also continued. New complications arose with the enforcement of the provisions of the Convention of 1818. The maritime provinces of Great Britain were insistent in interpreting the new provisions always from their point of view. The result was repeated quarrels, frequent seizures of American fishing vessels, and curtailment of privileges, if not of rights. After a trial of thirty-six years, stormy and unsatisfactory to both sides, the terms of the Convention of 1818 gave place to the Reciprocity Treaty of 1854, which was destined to last only one-third as long as its predecessor before it was discarded with no more ceremony than its own provisions necessitated.

The year 1819 was of importance in the history of the fisheries not only because it was the first in which the provisions of the Convention of 1818 went into effect but also because it was marked by the passage by Congress of a law that fixed the "bounty," or allowance to vessels employed in the fisheries, at a higher rate than had previously existed. The act, which continued in operation for forty-seven years, provided for bounties to the owners of fishing vessels employed at sea four calendar months exclusively in the codfishery under the following regulations.

Compensation to fishermen for their service must be by division of fish, or share in the proceeds of the sale of the fish; no person except the cook could receive wages. The master and three-fourths of the crew must be citizens of the United States. Fishing vessels had to be examined by an inspector as to their sea-worthiness, their equipment, and the number and nationality of the crew before sailing on a voyage on which allowances were to be paid. A regular log-book had to be kept on board day by day, setting forth the principal events of the voyage, which later was submitted to the collector of the home port. Arrivals and departures had to be recorded by the master or owner with the proper officers, but the required time of four months at sea did not have to be in continuous voyages. Vessels could engage in the mackerel or other fishery in their season if the masters so wished; these voyages, however, had to be distinct from the codfishery voyages. Allowances were paid as follows, according to the size of the vessels:—

If measuring more than five tons, and not exceeding thirty tons, three and one-half dollars per ton.

If measuring more than thirty tons, four dollars per ton.

If measuring more than thirty tons, with crew of not less

than ten persons, and having been employed at sea exclusively in the codfishery three and one-half calendar months, three and one-half dollars per ton.

The allowance to one vessel during the season, regardless of her tonnage, should not exceed three hundred and sixty dollars.[1]

Numerous attempts were made to have the act repealed, but none was successful until July, 1866. Since that date no allowances have been granted to vessels engaged in the fisheries. The subject of the fisheries came up before Congress in 1852 during the discussions that preceded the ratification of the Reciprocity Treaty of 1854. At the time it was shown by those who were friendly towards the fishing interests that there was need of bounties in supporting the industry not only for its own sake but also because this was the best means for supplying the navy and merchant marine of the nation with capable men and officers. In a series of speeches made before the House of Representatives the Hon. Zeno Scudder, congressman from Massachusetts stated, "the law of 1819 has continued to this time, and under its protection there has been a gradual increase in the business. It has, however, been smaller in comparison with that of other departments of commerce; showing that with all the aid which the government has rendered, it is still too poor in its returns to be followed to a great extent." Opposition to the granting of yearly allowances to the fisheries was speedily quelled when it was shown by the speaker that there were more than twenty thousand fishermen in New England whose principal income was only $76.89 per annum.[2] As five-eighths of the annual allowance went to the crew of the vessel and only the remaining three-eighths to the owners of the vessel, it is readily seen that the law was intended primarily, and continued

1 Sabine, pp. 166–169.

2 32nd Cong., 1st Sess., Vol. 29, pp. 919–920. (Aug. 12, 1852.)

throughout its operation, for the benefit of individual rather than corporate interests; it needed but little argument by the representative from Massachusetts to show "that the prime, medium, and final object of this series of statutes was, and is, the cultivation of seamen by the encouragement of a business which, in itself, was, and is, too poor to be sustained."

The amount of allowances paid fishing vessels under the act of 1819 for the forty years between the first of January, 1820 and the thirtieth of June, 1859, was $10,626,201.13, which gives an average of $265,655 per annum. The highest amount paid in one year was in 1857, when the allowances reached $464,178.[1] The amount of allowances granted to fishing vessels under the several acts before 1820 was $2,328,517.68, and the estimated amount granted from June 30th, 1859 to the repeal of the Act of 1819 in 1866 was $2,500,000. The total amount of allowances granted from the enactment of the first act in 1792 to the repeal of the last act in 1866 was between fourteen and fifteen million dollars.[2]

During the period from 1818 to 1866, duties upon fish imported into the country were levied by several tariff acts. By the Act of 1816, duties were levied on foreign caught fish at one dollar per quintal, on mackerel at one and one-half dollars per barrel, on salmon at two dollars per barrel,

1 36th Cong., 1st Sess., Sen. Docs., Vol. I, No. 41, pp. 14–15.

2 The geographical distribution of allowances to vessels engaged in the codfishery may be seen in the following statement, which shows the amount paid to each State from the commencement of the government to the 30th of June, 1859.

Maine	$ 4,175,050
New Hampshire	563,134
Massachusetts	7,926,273
Rhode Island	182,853
New York	78,890
Virginia	479
Total	$12,944,998

and on all other pickled fish at one dollar per barrel. The Act of 1842 continued these duties with additional rates of twenty per cent ad valorem on sardines and other fish preserved in oil; but fresh caught fish, brought in for daily consumption were exempt from duty. By the Act of 1846 a forty per cent ad valorem duty was placed on anchovies, sardines and all other fish preserved in oil, and a twenty per cent ad valorem duty on foreign fish, whether fresh, smoked, salted, dried or pickled, not otherwise provided for.[1]

The duty on salt, which formed the basis of the bounty, has been in the several tariffs as follows:—for 1824, twenty cents per bushel of 56 pounds; that of 1828, the same; that of 1832, ten cents per bushel; that of 1842, eight cents per bushel; that of 1846, twenty per cent ad valorem.[2]

The change of the schedule of 1846 to a twenty per cent ad valorem scale on salt and fish was of undoubted advantage to the foreign importer, but the effect was depressing upon American fishermen. The French fishermen and those of the British provinces were brought immediately into more active competition with ours. They possessed the advantage of proximity to the great fishing grounds, the French government paid its fishermen a bounty of about one and one-half dollars per quintal for the fish they caught, the price of labor was cheaper with both classes of people than in the United States, the act of 1846 was favorable to the importation of their fish into this country, consequently they became formidable rivals of our fishermen in our own markets. The imports of foreign fish increased at a rate that filled New England fishermen with the gravest fears for their industry. Their apprehension was well founded, for under the tariff of 1846 the quantity of codfish and

[1] Tariff Acts of the U. S. Acts of April 27, 1816; Aug. 30, 1842; July 30, 1846.

[2] Hunt, Merchant's Magazine, V, p. 26.

mackerel imported into the country increased to several times the amount imported previous to its passage. During the four fiscal years ending June 30, 1846,—previous to the introduction of the new schedule of duties—the imports of dried fish were 1,358 quintals, valued at $10,120; the imports of pickled fish were 74,634 barrels, valued at $561,593. For the four years from 1847 to 1850 the same items of import were: 42,332 quintals of dried fish, valued at $88,781, and 204,358 barrels of pickled fish, valued at above $1,000,000.[1]

The estimated amount of duty collected on imports of fish into the United States from the British North American possessions from 1850 to 1855 was $884,974.20, which represented a market value of more than $4,400,000 worth of fish.[2] Our fishermen became thoroughly aroused at the state of affairs, measures for remedying this condition of the fisheries were advocated, and the general discussion of the question that followed was one of the principal causes of the Reciprocity treaty of 1854.

The course of trade of the dried, smoked and pickled fish exported during this period shows an increasing proportion being sent to the West Indies and a falling off in exports to European countries until the latter amounted to almost nothing. The total exports of dried and smoked fish for the year 1821 were 254,947 quintals. The West Indies took 214,018 quintals, or 84 per cent of the total exports. In 1825 they took 251,034 quintals, or 88 per cent of the total exports; and in 1832 the amount was 233,247 quintals, which was 96 per cent of this kind of fish exported from the country. The export trade to Europe diminished from 21,184 quintals in 1821 to 3,042 quintals in 1825; in 1832 the quantity of exports had fallen to only 430 quintals, while the following year there were no exports to Europe.

1 Andrews' Report, p. 636.

2 Elliott, The U. S. and the Northeastern Fisheries, p. 148.

The French, Dutch, and Danish possessions of the West Indies, Cuba and Hayti took the bulk of fish that were sent to the West Indies.[1]

The quantity of dried fish sent to the West Indies during the years 1821, 1825 and 1832 averages nearly the same as was sent there in 1800, which was 244,352 quintals. The average annual exports of this kind of fish for the thirty-three years from 1819 to 1851 were 242,697 quintals, valued at $673,723 annually. This amount is slightly below the average for the period of 1783 to 1818. The tonnage employed in the fisheries during the latter period was about double that of the former, and the natural inference would be that exports should show a corresponding increase. As a matter of fact, European exports fell off until they need hardly be considered. Our exports to the British West Indies were suspended after the opening of the war of 1812, and when peaceful relations were resumed this trade was not recovered by the New England shippers, a trade which in 1800 amounted to 141,000 quintals of dried fish. The loss of the trade with the British West Indies was due to the successful competition of the fishermen of the British North American provinces. The value of the cod and mackerel fisheries of New England was in excess of former periods by a wide margin; the loss of trade with Europe and the absence of an increase of exports to the West Indies do not indicate, necessarily, that our fisheries were in a low state or that the decline of export trade was a calamity to the industry. The explanation is found in the increasing demand for American fish in our own markets, the demand at home being for the best quality of cured fish.

A writer of the times describes the change that took place in the course of our export trade as follows:—"Of late years (between 1830 and 1840) an entire change of markets for the products of the fishery, so far as it respects the

[1] Pitkin, p. 88.

large sized fish, has taken place. Since the opening of the Erie Canal and the increase of population and business consequent thereupon, an increasing demand for this article has grown up in that quarter; so that the New York and Albany markets, which previously required only a few thousand quintals for their annual supply, now afford a demand for nearly 150,000 quintals. The foreign export has diminished in a ratio proportionate to the increase of the domestic demand."[1]

The total amount of exports of pickled fish from 1819 to 1851 was 1,830,353 barrels and 139,557 kegs. The aggregate value of these exports for the years named was $7,289,-783, an annual average of $220,902. The average annual value of exports of dried, smoked and pickled fish during the same period was $894,624, and the aggregate value of all kinds of fish exported for that period reached a total of $29,522,628.[2] A comparison of these figures with the total value of our exports of fish for the thirty years preceding 1819 shows a falling off of about fifty per cent from the earlier totals, a decline that was due, as indicated above, to an increased demand for fish in our domestic markets.

The amount and value of the fish used for the home consumption can not be given for any number of years with accuracy. The statistics for the year 1840 afford a basis for a close estimate of fish products consumed at home. In that year the fisheries of the country produced 773,947 quintals of dried and smoked fish, and 472,360 barrels of pickled fish.[3] Of the exports, the 211,425 quintals of dried fish sold at $2.55 per quintal, and the 43,400 barrels of pickled fish sold at $4.12 per barrel.[4] Estimated at the same prices, the total value of the fisheries for 1840 was

[1] Niles' Weekly Register, Vol. 58, pp. 69–70.
[2] Compiled from Sabine's Report, pp. 176–178.
[3] Hunt, Merchant's Magazine, VI, p. 364.
[4] Ibid, IX, pp. 87–88.

$3,915,786, made up of $1,973,565 worth of dried fish and $1,942,221 worth of pickled fish. The value of fish used for domestic consumption, according to this estimate, was $3,201,543, made up of $1,434,431 worth of dried fish and $1,767,112 worth of pickled fish. The tonnage engaged in the codfishery and the amount of exports for 1840 were each nine per cent less than the average for this period of years; so that it is not an unsafe estimate to place the annual value of the fisheries between 1819 and 1851 near $4,000,000, with the home consumption considerably above $3,000,000 worth of fish.

The cod retained the distinction of being the principal food fish of the American seas down to the period of the Civil War. While there were years of depression in the codfishery from 1818 to 1866, the period as a whole was one of general prosperity and substantial gain in all phases of the industry. A survey of the Massachusetts towns engaged in the codfishery reveals, at first glance, a marked depression in this fishery; but where the tonnage of the State falls off in the codfishery it can generally be found to crop out again in some other new form of deep-sea enterprise, particularly in the mackerel fishery. For example, in 1851 the codfishery of Maine was above the 45,000 tonnage mark,—more than 5,000 tons ahead of the Massachusetts codfishery tonnage. But, on the other hand, Massachusetts had about 40,000 tons employed in the mackerel fishery, the tonnage for the two industries being almost alike for that year, while the mackerel tonnage of Maine was below 10,000 tons.

The amount of tonnage employed in the codfishery of New England during the period of forty-eight years from 1818 to 1866 reached an annual average double the tonnage employed during the first twenty-nine years of our Federal existence. The average tonnage for the period, and for different terms of years, are shown in the following table:

From 1819 to 1829, eleven years,	....68,700 tons.[1]	
" 1830 to 1839, ten "	....65,100 "	
" 1840 to 1849, " "	....68,200 "	
" 1850 to 1859, " "	...101,300 "	
" 1860 to 1866, seven "	....95,454 "	

From 1819 to 1866, forty-eight years, 79,200 tons.

In 1829 the tonnage was above 100,000 tons,[2] the only time during the period previous to the year 1852. From the latter year to 1864 the annual amount employed in the codfishery was above 100,000 tons, except in 1856. The average annual tonnage employed from 1852 to 1864—the most prosperous years of the period—was 112,700 tons. The highest tonnage employed was 136,654 tons in 1860.[3] Between 1850 and 1860 there were employed annually in the codfishery an average of 2,084 vessels, of about fifty tons each, carrying 14,570 men in the crew, or an average of seven men for each vessel.[4]

At the period of its greatest prosperity—the year 1859—the codfishery of the country was carried on in 2,593 vessels, of a total tonnage of 129,637 tons, carrying crews that aggregated 18,151 men. The status of the fishery in the several states is given in the table on the following page. The State of Maine led all others in the extent of the fishery; in fact, the codfishery of Maine was about equal to the combined fishery of the other states. The codfishery of Massachusetts, too, was so important that the

1 Previous to 1830 the tonnages employed in the cod and mackerel fisheries were listed together. Probably the tonnage employed in the codfishery previous to 1830 did not average above 50,000 tons annually.

2 This amount, however, included the mackerel tonnage for that year.

3 Compiled from Report of Register of Treasury, Jan. 23, 1868.

4 36th Cong., 1st Sess., Senate Doc., I, No. 41, pp. 14-15.

industry of the country may be said to have been embraced in these two states.

Turning now to the mackerel fishery we find that its development was immediate and very rapid after the treaty of 1818 went into operation. The catch of mackerel of the country was unimportant previous to 1819. For fifteen years preceding 1819, the total recorded catch of mackerel by the fishermen of Maine was 6,553 barrels; of Massachusetts, 231,085 barrels. The catch for Maine in the year 1819 was 5,322 barrels, and for Massachusetts, 100,111 barrels. From 1819 there was an almost unbroken line of development in the mackerel industry to the year of greatest prosperity, 1831, when 450,000 barrels of mackerel were salted in Maine, New Hampshire and Massachusetts. Of this amount, the fishermen of Massachusetts laid claim to 383,549 barrels, valued at $1,589,936. The catch of mackerel for 1831 is all the more remarkable when it is remembered that the fish were caught by hook and line, each fish being pulled in individually, where to-day by the use of the purse-seine several hundred barrels may be taken at a single haul.

For the decade following 1831 there was a sharp decline in the fishery until the low water mark was reached in 1840, when only 50,492 barrels were taken in Massachusetts.

THE STATUS OF THE CODFISHERY OF THE COUNTRY IN 1859.

State.	Estimated No. of Vessels.	Estimated Tonnage.	Estimated No. of Crew.
Maine	1,269	63,477	8,883
New Hampshire	43	2,137	301
Massachusetts	1,138	56,919	7,966
Rhode Island	10	475	70
Connecticut	125	6,228	875
New York	8	401	56
Total	2,593	129,637	18,151

The rise of the industry was generally steady from 1840 to 1851, after which there was a period of decline for eight years, except in the year 1855.[1] During the last three years of the Civil War there was a slight falling off in the catch of mackerel as compared with that of the early sixties, probably due to the lack of men to engage in the business, as many fishermen had enlisted for service in the war. But the value of the fishery was greatest during these years. The total value of the mackerel of the country for the three years 1864, 1865, and 1866 was $17,893,211. In these three years mackerel of the first quality sold in Massachusetts at from $22 to $30 a barrel. The cause of high prices appears to have been due to a steady demand for the fish; when the markets became empty of mackerel the price of the fish advanced. The most profitable year was in 1864 when the country's catch of 324,455 barrels of mackerel was valued at $7,001,098.[2] Nearly one-half of this total went to the fishermen from the town of Gloucester. The mackerel was king of the sea in war times.

During this period more than four-fifths of the mackerel industry was carried on by citizens of Massachusetts. From 1819 to 1866 the total product of the business in Massachusetts was 9,073,510 barrels of pickled mackerel, valued at $61,815,907, an average value of $8.60 a barrel. The average yearly product was 189,239 barrels, worth $1,671,240. The Maine catch of mackerel was less certain. During eleven odd years between 1820 and 1866 there was a total catch of 336,153 barrels; in 1865, the catch was 54,216 barrels. In New Hampshire the business was small, from 1830 to 1852 the total pack of salted mackerel being 153,370 barrels. During the Civil War the New Hampshire mackerel fishery was practically suspended, as only 722 barrels were packed from 1861 to 1866. It would appear

[1] U. S. Fish Com. Report, 1881, p. 280.

[2] U. S. Fish Com. Report, 1881, p. 299.

from the statistics of the fishery industry of New England from 1818 to 1866 that there was an average annual catch of 225,000 barrels of mackerel by our fishermen, valued at $1,935,000 annually.[1] Between 1850 and 1860 there were annually employed in the mackerel fishery of New England 662 vessels, carrying 5,252 men in the crews. The average size of each vessel was 65 tons; of the crew, eight men.[2]

Accounts of the extent and value of the New England fisheries for the year 1851 show that the number of vessels employed in the cod and mackerel fisheries was 2,691, valued at $12,000,000, including their outfits. The tonnage of the vessels engaged in the codfishery was 95,617 tons, in the mackerel fishery, 50,539 tons. The average codtonnage for the ten years previous to 1851 was 79,251 tons; nine quintals of codfish were taken per ton, worth $2.60 a quintal. There were 11,321 men in the codfishery for the ten years.

The status of the cod and mackerel fisheries of New England for the year 1859 is given on the following page, which shows the vessels, men and tonnage in the fisheries belonging to each State. Outside of the New England states, the deep-sea fisheries were unimportant and need not be considered in making the totals for the country's fisheries.[3]

The value of the products of the New England deep-sea

1 Statistics compiled from Sabine's Report, Hunt's Merchant's Magazine, U. S. Fish Commission Report for 1881, Boston Fish Bureau Reports.

2 From a careful estimate of both the cod and mackerel fisheries of Massachusetts for 1837 it is learned that in that year the catch of cod amounted to 510,554 quintals, valued at $1,569,517; and of mackerel, 234,059 barrels, valued at $1,639,049. The total value of the fisheries products of the State was $3,208,866. The number of fishermen was 11,146, with about as many more employés on land engaged in curing the fish. (32nd Cong., 1st Sess., Vol. 29, p. 904.)

3 36th Cong., 1st Session, Sen. Docs., Vol. I, No. 41.

fisheries for the year 1859 may be estimated conservatively at $1,650,000 from the mackerel fishery and $3,025,000 from the codfishery, a total of $4,675,000 from both industries.

Whether considered during periods of "lean" years, or when the prosperity of the fisheries is at its flood, the fisherman's share of the catch seems inadequate not only as an incentive for him to continue in the fishing business but also to support him and his family. How can they live on so small an income? The share of the cod fisherman for the decade between 1840 and 1850 was $62.31 a year on an average; his part of the government allowance was $14.58 more, making his total income from the codfishery $76.89. The income seems incredibly low, but it is a larger share than the mackerel fishermen received from the sea. For the five years previous to 1851, according to Congressman Scudder of Massachusetts, the 8,879 fishermen who were engaged in the mackerel fishery received an annual income of $64.04 derived from the sea.[1]

The fishermen themselves might manage to live very well on the scanty income because food and shelter were provided him when at sea and there was no opportunity, or need, of his spending money when aboard the vessel. His expenses for clothing, boots, oil-clothes and tobacco were

NEW ENGLAND COD AND MACKEREL FISHERIES IN 1859.

State.	Vessels.	Tonnage.	Crew.
Maine	1,432	73,291	10,187
New Hampshire	47	2,355	333
Massachusetts	1,422	74,957	10,238
Rhode Island	10	475	70
Connecticut	125	6,288	875
Total	3,036	157,366	22,703

[1] 32nd Cong., 1st Sess., Vol. 29, pp. 919–920.

not large. At the close of the fishing season he usually found employment ashore along the water-front or in curing the codfish preparatory for the markets. In addition, it should be remembered that often the members of his family were engaged in some gainful occupation. Previous to the Civil War and before the invention of machinery for the manufacture of nets hundreds of women and girls were employed in their own homes in making nets for fishermen of their own household, or for others. In general, however, women were not employed in curing the codfish on shore as they were in the early history of the industry, or as they are in Newfoundland to-day.

CHAPTER X

THE GROWTH ALONG THE COAST

A geographical review of the New England fisheries from 1818 to 1866 covers a wider range than has been considered previously, due in part to the extension of the fisheries into places already established, in part to the opening of new territory and the consequent development of the natural resources of those regions. In many places the records are few, leaving some doubt as to whether the fishing industry of a particular place was spasmodically pursued in connection with another industry, or whether only unusual cases of prosperous seasons raise the industry for a time to a place of importance great enough to receive mention in local records and newspapers. In the case of many towns where the fisheries were carried on extensively for a number of years the accounts of the industry are so scattering that an attempt to place them in a connected narrative would be a hopeless task.

The fisheries of Maine and Massachusetts receive greater consideration than other New England states. The fishing interests of Connecticut and Rhode Island were largely inshore. However, the whale fishery was carried on with considerable vigor from New London. The town early became interested in the fur-seal and sea-elephant fisheries in Antarctic waters. Vessels from here were the first American sealers to visit Desolation Island and Heard's Island in the South Indian Ocean, and large cargoes of sea-elephant oil were obtained annually from these islands for many years. The sealing fleet of New London in 1853 numbered eight sail. In 1858 it had increased to twelve sail, and for a score of years following it numbered annually from five to ten vessels.

It has already been noted how the fisheries of Maine assumed considerable importance before the second war with Great Britain. In 1820, Maine was admitted as a State into the Union, and was soon rising into a place of importance on account of the fisheries carried on by her citizens. She shortly outrivaled her mother state, Massachusetts, in the codfishery, and during this period was second to the Bay State in the mackerel fishery, as well as second in the Union in the extent of fisheries. From Eastport to Portland there was scarcely a place to be found, whether a village or a hamlet, on the coast itself or on the innumerable islands adjacent to the shore, where fishing industries were not pursued previous to the Civil War. A history of Maine fisheries is to be found very largely in a review of the local industries of her seaport towns for this period.

As early as 1820 the merchants of Eastport were extensively interested in the mackerel fishery. By 1830 the industry was at its height, there being fully forty vessels, of sixty to seventy tons, that fitted out at Eastport and sold the products of their catch there. These vessels carried a total of six hundred men and had an average catch varying from seven hundred to one thousand barrels of mackerel a season. The fishermen used to be engaged in the codfishery on the outer banks, or in the Bay of Fundy, before the summer mackerel season began. Several vessels fitted for the codfishery on the coast of Labrador by 1820, but the industry declined after 1830 and was wholly neglected after 1855. The Magdalen herring fishery began at Eastport by 1830, and continued to thrive until 1868.[1]

The town of Eastport claims the honor of having put up

[1] Goode, Geographical Review of the Fisheries, p. 16.

the first can of sealed goods of any kind within the limits of our country. The business began in 1843 with the canning of lobsters for market, an industry that has since risen to the rank of the highest importance in the State. In 1850 there were seven firms at Eastport engaged in the fish trade. These firms "employed 238 men; used 18,900 bushels of salt; cured 18,000 quintals of fish and 3,500 boxes of smoked herring; put up 12,000 barrels of pickled herring, 300 barrels of mackerel, and 3,503 barrels of other fish, . . . in addition to 450 barrels of oil and a quantity of canned goods, the whole having a value of $85,000."

The neighboring town of Lubec rivaled Eastport in the business of catching and smoking herring. By 1821, there were twenty smoke-houses in the place, each house having an annual output of from 2,500 to 3,000 boxes of herring. As early as 1830 the merchants of Lubec were sending vessels to the Magdalen Islands for additional herring to smoke and pickle. This herring fleet consisted of eleven vessels in 1860, each returning with cargoes of 700 to 800 barrels of fish. The smoked herring business was at its height between 1845 and 1865, there being from 400,000 to 500,000 boxes of herring smoked and packed annually.

Millbridge, located about midway between Machias Bay and Frenchman's Bay, became interested in the herring fishery in 1820. The business increased slowly until 1850, when people from Lubec built smoke-houses and presses for utilizing the catch of fish. The most prosperous period of the industry was between 1858 and 1863 when 75,000 to 100,000 boxes of herring were packed yearly. Lubec fishermen also visited Steuben, adjacent to Millbridge on the west, about the year 1850 to secure herring for smoking, and the business thrived at about the same time as for Millbridge, declining during the period of the war.

The towns bordering on the Frenchman's Bay district, which extends from Gouldsboro on the east as far west as Blue Hill, became engaged with varying success in different kinds of fishing enterprises. As early as 1810, settlers came to Gouldsboro to engage in the whole fishery, carrying it on from the shore to a limited extent. A large fleet of small vessels was employed in the codfishery for several years in the Bay of Fundy. With the decline of the business, the fishermen turned their attention to the hake fishery in Frenchman's Bay, which assumed a place of considerable importance after 1840. In some seasons as many as one hundred vessels from other parts of New England resorted to Frenchman's Bay for hake, which found a market in Portland and Boston. This industry, also, declined during the war.

Before the hake industry had begun to decline another branch of fishing was rising to take its place. For several years in the early part of the period in which hake fishing was carried on so exclusively, menhaden were caught for bait. In 1850 a Blue Hill woman discovered that marketable oil could be obtained from menhaden, and the business of catching the fish for their oil increased enormously in the Frenchman's Bay region. The best years for this industry appear to have been 1863, 1864, and 1865, although the business was continued with decreasing returns into the early part of the next decade. When the industry was at its height it is estimated that not less than one hundred try-houses, with two to four kettles each, were in operation between Lamoine and Gouldsboro. The yearly product of these houses has been estimated at fifty casks of oil, each holding forty gallons, worth a dollar per gallon, a total of $200,000 a year for the farmer-fishermen of Frenchman's Bay when the business was most prosperous.

The towns of Lamoine and Hancock were leaders in the fisheries of Frenchman's Bay. The people of Hancock became interested in the fisheries of the Bay of Fundy in 1845. In 1852 vessels were sent to the Western Bank for cod, and eight years later marked the beginning of sending vessels to the Grand Bank of Newfoundland, and others to the Magdalen Islands for herring. When the business of smoking herring was at its height in Hancock, from 30,000 to 40,000 boxes were packed yearly for the market at Boston and the West Indies.

About 1835 the people of Lamoine began sending pinkies of thirty or forty tons to fish for cod in the Bay of Fundy, and in 1848 the fleet numbered about twenty-five sail with six or eight men each. The fishery began to decline in 1850; the smaller boats were sold in other ports, and the larger vessels were fitted out for trips to the Grand Bank. This fishery, which began in 1857, was prosecuted from Lamoine with success for several years. When the menhaden business began to be profitable the farmers of the town left their fields to engage in the industry. There were scores of them who bought nets, boats and kettles and set up in business for themselves on the shores of the Bay. Along with the other fishery industries, that of smoking herring was extensively carried on in the place. The first schooner was sent to the Magdalen Islands for an additional supply of herring for smoking in 1855. The annual output of this industry averaged between 30,000 and 40,000 boxes; in 1865 or 1866 the number reached 125,000 boxes.

About 1825 Castine became a center for fitting out fishing vessels. Salt was imported direct from Cadiz and Liverpool to supply the demands of fishermen of Central and Eastern Maine. By 1850 five hundred vessels were fitted out annually at Castine. The fisheries of Deer Isle were carried on in small boats until 1830 when twelve large vessels were sent to offshore banks for cod; forty smaller ones fished inshore at the time. In ten years the number of vessels had increased to thirty and the boats to fifty. The height of the fishing business for the island was during the years of the Civil War. The larger schooners, to the number of thirty-five, were engaged almost exclusively in the mackerel fishery, during its season, in the Gulf of Saint Lawrence. After the close of the fishing season many of them were employed in the coasting trade between Boston or New York and the coast towns of Maine until the fishing season opened the following summer.

As early as 1825 there were forty boats engaged in the shore fisheries of the Isle au Haut. Some were engaged in the herring fishery; later vessels were sent to the Magdalen Islands for the fish. Bucksport, on the Penobscot River, had vessels engaged in offshore codfishery by 1825. The business increased steadily until there were twenty vessels of from fifty to one hundred and twenty tons landing 20,-000 quintals of fish in 1855. The Bucksport fishery began to decline after 1858.

Vinal Haven and North Haven have always occupied prominent places in the fisheries of Maine. The Labrador fisheries of Vinal Haven which began in 1804 were continued until 1840. The Magdalen herring industry began in 1830 and continued without interruption for twenty-eight years. Vinal Haven marketed $70,000 worth of dried fish in 1855. Three years previous to the opening of the Civil War from ninety to one hundred sail of vessels were owned in this place, and about forty at North Haven. About 1850, larger vessels were introduced into the fisheries of North Haven which gave an impetus to the fisheries of the town so that they soon outrivaled those of Vinal Haven. Some of their vessels were engaged in the codfishery in the spring and in the mackerel fishery during the summer. By 1861 their vessels followed the mackerel fishery during the entire season, going South in the spring and following the fish as they migrated into northern waters.

The shore towns from Matinicus to Portland were active in pursuing the industries of the sea. Matinicus was a favorite resort for herring for years. By 1840, ten thousand boxes of herring were being packed yearly for the Boston market. Bristol had twenty-five vessels engaged in fishing in 1830. Later the place became the center of the menhaden industry of Maine, the first oil and guano factory for the use of menhaden being built at Bristol in 1864. Between 1830 and 1840 Monhegan cured annually 9,000 quintals of fish. The mackerel fishery was carried on from the place through the Civil War. The smoking of Magdalen herring was carried on at Damariscotta from 1830 to 1845. North Boothbay sent eight or ten vessels to the cod-fishing grounds of Labrador in 1844. The fishing business began at Wiscasset in 1822, and was successfully prosecuted until the Civil War. At that time from 30 to 35 bankers and an equal number of shore boats were fitted at Wiscasset. The citizens of Southport successively tried their fortunes in the herring, cod and mackerel fisheries. Previous to the war of 1812 Georgetown, an island on the eastern boundary of the Kennebec River, had twenty-five vessels employed in the bank fishery. The town quickly recovered from the effects of the war. In 1843, between 25,000 and 30,000 quintals of fish were cured here. Both the inshore and the deep-sea fisheries increased in importance during this period. The merchants of the town bought fish from the neighboring towns of Westport, Woolwich and Phippsburg. In 1868, Georgetown handled $250,000 worth of fishery products.

Portland early became a center for the fisheries as well as for other maritime commerce. Shore and deep-sea fishing were carried on by her fishermen; vessels from other places resorted here to be fitted out for the sea; cargoes of fish were brought to her merchants from the shore towns; the shipping trade between Portland and the West Indies increased during this century, especially as more cod was packed in "drums" for southern markets. Portland was a trade center, as it is now, for the fishermen of Maine. In a single year previous to 1841, 45,000 barrels of mackerel were packed here. In the year 1864 there were 27,766 barrels packed at Portland, a quantity second only to the output of Gloucester.[1]

The state of the fisheries of Maine in 1840 shows that the business was carried on most extensively in Lincoln County, on the middle coast, and in Washington, the easternmost county. The counties of Hancock, Waldo and York ranked next in importance. There were in all 3,610 men employed in the fisheries; capital was invested to the amount of $526,967; the products of the fisheries consisted of 279,156 quintals of dried and smoked fish, 54,071 barrels of pickled fish, 118,851 gallons of oil, whalebone and other products amounting to the value of $2,351.[2]

In 1850 there were 37,218 tons employed in the codfishery of Maine, and 12,046 tons employed in the mackerel fishery.[3] Of the total 95,616 tons employed in the codfishery of the country in 1851, the amount of the State of Maine was 45,528 tons. The condition of the fisheries of the State for the year 1859 shows that in the codfishery there were 1,269 vessels of an aggregate tonnage of 63,477, carrying crews to the number of 8,883 men, and in the mackerel fishery there were 163 vessels of an aggregate tonnage of 9,814 tons, carrying 1,304 men. The fisheries of the State in 1859 had a total of 1,432 vessels, of a tonnage of 73,291 tons, carrying 10,187 men.[4]

At the close of the Civil War the fisheries were prosperous in the State of Maine, probably beyond any other period to that date. But the principal branches of fishing—cod, mackerel, menhaden and herring—were on the verge of

[1] U. S. Fish Com. Report, 1881, p. 287.
[2] Hunt, XI, pp. 322–23.
[3] Cotton, The Progress of the United States, p. 129.
[4] 36th Cong., 1st Sess., Sen. Docs. No. 41.

a change,—cod and mackerel fishing to decrease materially, herring and menhaden fishing to disappear almost entirely. The new industry, that of canning lobsters and other products of the sea, had already taken the initial step towards developing in prominence and economic value.

The fisheries of New Hampshire do not appear to have risen to any place of prominence after the war of 1812. In 1840, the quantity of fish caught, and smoked or dried in the State was 28,257 quintals, and of pickled fish, 1,715 barrels. There were 399 men employed in the fisheries and capital invested to the amount of $59,680.[1] The records of the value of the fishery products of New Hampshire in the Portsmouth custom house are wanting prior to 1867. The custom returns for that year, which are the most accessible for the period, place the total value of all fish products of New Hampshire for 1867 at $73,853.[2]

A review of the fisheries of Massachusetts towns shows that, like the State of Maine, almost every place bordering on the water front of the state was actively engaged in some branch of the industry during a part, if not all, of the period from 1818 to 1866. Newburyport held a prominent place in the cod and mackerel fisheries throughout the period. Until 1820, it was second only to Boston in the extent of the mackerel fishery, and for twenty years after it occupied third place among Massachusetts fishing towns. The number of barrels of mackerel packed in the town in 1831 was 36,424. During the Civil War there were 7,500 barrels packed annually. In 1835, Newburyport had 41 vessels of about fifty tons each in the codfishery, and 125 vessels engaged in mackereling; the latter number had fallen off to ninety in 1851. There were between 40 and 50 vessels engaged in the Labrador fishery in 1850, and 60 vessels there in 1860. Fishermen from this town used

1 Sabine, p. 175.
2 Hunt, Vol. XII, p. 96.

the water telescope for locating schools of cod on the shallow bottoms and a seine for catching the fish after locating them.[1]

The fishermen of Beverly centered their attention on the codfishery. The mackerel fishery never rose to a place of importance. As early as 1832 there were between 40 and 45 fishing vessels belonging to the town. Seven years later the number had risen to 49, averaging seventy-four tons each, the aggregate value, excluding the outfit, being $100,000. These vessels employed 306 Beverly hands and 112 others. The aggregate bounty on vessels owned in Beverly in 1839 was $17,040.[2] The value of fishery products for the year 1845 was $85,424. At the middle of the century there were 75 vessels from Beverly engaged in fishing, manned by 1,200 to 1,400 fishermen, mostly of New England birth. Subsequently, the fishery declined slowly, except for a short period of revival during the last three years of the war. In 1869, twenty-seven schooners employed 350 men, who caught 32,000 quintals of cod and halibut, valued at $200,000.

Salem played an unimportant part in the fisheries. The town was fifth in the State in the amount of mackerel packed in 1820, and in 1825, the amount for the latter year being 11,460 barrels. In 1836, there were only fourteen vessels, carrying 130 men, engaged in the fisheries. The value of the product of the codfishery for the year was $16,552, and of mackerel, $21,450. Only three vessels were engaged in the codfishery in 1845.[3]

Marblehead, ranking first for years previous to the Revolution, took a less prominent place in the fisheries after 1818. In 1829, about fifty vessels averaging sixty tons fitted for bank fishing. Fifty-seven vessels in the cod and

[1] Goode, Sec. II, p. 135.
[2] Hunt, Vol. I, p. 182.
[3] U. S. Fish Com., Report, 1881, pp. 259, 260.

mackerel fisheries are recorded for 1831, carrying 412 men. The value of the catch was $160,490. A writer of the time estimates that this amount gave to each of the men on an average, after all expenses are deducted, $214 for eight months' fishing. The most prosperous season for Marblehead was in 1839 when there were 98 fishing vessels, only three of which were under fifty tons burden. Six years later, when the period of decline had set in, the 65 fishing vessels of the town had a tonnage of 5,039 tons and 463 hands employed. These vessels landed 40,500 quintals of cod, which with other products of the sea had a total value of $123,256.[1] The number of vessels had fallen off to 48 in 1856; but the value of the catch, which was smaller than that of 1845, was $163,656. During the war the number of vessels in the codfishery declined from sixty-one in 1862 to twenty-five in 1866.

Since the Revolution, Boston has held high rank as a center for the commerce of the fishing industries. Down to the middle of the century, at least, Boston was the chief mart for the sale of dried fish, and a resort for fishermen of all classes for outfits. A large portion of the imports of fishery products into the country centered in Boston.[2] Between 1810 and 1826, more than half of the total mackerel catch of the State was brought to Boston to be sold. From 1804 to 1840, excepting one year when Gloucester took the lead, Boston held first place in Massachusetts in the number of barrels of mackerel inspected yearly. During the thirty years from 1821 to 1851, the imports of dried and pickled fish into the port of Boston consisted of 47,782 quintals of dried fish, valued at $111,643, and 379,587 barrels of pickled fish, valued at $2,126,128. For the eight years from 1843 to 1851, the exports of American caught fish from Boston consisted of

[1] Niles' National Register, Vol. lxx, p. 21.

[2] Hunt, Vol. xxiii, pp. 487–88.

990,489 quintals of dried fish valued at $2,616,845, and 90,321 barrels of pickled fish valued at $370,907. The total value of exports of both American caught and foreign caught fish from Boston for the eight years was $3,453,063.[1]

Information regarding the fishing fleet of Boston is incomplete. For the year 1855 Boston ranked third in the State in the extent to which its citizens pursued the mackerel fisheries, there being in that year 89 vessels, of a tonnage of 7,100, engaged in fishing, with 1,000 men in the crews. The amount of capital invested was $260,000, and the value of the product was $317,000.[2] No vessels were engaged in the codfishery.

The towns of Hingham and Cohasset carried on a thriving business in the mackerel fishery during most of this period. Part of the time Hingham ranked third in the State in the amount of the catch. Her fishing interests were carried on, in 1851, by 500 of her citizens in 37 vessels. The Plymouth district, which includes Scituate, Duxbury, Kingston and Plymouth, had, on an average, sixty-eight vessels in the codfishery from 1818 to 1866. The smallest number was 48 in 1826, when 24,000 quintals of cod were taken; the largest number was 94, in 1839, when 63,763 quintals of cod were taken. The mackerel fishermen of Plymouth caught 5,520 barrels of mackerel in 1830. This branch of fishery declined after 1838 until it became of no consequence a decade later.

Wellfleet takes high rank among the communities whose people pursued the mackerel fishery. The business began in 1826, and its increase was rapid and steady. Some years as many as 100 schooners were in the fleet. In 1833, there were 12,811 barrels of mackerel inspected; in 1845, the number was 19,900 barrels; in 1848, it reached 28,219 bar-

[1] Andrews' Report, pp. 636, 637.

[2] 36th Cong., 1st Sess., Sen. Docs. No. 41, quoting from "Industry of Massachusetts."

rels. From 1845 to 1865, Wellfleet was generally next to Gloucester and Boston in the rank of mackerel fisheries. In 1860, the 75 schooners comprising her fleet were valued at $375,000. During the war the industry was prosecuted with excellent returns, the number of barrels of mackerel inspected being 111,944.[1]

Freeman, in his History of Cape Cod, says, "In 1860, Provincetown might be pronounced beyond contradiction one of the most enterprising and flourishing towns in the country. The fisheries now, as ever, command much attention, and employ a great number of men and a very large amount of capital. These fisheries, it may be said, train a large number of the most experienced and intrepid mariners in the world." The fisheries were principally whale, cod, and mackerel. The character of the codfishery cannot be ascertained for the period of 1818 to 1866, it is, however, known to have been extensive and prosperous. In 1862, the number of vessels was about seventy-six, below what the former tonnage had been, and the aggregate of the catch was about 65,000 quintals. During the progress of the war the number of vessels increased; in 1867, the fleet numbered ninety-one, carrying 988 men. The catch for that season was 78,500 quintals of cod and 15,156 quintals of halibut. The mackerel fishery did not assume a position of great importance at Provincetown until between 1845 and 1850. The best year was in 1848, when 31,049 barrels were inspected. For a decade after 1859 Provincetown held third or fourth place among Massachusetts towns in the number of barrels of mackerel inspected.

Chatham had 21 schooners employed in the fisheries in 1837, yielding 15,500 quintals of cod, valued at $46,500. Prior to 1845, the greater part of the fleet was engaged in codfishing. After this date the codfishery declined on

[1] U. S. Fish Com. Report, 1881, pp. 258–274.

account of the harbor filling with sand, which prevented the larger vessels from passing in and out easily. The mackerel fishery, employing smaller craft, increased as the cod-fishery fell off. The people of Dennis began to catch mackerel in 1836. Nine years later the town had from 50 to 70 vessels in the fishery. The fisheries increased and prospered year after year. Harwich carried on a prosperous business in the mackerel fishery from 1846 to 1866. Other towns on the south coast of the State were engaged in the fisheries, notably Nantucket and New Bedford in the whale fishery. Cod, mackerel, shad and oyster fisheries were carried on in this section, but their importance was not considered seriously by the people until the whaling industry began to decline.[1]

Gloucester has been left until the last among Massachusetts fishing towns in order that its importance as a fishing port might be emphasized. In 1818, its fisheries were of minor consideration in comparison with some other towns. In 1866, Gloucester led all ports in the New World in the extent and importance of its fishing interests. This distinction has been held by the town ever since, notwithstanding the fact that an immense trade in fresh fish is carried on from Boston, and no account of American sea fisheries for the last fifty years is complete without frequent reference to the extent and importance of Gloucester's fishing interests.

The story of the cod far antedates that of the mackerel; but none has a more absorbing interest than the mackerel fishery both for the importance of the industry and the methods pursued. The first trip for mackerel to salt was made by the schooner *President,* Captain Simeon Burnham, in 1818. Previous to this the fish was used principally as bait. Down to 1818, the Gloucester catch of mackerel was small and remained so until 1821, when 2,177 barrels of

[1] Goode, Sec. II, pp. 729–734.

the fish were inspected in that port. The total amount of mackerel previously inspected there was only 1,272 barrels. In 1822, Gloucester schooners first went to George's for mackerel, and, in 1830, the first vessel went to the Gulf of Saint Lawrence for the same purpose.

Mention is made of an immense school of mackerel which suddenly appeared on Middle Bank in 1825. For three days a fleet of about 200 vessels fished as continuously and as fast as nature would allow; at the end of the third day the fish disappeared as mysteriously and as suddenly as they had come.[1] A single jigger in that year, with a crew of eight men, caught 1,300 barrels of mackerel.[2] The fish were in abundance again the following year and so continued until 1831, which was a record-breaking season. During this period of great plenty, the vessels of the fleet averaged 800 barrels of fish a season. The boats at first were from 40 to 50 tons burden. As the industry became more assured in its permanency and was extended profitably to George's Bank and the Gulf of Saint Lawrence, the vessels increased in tonnage and the fleet in size.

Gloucester surpassed Boston in the extent of its mackerel fishery in 1840, and has held first rank ever since. During the period of the Reciprocity Treaty, 1854 to 1866, a very prosperous business was developed by Gloucester fishermen in the Gulf of Saint Lawrence, as the provisions of the treaty allowed American vessels to catch fish close inshore. Several hundred vessels were annually fitted out for the Bay, as the Gulf of Saint Lawrence is popularly designated. Often the mackerel were shipped home from there in Provincial vessels, thus allowing our fishing vessels to take two or more fares a season.[3] The method of catching was still by hook and line, or jigging, as it was

[1] Fisherman's Own Book, p. 197.
[2] Fisheries of Gloucester, p. 36.
[3] Goode, Sec. II, p. 152.

called. Bait was thrown from the vessel to draw the fish from the bottom and to attract them to the reach of the lines of the crew, who fished from the vessel's deck. Many vessels carried as many as 75 barrels of menhaden slivers, which were ground fine in mills for bait to be cast overboard to attract the fish.[1]

The catch of the Gloucester mackerel fleet for 1831 was 69,756 barrels; for 1851, it was 81,627 barrels. The average catch for the years of the war, 1861 to 1865, was 131,432 barrels. The record year was 1864 when 154,938 barrels of mackerel were taken. The value of this catch for Gloucester was in the millions. There were 68,061 barrels of No. 1 mackerel, quoted at $30 per barrel, and 73,002 barrels of No. 2 mackerel worth $20 per barrel. Based on these prices the value to the town of the mackerel industry for 1864 alone was in excess of $3,500,000.[2] The high-line stock for a vessel at the Gulf of Saint Lawrence in 1865 was made by the schooner *Colonel Ellsworth,* her net stock for a trip of five months being $13,728. This was the highest stock made by a schooner to that date. Among the crew, the high-liner's share was $558.[3]

The Gloucester fishery for cod on the Grand Bank of Newfoundland was at a very low mark from 1804 to 1819. In the latter year a company was formed with $50,000 stock for the purpose of reviving the industry, but after operating unsuccessfully for about three years the attempt was abandoned. The Newfoundland codfishery thereafter did not become of marked importance. In 1821, four schooners from the town made initial trips for cod to the Gulf of Saint Lawrence, returning with full fares near the first of July, after an absence of about two months.

While these pioneers were in quest of cod in the Gulf,

1 Fishermen's Memorial and Record Book, p. 63.
2 U. S. Fish Com. Report, 1881, pp. 274, 300.
3 Memorial and Record Book, p. 86.

three other schooners were making attempts on George's Bank. As early as the middle of the previous century, Marblehead fishing vessels were accustomed to make trips to George's for cod. The vessels did not anchor, but drifted about on the fishing grounds. It was the prevailing belief that no vessel could anchor safely on George's for fear of being overwhelmed and sunk at anchor by the strong current that is found on the Bank. The three Gloucester schooners kept close together for mutual assistance if any danger arose. Finally the crew of one of the vessels got up courage to cast anchor, only to weigh it again immediately.

For a number of years the codfishery did not progress. In 1827, the product of the fishery was 66,133 quintals and 2,204 barrels of oil. The offshore codfishery of 1829 was in a depressed state. The business had become of little value to the owners of the fishing vessels, due largely to competition of foreign fishermen and higher rates of bounty paid to them. The falling off of foreign trade also meant the loss of profitable returns formerly made on cargoes of sugar, wine, and other imports, brought by vessels on their return trips from selling their fish abroad. Then, too, the remarkable development of the mackerel fishery turned much capital into the newer and more remunerative occupation. The period of decline in the codfishery did not change for the better before 1841.

In the meantime, other kinds of fishing were profitably pursued. The shore fishery of Gloucester was of considerable importance about 1832, when 799 men were employed. The catch of 63,112 quintals of cod was valued at $157,780, and there was a government bounty of $25,172. As early as 1819, halibut had been found on Middle Bank.[1] The first trip to George's for halibut was made by the schooner

[1] Memorial and Record Book, p. 67.

Nautilus in 1830.[1] The codfishery revived somewhat, and the halibut fishery became of importance as a permanent business about 1835. The value of cod for 1837 was $186,516. Seven years later the fisheries of Gloucester employed 249 vessels, carrying over 1,200 men, and secured 86,315 quintals of fish. In 1846, there were 220 vessels in the fleet, and 1,850 fishermen.

The railroad connecting Gloucester with Boston was completed in 1846. This means of communication was of immediate and lasting value to the fisheries of Cape Ann. Previous to this time Boston had been the market for fresh fish. Vessels did not then carry ice at sea to preserve the fish, but some of them were fitted with water-tight compartments amidship. By boring holes through the bottom of the vessel into the compartment the free access of water was secured, in which the halibut were kept alive until the vessel reached market. In 1848, a company was formed at Gloucester for the purpose of making the town a shipping port for fresh halibut. The company failed the first year; they had agreed to buy all halibut furnished by the fishermen; the season's catch proved to be exceptionally large, so that markets could not be found to take the product that was forced upon the company.

The total value of the fishery products of Gloucester in 1847 was $589,354. The number of vessels employed was 287, of which 126 were less than forty tons burden; the total tonnage was 12,354 tons; the number of men employed was 1,681 and 186 boys.[2] During the twenty years following 1847, the fisheries of Gloucester developed rapidly. More than 2,000 men were employed in the industry in 1851. The value of the halibut catch that year was $120,000. The introduction of frozen herring from

[1] Ibid, p. 77.

[2] Babson, History of Gloucester, p. 599.

Newfoundland in 1856 for bait in the George's codfishery was a great stimulus to the prosecution of that fishery. The herring furnished excellent bait for the school of cod that appears on George's in winter. There were twenty firms in the town that owned schooners engaged in the mackerel fishery or fitted out other schooners for the business. In 1859, three hundred and one schooners formed the fishing fleet of the place, manned by 3,434 men and 134 boys. The value of the products of the fishery that year was $1,276,704, exclusive of the value of the herring trade with Newfoundland, which was worth $250,000 more.[1]

According to the census report of Massachusetts for the year 1865, the fisheries of Gloucester employed 358 vessels with an aggregate tonnage of 25,670 tons, and the value of the products was $3,319,458. This is a remarkable record of growth, for the value of the fishing industry of Gloucester had increased fivefold, or more, within eighteen years. The products of the fishery were sold principally in Boston, New York and Philadelphia.[2]

The status of the Massachusetts cod and mackerel fisheries for 1855, and the relative standing of the principal fishing towns show that the number of vessels engaged in the fisheries of the State that year was 1,050, of a tonnage of 71,372, manned by 9,756 men. The amount of capital invested in the industry was $3,638,041, the value of the production for 1855 was $2,753,535, and the gross earnings for the year were nearly 76 per cent of the capital invested. The extent and value of the fisheries of Gloucester surpassed the combined products of Boston and Provincetown, the next towns in importance in the State.[3]

In 1859, Massachusetts had 1,138 vessels employed in the

[1] Goode, Sec. II, p. 692.

[2] Hunt, Vol. xliv, p. 229.

[3] 36th Cong., 1st Sess., Sen. Docs., Vol. I, No. 41, quoting "Industry of Massachusetts."

codfishery, aggregating a tonnage of 56,919 tons, and carrying crews of 7,966 men. The mackerel fishery employed 284 vessels, with a tonnage of 17,038 tons, and crews of 2,272 men. The total number of vessels engaged in the cod and mackerel fishery was 1,422, of an aggregate tonnage of 74,957 tons, and carrying 10,238 fishermen. In addition, Massachusetts had a fleet of 514 vessels in the whale fishery, carrying 12,336 men.

The records of the fishery industries of the other states of New England are very unsatisfactory. The fishermen of Rhode Island carried on a business from time to time catching menhaden and developing oyster beds. Owing to the high price of paint oils in 1812, the inhabitants of the State began to use fish oil in place of the more costly material. The process of extracting oil from fish was improved in 1820 by first boiling the fish in kettles. Ten years later further progress was made by the inception of steam-cooking. Between 1835 and 1840, the refuse parts of menhaden, known as chum, became of value as fertilizer. The introduction of the purse-seine for taking the fish, probably before 1850, was revolutionary and stimulated the industry greatly. Between 1855 and 1860, presses for separating the oil and water from the chum came into use and were of additional economic importance.[1]

The practice of introducing oysters from Chesapeake Bay and laying them down in the shore waters of Rhode Island dates from the early part of this period. The oyster industry in Rhode Island flourished with increasing interest until the Civil War, when it decreased principally because the southern supply of oysters for planting was cut off by the opening of hostilities. The value of the oyster fishery of the State for 1860 is placed at $382,170, by Prof. Goode. The general fisheries of the State for that year

[1] Goode, Sec. V, Vol. I, pp. 366–68.

yielded 118,611 barrels of menhaden and other fish for fertilizer, worth $27,817; about $25,000 worth of food fish; and $11,692 worth of clams and other shell fish. According to the State census for 1865, the product of the fisheries of Rhode Island that year was as follows: fish seined for manure and oil, 154,468 barrels, worth $126,035; fish caught for food, 2,462,360 pounds, worth $121,094; 31,697 bushels of clams, 72,895 bushels of oysters, and 42,900 pounds of lobsters, having a total value of $118,655. The aggregate value of the fish products of Rhode Island for 1865 was $365,784.

The Connecticut River seems always to have been famous for its shad fishery, which was pursued with profit as far up its course as Hadley, Massachusetts. It is reported that in 1801 there were as many as fourteen wharves at South Hadley, where shad were taken by means of scoop nets and seines, sometimes as many as 1,200 at a single haul. In 1848, it was not an uncommon thing for a man to take from 2,000 to 3,000 in a day. The method of pound fishing was introduced in 1849, after which the fishery increased all along the coast.

Menhaden were caught and the oil extracted as early as 1850 or 1852 at an establishment at Fort Hale, New Haven Harbor. The discovery of the process of extracting the oil by steam was claimed by a Connecticut man as early as 1852 or 1853.[1] In 1840, Connecticut was second only to Massachusetts in the amount of capital invested in the fisheries. It is probable, however, that a large part of the capital was employed in the whale fishery, which was carried on principally from New London.

The catch of menhaden for Connecticut can not be ascertained for any town or for any period of years. In 1851, five million of the fish were caught at Westbrook, but the industry subsequently declined. In the earlier half of the

[1] Goode, Sec. V, Vol. I, p. 389.

century, salmon fisheries were carried on in the rivers of the State, but to a less extent than either the shad or the menhaden fisheries.

Fair Haven, Connecticut, was one of the first places in New England to import oysters from New Jersey, and later from Virginia, to be transplanted in northern waters for additional growth. The Virginia trade began between 1830 and 1840, and there was a rapid development of the industry. The oyster establishments of Fair Haven had branch houses in the principal inland cities as far west as Chicago and St. Louis. In 1857–58, from 200 to 250 schooners were employed in supplying the establishments of Connecticut with clams from the Chesapeake. In 1850, one of the more enterprising merchants of Fair Haven transferred his business to warehouses nearer the source of supply of oysters and opened branch houses at Baltimore. Others followed the lead, so that it came about, according to Ernest Ingersoll, that all the great Baltimore firms of old standing originated in Fair Haven. The result was that the oyster trade retrograded; new changes in method came into use, and different results followed.

CHAPTER XI

INSHORE FISHERIES—THE HERRING FAMILY

The fisheries of the herring family most common to the New England coast are the common herring, the shad, the alewife, and the menhaden. Other species of these occur, as the branch herring, the glut herring, and the hickory shad or tailor herring; but for our purpose it will be sufficient to consider the more common kinds. The herring family is a fish of great importance commercially. This value is due to the wide range of the fish, the great abundance of the supply, the small amount of capital necessary for catching the fish, and to the great variety of food values and commercial purposes for which it is used. It has been for ages the poor man's food. For centuries the people of Europe have drawn immense quantities of the fish from the sea with no apparent diminution of the supply. As food for man it is used fresh, pickled, smoked and canned, being labeled as sardines, mackerel, and trout; it is frozen to be used as bait on the deep-sea banks making possible the prosperity of the bank fishery of our country; thousands of tons of fertilizer have been used in restoring worn-out farm lands; and millions of gallons of oil have been used by painters as the basis of their paints.

The different species of the fish are found on the banks in immense schools; others frequent our bays and harbors, entering into almost every water area along the Atlantic seaboard as far north as Labrador; still others ascend rivers for hundreds of miles, bringing to the very doors of inland people the fresh products of the sea. Not only do the last named fish increase the ease by which they may be taken, but their very presence in inshore waters attracts other fish of a deep-sea variety and renders possible the pursuit of bank fishery by fishermen of small means.

In our country, all varieties of the herring family have been caught from the earliest days. At first their chief importance was for fertilizer and for fresh food. After the middle of the last century, their commercial importance was greatly increased by the discovery of the value of the oil, by their use as bait for deep-sea fisheries, and subsequently for their value in the sardine canning industry. The records of the fishery are intermittent, as no accurate attempt at recording their catch and value was begun until after the close of the Civil War. There is little doubt but that this family of fish will continue for generations to be of importance in our fisheries, for the demand is greater than the supply, and the supply is of such a kind that there is little likelihood of the abundance being diminished materially by the annual catch of fish.

THE HERRING.

The true herring, sometimes called the sea herring, or the English herring, ranks among the foremost of the world's food-fishes. They occur on the Atlantic coast from Labrador on the north to Cape Cod, occasionally even to Cape Hatteras, on the south. It is probable that these schools resort to inshore grounds for the purpose of spawning. The greater abundance of herring north of Cape Cod makes it essentially a northern fish. On the east coast of Maine young herring are canned and sold as sardines.

In America, as in Europe, the herring spawn at different seasons of the year. According to Mr. Earll,[1] they spawn

[1] The Herring Fishery and the Sardine Industry, Goode, Sec. V, Vol. I, p. 402.

on the south coast of Newfoundland between the middle of May and the first of July. In speaking of their movements further, he says:

"They usually approach the shores of the Magdalen Islands, in the Gulf of Saint Lawrence, during the last week in April. They visit this locality wholly for the purpose of spawning, and leave as soon as the eggs have been deposited, which usually requires from three to five weeks.

"About the time of their departure from the Magdalens, schools of spawning fish make their appearance at the western end of Cape Breton Island. They are first seen in the vicinity of Port Hood, and from this point they gradually work toward the Strait of Canso. These remain but a short time, frequently leaving in less than two weeks from the time they are first seen.

"At Grand Manan Island, New Brunswick, the herring strike in, in July, and the spawning season continues from the first of July to the middle of September. . . .

"At Wood Island, Maine, and at Cape Ann, Massachusetts, they usually arrive about the twentieth of September, and the spawning is at its height from that time till the middle of October. According to Professor Baird, the spawning season occurs even later as we proceed southward.

"It is claimed, however, and is doubtless true, that the spawning season for the winter schools in the vicinity of Eastport is in March and April, and that they frequently spawn at St. Andrews Bay as late as the middle of May."

The herring spawn in shoal water, never entering rivers for that purpose. In New England, they are found at times in almost every bay and cove from Cape Cod to Eastport. Young and old go in schools by themselves. The region of Passamaquoddy Bay is a famous resort for herring, and the principal locality where young fish are taken in abundance. Other places resorted to by American fishermen for herring are the coast of Labrador, the southwestern coasts of Newfoundland, the Magdalen Islands,

Anticosti, Nova Scotia and New Brunswick. There are several distinct branches of industry arising from the herring fishery, according to the use that is made of them. These will be treated as the frozen herring industry, the smoked and pickled herring industry, and the sardine canning industry.

The frozen herring industry of New England, which has laid the foundation for the prosperity of the deep-sea fisheries of the country of late years, was begun by Capt. H. O. Smith, of Gloucester, in the winter of 1854 and 1855. The first trip was made from Newfoundland, consisting of a mixed cargo of codfish, halibut, and 80,000 frozen herring. This venture to Newfoundland in the middle of the winter season was in the nature of an experiment; but the business proved to be so successful from the first that it has been followed more or less successfully ever since. The first dozen years from its inception the industry was confined to Newfoundland herring. The second year of the trade, four vessels made trips aggregating 730,000 fish that were brought back. Year by year the number of vessels engaged in this branch of fisheries kept increasing until the winter of 1863 and 1864, when a fleet of 39 vessels brought 10,700,000 frozen herring from Newfoundland. The total number of herring brought from Newfoundland down to 1866 and 1867 was about 60,000,000.[1]

During that season there were 45 vessels comprising the frozen herring fleet, one of them going for the first time from Gloucester to the herring grounds of Grand Manan, in the Bay of Fundy. At that time the Gloucester fishermen hesitated to visit that region on account of their prejudice against net herring, the higher price demanded for the fish, and the smaller size of the herring when compared with those of Newfoundland. But the fleet soon found it more profitable to make short, quick trips to the

[1] Goode, History and Methods, I, p. 458.

Bay of Fundy than the longer and more dangerous ones to Newfoundland. The supply of herring at Newfoundland, too, did not always prove adequate to furnish every vessel with a cargo of fish, while there was no scarcity of the fish at this time at Grand Manan. From 1866 to 1867 the number of vessels that went to Newfoundland kept diminishing yearly, while the Grand Manan fleet grew larger. In 1870 and 1871, twenty-three of the fleet went to Newfoundland, thirty-six went to the Passamaquoddy region. In ten years more the Newfoundland industry in frozen herring had dwindled to almost nothing, as the fleet numbered but three vessels. The same winter, 118 vessels went to Nova Scotia, returning with more than 32,000,000 fish.

From the winter of 1866 and 1867 to that of 1880 and 1881, inclusive, about 110,000,000 of herring were brought back by the Newfoundland fleet, against 185,000,000 secured by the fleet going to Nova Scotia waters. The average price paid to fishermen was 30 cents per 100. The cost of barrels and handling gave an added value of about one-tenth to the cost of the herring. During the winter of 1880 and 1881, there were shipped from Eastport and the vicinity 32,630,000 herring, estimated to be worth $98,700 to the producers.[1]

The following decade witnessed a rather remarkable reversal of the frozen herring industry. The Nova Scotia trade declined until it failed entirely. In the meantime, the Newfoundland industry revived to a prosperous condition, far ahead of previous successful years. In 1885 and 1886, there went to Grand Manan only 13 vessels, in place of the 118 that made successful trips there only five years before. The Newfoundland fleet had increased from three to twenty vessels and brought back 7,540,000 fish,—about 1,000,000 in excess of the Grand Manan supply for that

[1] Goode, Sec. V. Vol. I, p. 458.

year.[1] Five years later, 1890 and 1891, the Passamaquoddy fishery proved to be a failure. The Newfoundland fleet numbered seventy vessels and made profitable voyages. Many of the fleet failed to secure cargoes of frozen herring on account of the mild weather; they brought back salt herring instead, which were used in making smoked bloaters. This practice has been followed considerably since that time. The fleet for the next winter—1891-1892—was a record-breaker, there being 100 sail that went to Newfoundland. During the remainder of the century the number of vessels and the quantity of the catch varied from year to year. There were 48 arrivals in the winter of 1898-1899, and 38,850 barrels of frozen herring were brought back, selling at from two to three cents from the vessel. Thirty arrivals the following winter brought in about 15,000 barrels. The next three years were not up to the usual standard, the supply being short. From 1903 to 1906, about 23,000 barrels were landed each year.[2] *The Fishing Gazette* summarizes the Newfoundland Herring Fishery for 1906 and 1907 as follows:

"According to official reports from Bay of Islands, at the conclusion of the herring fishery this year the total catch amounted to 116,236 barrels, valued at $452,144. Of this 65 American vessels took away 70,346 barrels, valued at $268,596, while 43 colonial vessels gathered in 45,887 barrels valued at $183,548. These figures show that the fishery this year was the largest in quantity and value in the history of the West Coast herring industry, and that the colonial vessels secured a larger share of the haul and profits than ever before. Five American and five colonial vessels, nearly all of them with cargoes on board, were wrecked during the season." [3]

[1] Fish Commissioner's Report, 1886.
[2] Boston Fish Bureau Reports.
[3] The Fishing Gazette, January 26, 1907.

The Newfoundland herring fishery has been the cause for much misunderstanding and considerable ill-feeling between the fishermen of New England and the residents of the Island. The most famous collision between the two classes of fishermen occurred in the winter of 1877–1878, and is known as the Fortune Bay riot. By the terms of the treaty of Washington, in force from 1873 to 1885, the Americans were given rights on certain parts of the Newfoundland coasts in common with the British subjects, as had been granted by the Treaty of 1818. The award of the Halifax Commission of 1877 made it incumbent upon the government of the United States to pay Great Britain for the excess of privileges granted under the new treaty. This amount was set at $5,500,000, and was paid by the United States without controversy.

"The fishermen of Newfoundland were especially restive under what they regarded as the ceding away of their rights by the imperial government. They looked upon the shore herring fishery of their island as peculiarly their own property. It was their principal source of revenue, on which themselves and their families were dependent for bread; and if this fishery were taken away from them, or its value destroyed, starvation would stare them in the face. Accordingly, when a large American fleet appeared upon their coast, in the winter of 1877 and 1878 with improved apparatus for the catching of herring, instead of purchasing a supply of the native fishermen, as had previously been the practice, they were greatly incensed, and resolved to defend what they felt to be their rights." [1]

The Gloucester fleet consisting of twenty-two vessels arrived at Newfoundland about two weeks ahead of the appearance of the school of fish.

"The herring struck into Fortune Bay on Sunday, Jan. 6, 1878, in such quantities that a few hours' seining would have

[1] Fishermen's Own Book, p. 109.

sufficed to load the entire fleet. The schooners *New England*, Capt. Peter McAuley, and *Ontario*, Capt. John Dago, joined and set their seines, making a double seine about 2,400 feet long and 150 feet deep, which soon filled with herring, the catch being estimated at fully 2,000 barrels. This, with the operations of the other Gloucester schooners, maddened the Newfoundland fishermen, who had gathered in the vicinity to the number of two hundred or more, and they commenced to put off in boats, making warlike demonstrations and using threatening language, and commanding the American fishermen to desist from fishing. The seines of the *Ontario* and the *New England* were seized by the mob, the fish let out, and the seines torn in pieces and carried away.

"The rioters next turned their attention to the schooner *Moses Adams*, Capt. Solomon Jacobs, whose seine had been set and filled, and whose crew were busily engaged in scooping the herring into boats and transferring them to their vessel. Three separate attempts were made to seize the seines, but Capt. Jacobs and his crew were provided with loaded revolvers, and by threatening to shoot the first man who dared to interfere with them, succeeded in keeping the mob at bay until a partial cargo had been secured, when the Newfoundlanders tripped the seine and allowed the remainder of the herring to escape. Had they been allowed to continue fishing, they could have loaded their vessels and helped in supplying others.

"In the evening the rioters had a jubilee, blowing horns, firing guns and shouting, as if celebrating a victory. This was taken as an indication that the same course would be pursued if further attempts were made to fish, and the other vessels were deterred from making the attempt. Next day the herring struck off shore, and finding it difficult to purchase cargoes, the fleet soon set sail for home, mostly in ballast, although a portion of the fleet secured partial cargoes, mostly by purchase." [1]

In July, 1879, two American vessels appeared at Aspee Bay, Cape Breton Island, to engage in the taking of squid

[1] Fishermen's Own Book, pp. 110–111.

for bait with purse seines. The threats of the local fishermen were so serious and determined that the Americans were obliged to refrain from fishing, and returned home without fares. In August, of the same year, the schooner *Howard Holbrook* put into Trinity Bay, Newfoundland, for the purpose of seining bait. The fishermen were opposed by about thirty natives, who threatened to destroy seine and dory if used within the three-mile limit, the spokesman of the party declaring that no one should set a seine for squid within three miles of the shore and live to haul it. In the summer of 1880, other New England fishing vessels were interfered with while in Newfoundland waters. Negotiations were begun between this government and that of Great Britain to recover damages for injuries caused at Fortune Bay and other places. As evidence was produced, very largely from testimony and proceedings of the Halifax Commission, showing the undoubted right of American fishermen to catch fish as they pleased in provincial waters, a settlement was effected in the summer of 1881 by which the United States received from Great Britain the sum of £15,111. The Fortune Bay fleet put in claims amounting to $100,247.58. The amount actually disbursed to them was $52,977.26, the payment being made on the basis of actual expenses, less the value of cargo secured, with seventeen and one-half per cent interest, or at the rate of five per cent per annum.[1]

The Newfoundland herring fishery has been, in recent years, also, the source of complaints of the fishermen of New England and the natives of the island. Attempts at satisfactory adjustment of differences have resulted in no agreement acceptable to both parties. The present outlook is that an amicable settlement will be effected by leaving the question to the Hague Tribunal for its decision.[2]

[1] Fishermen's Own Book, p. 112.

[2] See Ch. XIX.

The smoked and pickled herring industries are closely related, separate treatment being used in curing herring for the market. For seven or eight years previous to 1866, the herring trade was carried on by American fishermen on the coast of Maine from Deer Island eastward to Eastport. The cargoes of fish were secured at the Magdalen Islands and on the Maine coast. About fifty vessels went to the Islands in the spring bringing home full fares, which were sold to good advantage. Since 1866, herring have failed to visit the Magdalen Islands regularly and many vessels resorting to the islands for the fish have returned with broken fares.[1]

Many vessels have resorted to the fishing grounds of Newfoundland for herring after failure to secure a fare at the Magdalen Islands. The principal bays and harbors of the coast of Newfoundland visited by American fishermen for herring are Fortune Bay on the southern side, Bonne Bay and Bay of Islands on the western side. The principal fishing season for Americans occurs in the fall, from October to the last of December. The western bays are visited by the pickled herring fleet: for the past five years nearly all the frozen herring have come from the Bay of Islands, on the northwest coast, having been taken by gill-nets. Ice forms early in the western bays so that vessels are sometimes caught and frozen in, where they have to remain until spring.

The capture of herring at Magdalen Islands used to be by gill-nets and haul-nets. Purse-seines were first carried by the Gloucester fishermen about 1865, and for seven years the larger part of the herring was taken in that way. For many years the seine fishing was profitably carried on by vessels from Lubec, Eastport, Lamoine, and other towns on the Maine coast. The number of vessels visiting the Magdalen Islands decreased after 1870. In 1869, this fish-

[1] Goode, Sec. V, Vol. I.

ing ground yielded 75,000 barrels of herring. In 1870, the yield was almost an entire failure. One hundred and nineteen schooners from New England visited the islands in 1870. The catch for the American and the provincial fleet was only 2,100 barrels. Lamoine, Maine, had ten vessels in this fishery, all of which together did not secure enough to make a single full fare. In 1872, 14,806 barrels were caught by American fishermen. In 1873, the fishery was again a failure on account of the ice which prevented the fleet from reaching the islands. The fishery dwindled until 1880, when no American fleet went to the Magdalens for herring to smoke, although cod vessels resorted to the islands for herring for bait on their bank trips.

The quantity of pickled herring secured by the American vessels at Newfoundland during these years can not be ascertained. Mr. Goode estimates that from six to ten cargoes aggregating from 3,000 to 4,000 barrels were landed yearly by American vessels.[1] The average quantity of pickled herring imported into the United States from Newfoundland annually for the ten years ending with 1876 was 33,000 barrels, valued at $97,425 per year.

The practice of smoking herring in America doubtless dates from the earliest days, when that method of curing fish would be more convenient than curing by salt, which could not always be obtained. The smoking of herring was introduced into Maine at Eastport in 1808, and at Lubec in 1812. By the middle of the last century the business had reached a prosperous condition, being carried on most extensively at Lubec. For twenty years previous to 1865, not less than 500,000 boxes of herring were annually cured in the limits of the town. After the close of the war the demand fell off somewhat. During the period of the Treaty of Washington, from 1873 to 1885, the demand for smoked herring was so reduced by large impor-

[1] Goode, Sec. V, Vol. I, p. 470.

tations of smoked herring from the British Provinces, which, by the terms of the treaty, were admitted free of duty, that in 1880 the output of the entire State of Maine was only 370,615 boxes, valued at $99,973. In the meantime the imports increased from 1,029,095 pounds, valued at $34,670, in 1874, to 10,441,355 pounds, valued at $129,034 in 1885.[1]

In the last quarter century, the smoked herring industry has increased about four-fold in the quantity of herring used, and about three-fold in the value of the product.[2] The condition of the industry to-day is better than ever before. The business is carried on exclusively in Washington County in the vicinity of Eastport and Lubec. When herring for this purpose used to be secured in abundance from the Magdalen Islands, the business of smoking the fish was carried on in towns farther to the west. In 1880, for example, one-fifth of the business was carried on in counties west of Washington County, in the vicinity of Frenchman's Bay and Portland, principally.

1 Ansley Hall, The Herring Industry of Passamaquoddy Region, Maine.

2 The following table gives the total quantity and value of the herring prepared by fishermen, canners, and regular smokers, in Maine for each year from 1880 to 1906 for which statistics are available.

YEARS.	Pounds.	Value.	YEARS.	Pounds.	Value.
1880	4,434,111	$ 99,973	1892	10,151,695	$232,036
1887	3,419,485	100,488	1893	10,671,170	185,836
1888	4,360,435	140,154	1902	14,910,560	314,152*
1889	5,090,425	159,330	1905	16,495,618	261,536
				Boxes	
			1906	3,359,100	317,480

* Statistics of the Fisheries of the New England States for 1905, p. 39.

With the decay of Magdalen herring supply the western smoke-houses went out of business, and the industry was carried on in the Passamaquoddy region, where the supply of herring is more abundant and assured.

The name sardine (fish of Sardinia) is applied to a number of species of the Clupeidæ family, of which there are some sixty kinds. The true sardine of the Mediterranean and nearby waters is the *pomolobus* or *Clupea Pilchardus*, but the name now applies also to sprats, pilchards, and to several varieties of small herring, when packed in oil and enclosed in tin boxes.

The idea of packing small fish in oil under the name of sardines originated in France as early as 1850. The business increased enormously during the first decade of its existence and was taken up by other countries,—Spain, Portugal, Italy, Germany, Sweden, Norway and Japan. It was not until about 1865 that Americans learned that small fish, differing from the pilchard of France, were being packed by the different countries as sardines.[1]

The pioneer of the sardine industry in the United States was Mr. George Burnham, of Portland, who thus describes his experiences in investigating the sardine industry in Europe:

"The idea of using the small herring as a substitute for sardines occurred to the writer in 1865. It was well-known to me that myriads of small herring were annually caught at Eastport, Maine. These were too small to use for smoking or pickling, and I thought they might be used with profit as a substitute (for sardines), and if properly prepared they would be equally good."[2]

[1] Bureau of Industrial and Labor Statistics, State of Maine, Report for 1895, p. 142.

[2] Goode, Sec. V, Vol. I, pp. 489–490.

Mr. Burnham visited France to study the question on the scene of actual operations, and became familiar with the details of the French canneries. After this investigation he purchased a quantity of olive oil and other articles to be used in the packing of herring before his return to America. In 1867, he went to Eastport, secured a plant and began a series of experiments. Great difficulty was found in properly drying the fish on account of the dampness of the climate. After a loss of considerable time and money, he was obliged to abandon the business because he could not get rid of the herring-oil flavor. Subsequent developments of the industry show that Mr. Burnham stopped his experiments when on the very threshold of success. His work, however, had called the attention of others to the business, and he should be credited with being the originator of the sardine industry in America.

It had been the practice of a New York firm for about a decade before 1872 to import from Germany small herring under the name of "Russian sardines." These fish were packed in kegs, weighing four, seven, and eleven pounds respectively, the fish being placed in layers and sprinkled with whole spices, such as cloves, pepper, bay leaves, allspice, and others. When in the early part of the Franco-German war, in consequence of the blockade of the German ports by the French navy, the importation of the article from Hamburg had to be abandoned, the price of "Russian sardines" in New York advanced fifty per cent. This condition of affairs led to a consideration of the use of small herring as a substitute for "Russian sardines." Experiments were begun at once to make the Eastport herring serve the purpose. The quality of the Eastport product proved to be superior to the imported "Russian," and an important business was begun. The "home-made Russians" were packed in attractive oblong and square tin boxes, they were guaranteed to keep sound, and since they were also of a better quality than the imported "Russians" they succeeded in driving the German article from the American market.

While this industry was in the early stages of its progress important experiments were being made in putting up herring in oil. These efforts were undertaken by Mr. Henry Sellmann and Mr. Julius Wolff of New York who, in 1876, began to study the problem on a practical basis. Mr. Sellmann moved to Eastport in order to facilitate his experiments. In 1879 a new company was organized to carry on the sardine industry known as the American Sardine Company, and their first factory was built at Eastport. The firm devised a new and better way of preparing the fish for sardines, which differed from the French method, and secured a patent for it. Prior to 1880, Eastport was the only town in America engaged in canning sardines. At that time, five canneries were in operation there and the processes of packing were, as far as possible, kept secret.

The first cannery for sardines was built at Eastport, in 1875. During each of the succeeding years one new cannery was added to the number so that in 1879 there were five in operation. In the spring of 1880 eight more were built at Eastport, and one each at Robbinston, Lubec, Jonesport, Lamoine, and Camden, making eighteen in operation in the State. By 1886 there were thirty-two canneries in operation at Eastport and the neighboring places. Along the coast, scattered from Cutler westward, there were thirteen others in operation, making forty-five canneries in the State in 1886.

The following table shows the rapid growth of the canneries in Maine previous to 1886, and the amount and value of sardines packed at periods since the business began.

YEAR.	Sardine Canneries.	Cases Packed.	Value.	Average Value Per Case.
1875	1	600	$ 6,600	$11.00
1880	15	74,255	743,618	10.14
1886	32	337,553	1,343,723	4.00
1889	23	371,195	1,417,685	3.82
1892	32	461,552	1,618,960	3.50
1898	61	1,202,657	3,103,723	2.58
1902	52	1,203,970	3,631,035	3.02
1905	33	2,159,797	5,078,587	2.35 [1]

From the above table it will be seen that while the increase in the number of canneries was at times rapid, especially from 1875–1886, there has been no year in which the annual output and value of sardines have not increased with still greater rapidity. The number of sardine canneries in 1905 is little more than one-half the number operated in 1898. Within the past few years the sardine industry has undergone considerable change. In 1899 two companies were formed, known as the "Seacoast Packing Company" and the "Standard Sardine Company," which included most of the canneries in Washington and Hancock Counties. The Seacoast Packing Company eventually absorbed its younger rival, and a number of the more antiquated plants were discontinued. Some of the canneries were fitted with new and improved machinery and were thus rendered more effective than formerly. Eleven plants at Eastport, owned by the Seacoast Packing Company, were not operated in 1902. This company was reorganized in 1903, and a greater number of its canneries were sold.[2] The result of this consolidation of interests is that there

[1] Ansley Hall, p. 465; Fisheries of the New England States, 1902 and 1905; Maine Report, 1905 and 1906.

[2] Fisheries of the New England States for 1902, p. 273.

has been a much greater output from the thirty-three canneries of 1906 than from the sixty-one canneries of 1898.

An estimate was made in 1886 of the cost per case of quarter-oil sardines. The material cost $2.83 and the labor $1.17, making the total cost $4.00 per case at the factory. In 1895 the cost per case was only $2.34 at the factory, being made up of $1.42 for material and $.92 for labor. This reduction in cost was due more largely to cheapness in material than in labor. Of the total difference of $1.66 per case, 85% was in material and 15% in labor. In 1906, the value at the factory of all kinds of sardines was $2.35 per case, which differs little from the value ten years ago.[1]

The canning of other fish products is so closely connected with that of canning sardines that it is not out of place for mention to be made of it at this place. These canning establishments put up cod, herring, clams, mackerel, menhaden, and in some cases, when not preparing fishery products, they engage in the canning of berries and vegetables. Formerly the canning of lobsters was an important part of the business, but in the year 1895 the state enacted a law prohibiting the capture of lobsters under ten and one-half inches in length; and the larger lobsters are too valuable for canning purposes.

The number of canneries engaged in canning fish other than sardines was seventeen in 1898, twenty-three in 1902, and thirteen in 1905. These firms have been engaged for the most part in canning clams. In 1876 large herring were canned under the name of "sea-trout," and in the next year from the same kind of herring the canning industry evolved canned "brook-trout." Mackerel were added to the list of canned fish in 1880. The value of these three products of the canning industry in 1880 was

1 Ansley Hall, pp. 467–468.

brook-trout worth $6,500, sea-trout worth $22,058, and mackerel worth $16,400.[1] No mackerel were canned in 1889 but in 1898 the pack of canned mackerel was valued at $44,848. The output of canned clams in the same time increased from $43,050 in 1889 to $206,087 in 1898.

By 1902 packers began to can kippered herring, the output for that year being 1,750 cases, valued at $8,720. Mackerel, lobsters, menhaden, and cod have not been canned to any extent for the last half dozen years, the canning business being confined largely to sardines and the different products of clam, as canned clams, clam juice, clam chowder, and clam extracts.

In 1905, the canning industry of Maine was worth, in manufactured and secondary products, $5,342,062. This total was made up of the following products: sardines, in oil, mustard, and spices, $5,078,587; plain herring $7,200; mackerel, $340; cod, $8,931; Russian sardines, $1,750; herring, salted and smoked, $34,285; clam products, $192,479. Secondary products, oil, pomace, scrap and fertilizer, $18,-490. The number of persons engaged in the canning industry of Maine in 1905 was 7,017 and the wages paid $1,160,434.[2]

THE SHAD.

There is no food-fish on the Atlantic coast that can compare with the shad in its importance to so many persons. Other fishery industries are carried on with greater capital invested, at a greater risk of life and property, with more effort, and with greater returns. But the shad fishery is probably the most universal of the Atlantic coast fisheries. The fish occurs more or less abundantly along the whole coast from Florida to New Brunswick. Like the alewife,

[1] Goode, Sec. V, Vol. I, p. 521.

[2] Fisheries of the New England States for 1902 and 1905.

the shad is anadromous—ascending rivers for the purpose of spawning—when it may be taken by the farmer-fisherman at his very door, hundreds of miles from the coast.

Within the last half century, the abundance of shad in our rivers has been greatly affected by the agency of man, especially in the rivers where dams have been constructed and on which certain mills have been built. In the early part of the last century shad used to enter and ascend the rivers until they met with impassable falls or reached the head waters of the stream. They were taken at all points along the run. The most important agencies that have brought about a limitation of the range of shad in rivers are insurmountable dams, the pollution of water by manufacturing plants and the sewage of cities, agricultural operations that, carried on near the rivers, cause the waters to become muddy during the spawning seasons, and the extensive fisheries usually placed at the mouths of rivers. In a few rivers the development of the water power has resulted in completely exterminating the anadromous fishes, this being the case in the Thames, the Blackstone, the Saco, and the Merrimac. In the case of the Connecticut, the abundance of the shad has decreased greatly since the building of the dam at Holyoke in 1849. The fish were thus prevented from ascending the river above the dam and for several years were caught in greater abundance at points below the obstruction. But from the six years from 1865 to 1870 the annual catch averaged 4,482 shad, less than one-half the former yield. The record of the catch on the Connecticut from 1853 to 1896 shows that the total yield below the dam decreased from nearly 500,000 annually to an average of less than one-tenth of that number.[1]

[1] The Shad Fisheries of the Atlantic Coast, pp. 112–113.

How the range of shad in New England rivers has been cut down is apparent from the following tables, exhibiting the original and

The shad appear on the New England coast from about the first of April to the first of May. They immediately ascend the streams for the purpose of spawning. After remaining in the rivers for several months they disappear, in their annual migration to southern waters, for the winter. The fish is of value not only on account of its availability of capture and its commercial worth to mankind but indirectly because the large schools of young shad, when leaving their native streams, attract to the coast the deep-sea fish, as the cod, haddock, and other offshore species. Thus the larger kinds of fish may often be taken without the expense and risk of trips to the distant banks.

The shad fishery of the United States centers principally in New Jersey, Virginia, North Carolina, and Maryland. The industry in New England is inferior when compared with the fishery that is carried on in the other Atlantic states. The number of shad taken in the country in 1896 was above 13,000,000, valued at $1,651,443. The catch of the New England states was 490,000 fish, valued at $51,-696. The value of the catch for 1898 was $44,018, and for 1902, $58,564. The rivers of Maine and Connecticut furnish the larger part of the New England catch of shad. The

the present range of the fish. The present range of shad in these five rivers is less than 40% of the original range.

Rivers.	Original limit of shad run.		Present limit of shad run.	
	Locality.	Distance from coast line. Miles.	Locality.	Distance from coast line. Miles.
Housatonic	Falls Village	150	Birmingham	92
Connecticut	Bellows Falls, Vt.	204	Windsor Locks, Ct.	89
Merrimac	Winnespesaukee, N. H.	125	Lawrence, Mass.	20
Kennebec	Carritunk Falls	108	Augusta	44
Penobscot		90	Verona	35

catch and value of the shad fishery of Maine within recent years has been as follows:—1896, catch 366,738 fish, valued at $30,788; 1898, catch, 861,879 fish, valued at $19,752; 1902, 848,999 fish, valued at $28,959; 1905, 1,087,200 fish, valued at $54,286; 1906, 470,200 fish, valued at $7,716. The number of persons employed in 1905 was 285; in 1906, there were 350.

For thirty years or more the people of Connecticut have been endeavoring to keep alive the shad fishery of the Connecticut River. Of all rivers of New England this one, perhaps, has suffered the most from the destruction of its fish. The salmon began to fall off in numbers as soon as dams were erected. They disappeared entirely from the river more than a hundred years ago. The movement of restocking the Connecticut with salmon and shad had its beginning in a meeting of commissioners from New Hampshire, Vermont, Massachusetts, and Connecticut held at Boston on February 27, 1867. An agreement was entered into at this meeting whereby "New Hampshire was to procure and distribute impregnated ova of salmon and shad in the head waters of the river. Vermont and Massachusetts were to build suitable fishways for the passage of fish over the dams to their spawning grounds; and Connecticut was to abolish gill-nets, stake nets, and pounds in the river and on the sound."[1]

Connecticut immediately restricted the shad fishing to the period from March 15 to June 15. New Hampshire planted 20,000 salmon fry in the head waters of the Connecticut that year. Seth Green began the artificial propagation of shad at Holyoke. His attempts were not successful at first but the next year he succeeded in hatching several million of shad. In 1869 no attempts were made to hatch shad.

[1] Report of the Commissioners of Fisheries and Game, State of Connecticut for 1906, p. 8.

In 1870 shad appeared in unusual abundance, there being more than for twenty years past. The fishing continued uncommonly good throughout the season. Fifty-four million shad fry were hatched and turned into the river at Holyoke that season. The next year, also, shad were in abundance, and 65,000,000 shad fry were turned into the river at Holyoke, and in 1872 nearly 90,000,000 more.

In 1874 the United States Commissioner co-operated in hatching 44,000,000 shad fry. The four states in co-operation planted over one million salmon fry. The following year 460,000 salmon fry were planted. From 1867 to 1874 the operations of Seth Green were conducted jointly by Massachusetts and Connecticut.

From 1875 to 1884 only 12,000,000 shad fry were planted in the river by Connecticut. The artificial propagation of shad on the Housatonic, at Birmingham, was begun in 1884, and the place was operated until 1898, hatching out over 75,000,000 shad fry during that period. About that time the run of shad ceased on the Housatonic and the artificial propagation of shad fry was discontinued on the part of the State until 1904. In the meantime the whole supply was obtained from the United States Government. The Connecticut Commission established a hatchery on the Farmington River in 1904, which has been conducted with satisfactory results. It has been the belief of the Fish Commision of the State for years that unless natural reproduction be aided by artificial propagation it is a matter of a few years only before the shad will be exterminated.[1]

THE ALEWIFE.

Alewives, or river herring, are the most abundant river food-fish frequenting the rivers of the Atlantic coast of

[1] Report of the Commissioners of Fisheries and Game, Conn., 1906.

our country. Their range is from Florida to Maine. There are two species of the fish, the branch herring or alewife proper, and the glut herring or kyack. The branch herring usually appears in rivers earlier than the glut herring. The glut herring appears suddenly, about the middle of the shad season, in enormous schools. Spawning takes place at a shorter distance from the sea than in the case of the branch herring. The alewife may be readily distinguished from the glut herring by the pale lining of the abdominal cavity, this lining being black in the glut herring.

As a food-fish the alewife is generally regarded as superior to the sea herring, being larger and of a better flavor, but in the latter quality they are inferior to the shad. Large quantities are consumed fresh. They are also prepared for market by being salted and smoked. The fish is highly satisfactory among the fishermen of Maine and Massachusetts as bait for cod and other ground fisheries.

The alewife is a shore fish. The method of catching alewives is by pound nets, trap nets, gill nets, weirs, and seines. They are caught in much larger quantities than any other fish entering the fresh waters of the country, and among all fish in American waters are surpassed in the abundance of the catch only by two species, the sea herring and the menhaden. The leading states in which the alewife industry is pursued are Maryland, North Carolina, and Virginia. Important fisheries exist along the New England, New York, and New Jersey coasts. In 1896, the six states leading in the alewife industry were Maryland, North Carolina, Virginia, Massachusetts, Rhode Island, and Maine.

The value of the alewife industry in the country in 1880 was $526,546, in 1888 it was $500,713, and in 1896 it was $459,598. The fishery of the middle Atlantic States was between 40% and 50% of the total during these years.

The value of the alewife fishery in the several New England States is shown in the following tables:

	1880.	1888.	1896.	1898.	1902.
Maine	$35,823	$30,103	$25,336	$25,302	$21,732
New Hampshire	8,500	3,080	3,045	2,750	3,813
Massachusetts	35,802	83,530	45,124	31,288	40,979
Rhode Island	14,460	21,165	27,884	10,273	7,366
Connecticut	8,700	1,253	12,031	7,346	15,399
Total	103,285	139,131	113,420	76,959	89,289
Other States	423,261	361,582	346,178	......	......
Grand total	526,546	500,713	459,598	......	......

The number of alewives caught in New England in 1880 was a little below 10,000,000, in 1888 it was, in round numbers, 11,000,000, in 1896, 12,000,000. In 1902, the catch for some years had been on the increase but the value of the catch varied with different years. The average price for the fish in 1896 was nine cents per pound, in the New England States.[1]

Among the streams of New England in which alewives were taken the Damariscotta River in Maine had the largest catch in 1896, followed by the Connecticut, Taunton, Merrimac, St. George, and Penobscot. The alewife fishing industry of Connecticut has developed greatly during the present century. From the returns of this fishery for the biennial period ending October 1, 1906, it is found that there were 342 persons employed, using 216 boats valued at $9,538 and 636 nets valued at $19,026, showing the amount of $28,564 invested in this branch of the fishery and yielding 2,703,201 pounds of fish valued at $38,755.[1]

[1] Report of the Commissioners of Fish and Game, Conn., 1906, p. 16.

"In the early season, March and April, hundreds of barrels of these fish are shipped to Boston and other eastern points, furnishing our Atlantic fishing fleet with bait. Many barrels are also shipped to different parts of New England and New York for food. The greater part are sold fresh until warm weather and at a price that warrants the encouragement and protection of this industry. Many hundreds are salted when the weather is too warm to ship them fresh without great risk. The market for salt fish is mostly foreign. The quality of the Connecticut river fish, the great care in salting and packing, with the long experience of the fishermen make them rank second to none in price obtained, and they always find a ready market."

The value of the alewife fishery in Massachusetts for 1902 was $40,979. This value consisted of fresh alewives, 1,320,350 pounds worth $15,220; of salted alewives, 1,979,000 pounds worth $24,619; and of smoked alewives, 114,000 pounds worth $1,140. A considerable business is done, at Provincetown, in canning the species called "kyacks," which are labeled "mackerel" and "trout." The fish

THE CONDITION OF THE ALEWIFE FISHERY OF NEW ENGLAND FOR 1896.

State.	Men.	Weirs, etc.	Seine.	Value.	Boats.	Value.	Value of shore property.	Total value of investment.
Maine	247	143	10	$8,497	233	$3,495	$ 5,369	$17,583
N. H.	10	10	5	530	12	155	355	1,040
Mass.	388	10	223	3,225	88	2,974	12,958	19,470
R. I.	78	28	32	1,830	33	610	1,050	3,490
Conn.	113	3	110	2,475	31	683	470	3,648
Total	836	194	380	$16,557	387	$7,917	$20,202	$27,231

(Alewife Fisheries of the U. S. in 1896, H. W. Smith.)

are taken from the local weirs each day, packed in the best of condition, and are reputed to make a very desirable food for people of moderate means.

THE MENHADEN.

The name "menhaden" is given to a species of the herring family in southern New England, while north of Cape Cod the name "porgy" is almost universally applied; altogether, the fish is designated by about thirty different names. The range of menhaden is from Nova Scotia to Brazil and it is by far the most abundant fish on the eastern coast of the United States.[1] "The menhaden's place in nature," says Prof. Goode, "is not hard to surmise; swarming our waters in countless myriads, swimming in closely packed unwieldy masses, helpless as flocks of sheep, near the surface of the water and at the mercy of every enemy, destitute of means of defense or offense, their mission is undoubtedly to be eaten."

During the period of the Civil War and for several years thereafter the fish appeared in immense schools on the coast of the Gulf of Maine. Since 1879 the fish have not frequented the coast north of Cape Cod except in limited quantities. The uncertainty of the coming of the menhaden to the coast of Maine in recent years has restricted the menhaden industry to Lincoln County.

Menhaden approach the southern shores of New England the latter part of April or the first of May; they reach Cape Ann about the middle of May; and the coast of Maine the latter part of that month or the first of June. In September they begin to leave the Maine coast and disappear from the waters of Long Island in November or December. The fecundity of the menhaden is very great, exceeding that of the shad and the herring. More than 140,000 eggs have been taken from one fish.

[1] Jordan and Evermann, American Food and Game Fishes, p. 109.

Commercially the menhaden is of importance as a fertilizer of soils, as a bait-fish in the cod and other ground fisheries, and for the oil and scrap produced by cooking and pressing the fish. At a time when the hand-line mackerel fishery was at its height thousands of barrels of "slivered" menhaden were used annually for bait by the New England fishermen. As a producer of commercial oil the menhaden has taken high rank, both in the quantity and in the quality produced. In 1874, the yield of menhaden oil on the Atlantic coast was 3,373,000 gallons, nearly equal to the aggregate of all the whale, seal, and cod oil made in America.[1]

The value of menhaden for bait was taken up for consideration by the Halifax Commission. Then it was shown that for bait for mackerel, the menhaden was preferred. It was customary for American vessels to carry twenty or more barrels of "slivered" menhaden to the Bay to use for bait. The claim was made that American fishermen sold $8,000 or $10,000 worth annually to Canadian fishermen. The demand for bait can be seen when it is remembered that each vessel took 15 or 20 barrels of bait for a trip to George's Bank and each mackerel fisherman took 75 to 100 barrels or even more than that. Prof. Goode estimates the total consumption of menhaden for bait for 1877 at 80,000 barrels, or 26,000,000 fish, worth perhaps, $500,000. The entire amount used in the mackerel fishery yearly was near 8,000 or 9,000 barrels.[2]

The industry of extracting menhaden oil, or the development of the "porgy press," originated at Blue Hill, Maine, in 1850. During the Civil War the business was pursued by thousands of farmer-fishermen on the Maine coast, three or four of them uniting in partnership and having

1 Goode, Sec. V, Vol. I, p. 330.

2 Ibid, pp. 348–349.

their porgy stand or press near the shore. This was before factories were built by organized companies for extracting the menhaden oil.

The first factory in Maine was erected at Blue Hill by a Rhode Island company, in 1864. The same season another one was erected at Bristol. This business now entered upon a flourishing stage of activity. In 1866 eleven factories were built, all using steam to operate the hydraulic presses to separate the oil from the fish. Nearly every year for a decade saw the erection of one or more factories, the principal center being about Bristol, Bremen, Booth Bay and Southport. By 1876, eighteen factories had been built at a cost of $260,500 for buildings and equipment. Within three years, owing to the departure of menhaden from Maine shores all these factories were idle. By 1877, there were upwards of 60 menhaden factories on the New England coast, with an invested capital near $3,000,000. These factories gave employment to 1,197 men, 383 sailing vessels, and 29 steamers, besides numerous boats. The yearly consumption of the factories was computed at 1,191,100 barrels, or about 300,000,000 fish. The following table exhibits the state of the industry in 1876. It will be seen how important the industry had become in Maine, as compared with the other states, and that there was a greater commercial return for the output than in any other locality.

LOCALITY.	No. of Men.	No. of Vessels.	No. of Steamers.	Total Capital.	Bbls. Fish Used.	Bbls. Oil Mfd.
Other States	1,629	291	3	$1,767,000	826,886	848,729
Maine	1,129	29	43	983,000	709,000	2,143,273

The number of tons of crude guano manufactured in other states in 1876 was 29,831; in Maine it was 21,414 tons.

The most striking fact brought out is that Maine realized 71% of the oil from 46% of fish used. Further, the total Maine capital, $983,000, turned out a total product of $1,071,449, whereas the rest of the country realized only $637,600 from $1,767,000.[1]

The amount of capital invested in factories, vessels and gear in the Maine menhaden fishery increased from $613,500 in 1873 to $1,051,812 in 1878. During the period of these six years there were 57,456 barrels of fish sold for bait, and 3,570,041 barrels taken at the factories. There was also produced 108,063 tons of crude guano, and 9,673,843 gallons of oil. The average yield of oil was 2¾ gallons of oil from every barrel of fish used. The amount of oil taken from a barrel of menhaden in the Maine factories was always greater than at factories farther south even on the New England shore.

The value of guano at $11 per ton was $1,194,093. The fish sold for one dollar per barrel for bait; the oil, at 37 cts. per gallon, was worth $3,679,321.91. The total value of this industry for Maine for the six years, 1873 to 1878, was $4,930,870.91, a yearly value of $821,611.82. So rapid and disastrous was the decline of the fishery along the Maine shore that when the census of 1880 was taken the State had neither fishermen nor factory hands employed in the business. In 1889 there were three factories, valued at $22,200 in the menhaden industry of Maine. In 1898 there were four factories, valued at $190,000. The products of 1889 were 282,465 gallons of oil, valued at $62,405 and and 2,305 tons of scrap worth $24,735, a total value of $87,140.

The results of 1898 show there was a considerable increase in the Maine menhaden industry after 1889. The product increased to 765,000 gallons of oil made, valued at $191,250, and 9,120 tons of guano, valued at $91,200.

1 Goode, Ibid, p. 360.

The total value of the fishery, not including the bait sold to the fishermen, equaled $382,450. In 1899 these factories were not carried on at all as no fish appeared on the coast. In 1902, the fish were not abundant along the coast of Maine. No factories of the State were operated except at Boothbay Harbor. The reported value of fresh and salted menhaden was $1,872. In 1906, fishermen were employed in this industry from Maine, not in the State but in waters South of Cape Cod, coming into the Maine waters long enough only to ascertain whether or not the fish were schooling.

The condition of the New England menhaden fishery for different years since 1880 may be seen in the following table, which shows the value of the product for 1880, 1898 and 1902:[1]

In 1880 the total number of persons employed in the menhaden fishery in the manufacture of oil and guano in the United States was 3,635; the capital invested was $2,362,841; and the value of the product was $2,116,787. In New England there were 1171 fishermen and 339 factory hands, 173 vessels, 282 boats; the value of fishery and fixtures was $531,087, a little over one-half the factory value of the country.

It will be seen that there has been a very noticeable decline since 1880. There was a tendency toward decline in Rhode Island for a number of years previous to 1898.

[1] Goode, p. 361; Statistics of the Fisheries of New England, 1898 and 1902.

YEAR.	Maine.	Mass.	R. I.	Conn.	Total.
1880	......	$61,769	$221,748	$256,205	$539,732
1898	$20,706	10,544	7,591	26,334	65,175
1902	1,872	5,409	1,156	47,964	56,401

In 1892 the value of the fishery was $115,992. The great falling off for 1898 was due principally to the fact that in the early part of the year the menhaden industry of Rhode Island was consolidated under the control of a company having headquarters in New York City. Most of the vessels that were engaged in the Rhode Island menhaden industry were transferred to New York and therefore were credited with crews and catch to that State. There were two factories in Rhode Island in 1898 and the value of the product was actually $96,389. Nearly $90,000 of this amount was credited to New York. In 1902 there was but one factory in Rhode Island. This is located at Tiverton and is one of the largest on the coast. In 1902 it employed 195 hands, received about 115,000,000 menhaden and produced 897,188 gallons of oil worth $225,912, and 15,727 tons of scrap worth $203,906, the total value of the products being $429,818. As the vessels supplying this factory are owned in New York the State of Rhode Island gets credit for its menhaden for 1902 to the extent of $1,156 worth only.

The value of Connecticut menhaden in 1889 was $100,569. In 1898 this had decreased to $26,334, there being three factories employing 52 men in that year. There was an increase from 1898 to 1902 in the catch of menhaden from 11,000,000 to 16,800,000 pounds, due to the greater abundance of fish in the Long Island Sound.

An investigation of the menhaden fishery was ordered by Congress and made in 1894. This was to determine the kind and extent of food-fishes taken in prosecution of the menhaden fishery. "It was the contention of sportsmen and others that in menhaden fishing large quantities of game and other food-fish are taken: that these are usually landed at the factories, where they serve the same purpose as menhaden; that on account of the extensive menhaden fishing along the coast the supply of food-fish has been greatly reduced; that important fishing grounds for game fish have been ruined; that where food-fish are not actually caught in the purse seines they are driven off; that fish that frequent the bays and there undergo the spawning process are prevented from reaching the desired grounds by the presence of menhaden vessels at or near the mouths of the bays. Of the foregoing objections to the fishery, greater importance is laid on the first two points."[1]

Two vessels were used by the Fish Commission during the summer of 1894 in making investigations in the menhaden business. These two vessels made 1,078 seine hauls and took a total of 28,060,565 fish of all kinds. Of this total, there were 27,965,755 menhaden, and 94,810 other fish. Alewives and shad made up 89,043; so that the actual number of fish taken outside of the herring family was only 5,767 fish, in a total of over 28,000,000 fish taken. After sportsmen had determined that the amount of game fish destroyed in taking 28,000,000 menhaden was only about one-fiftieth of one per cent, they, and the public with them, realized how weak their claims had been that the catching of menhaden destroys many kinds of food-fish.

[1] Smith, Investigation of the Menhaden Fishery in 1894, p. 286.

CHAPTER XII

INSHORE FISHERIES—SHELLFISH

THE LOBSTER.

The lobster is found on the Atlantic coast from Delaware to Labrador. It occurs most abundantly on the coast of Maine and Nova Scotia. It is found in depths of 100 fathoms, and while more abundant on the shores it has been found in considerable quantities on the off-shore banks. In April or May the lobster on the Maine coast comes into comparatively shallow waters, where it remains until late in the fall, going back to the ocean or deeper waters of the bays in October or November. Its favorite resort is on rocky bottoms, very few being captured on sandy or muddy ground.

The enemies of the lobster are so numerous that one writer has said that "every predaceous fish which feeds upon the bottom may be looked upon in general as an enemy of the lobster."[1] The most destructive enemies are the cod, the sea eel and the dog-fish. The greatest enemy of the adult lobster is man. For several years past, the decline in the lobster fishery of New England has aroused the attention of all who are interested in preserving this valuable food-fish. Where inquiries are made into the cause of the decline of the industry it is found that, in addition to its enemies, the method of propagation of the lobster is a natural hindrance to its rapid increase.

[1] Francis H. Herrick, The American Lobster, Bulletin, United States Fish Commission for 1895, pp. 1 to 252.

The female lobster spawns but once in two years. The eggs, after being laid, are carried by the female in clusters under the tail for a period of ten or eleven months, before they hatch. The number of eggs laid varies with the size of the fish. The law of reproduction has been expressed as follows: The number of eggs produced at each reproductive period varies in a geometrical series, while the length of lobsters producing these eggs varies in an arithmetical series. Thus, an 8 inch lobster produces 5,000 eggs, a 10 inch lobster, 10,000 eggs, and one 12 inches long, 20,000 eggs.[1] The high rate of production is not maintained beyond the length of fourteen or sixteen inches. The average female lobster lays eggs for the first time in the summer of its sixth year. The largest number of eggs recorded for a female is 97,440.[2]

After being hatched the young lobsters cut loose from their mother, and rise to the surface where they live as larvæ, being about one-third of an inch long. Within a period of eight weeks the larvæ have passed through three stages, and have molted five or six times. By this time they are three-fifths of an inch long and sink to the bottom, seeking to find some sheltering piles of rocks near the shore where they can burrow. When winter approaches the young lobster is from one to three inches long. This short sketch of the method of reproduction of the lobster and its early life as larvæ is given that it may be more easily understood why the lobster needs ample protection from man if it is to persist as a food-fish on our shores.

Previous to the year 1880 there had been no attempt to write a history of the lobster fishery or to publish any extensive reports on the subject. Mention is made in the early records of the New England colonists of the abun-

[1] Commissioners of Inland Fisheries, Rhode Island, 1906.
[2] F. H. Herrick, The American Lobster.

dance and good qualities of the lobster, and it is very probable that it was regarded as an important food-fish. Its use, however, was confined to the people of the sea coast for a long time, and it was not until 1840, or later, that the lobster was used to any extent at any distance from its natural surroundings.

The lobster canning industry of the United States was first attempted at Eastport, Maine, in 1840, and it was made successful in 1843. The business was in the experimental stage for several years. Salmon, halibut, and vegetables were canned in addition to lobsters. By 1854, canned goods from Eastport were being shipped to California. In 1850, there were only three canneries in the United States engaged in hermetically sealing fish, meats, fruits and vegetables. The supply of lobsters for the Eastport canneries came at first from the westward, probably beyond Penobscot Bay; the lobster fishery was not introduced in that vicinity until 1853, as it was supposed there were no lobsters in the vicinity. About 1870, owing to the fact that there was a noticeable scarcity of lobsters on portions of the Maine coast, American capitalists interested in the lobster canning industry began to establish canneries on the coast of the British Provinces. By 1880 the amount of American capital so invested exceeded that invested in lobster canneries on the coast of Maine.[1] The canning of lobsters at Eastport continued to develop from 1855 to 1865, reaching its height about the latter year. Since that time it has declined in American waters, until it is no longer profitable to be carried on.

"As the decline in the supply was attributed to the canneries, a sentiment against them was gradually formed, and laws were enacted regulating the time in which they should operate and the size of the lobsters they should put up. Prior to 1879, they were permitted to pack lobsters at

[1] Goode, Sec. V, Vol. II, pp. 687–689.

any season of the year, but they usually operated only between April 1 and August 1, and again between the tenth or middle of September and the first of December, the length of the season depending very largely upon the weather and the abundance of lobsters. In 1879, it was enacted that no canning of lobsters should be allowed from August 1 to April 1 following. In 1883, it was made illegal to can lobsters less than nine inches in length. In 1885, the canning season was fixed from April 1 to July 15. In 1889, the season was fixed from May 1 to July 1, and the minimum length of lobsters to be canned placed at nine inches. In 1891, this act was so amended as to make the season from April 20 to June 1. After 1880, the number of canneries gradually declined, until in 1895 the last one suspended the canning of lobsters, owing to the passage of a law fixing the minimum size at ten and one-half inches. This law went into effect July 1, 1895. As they could not afford to pay the high price demanded for lobsters of this size they were compelled to give up the business."

The following table shows the number of factories in operation, the quantity and value of fresh lobsters used, and

	1880		1889		1892	
	No.	Value.	No.	Value.	No.	Value.
Number of canneries	23	....	20	....	11	....
Lobsters used fresh, lbs.	9,494,284	$ 95,000	5,752,654	$ 72,092	5,326,322	$ 78,720
Canned:						
One-pound cans	1,542,696	....	999,521	$126,577	1,228,944	$195,114
Two-pound cans	148,704	....	85,520	16,036	3,096	839
Other sizes	139,801	....	....	....	....	....
Total cans	1,831,201	$238,280	1,085,041	$142,613	1,232,040	$195,953

(Cobb, The Lobster Fishery of Maine, pp. 256 to 257.)

the number and value of cans of lobsters put up in the State of Maine, in the years 1880, 1889, and 1892.[1]

The lobster fishery was begun many years before the lobster canning industry arose, and has always been of greater importance. The coast of Maine always has been and is to-day the favorite locality in the United States for the pursuit of the lobster fishery. The present value of the fishery in New England is second only to the oyster fishery. Our account will be largely a sketch of the fishery as carried on in the State of Maine, which represents three-fourths of the New England lobster industry.

As early as 1830 vessels from Boston and Connecticut visited Harpswell and probably other places in Maine for supplies of fresh lobsters. The industry flourished and in about a decade the carrying trade was begun by a resident of the place, an account of which follows:—

"In 1841, Captain E. M. Oakes began to carry lobsters from Cundy's Harbor and Horse Island Harbor, Harpswell, to Mr. Eben Weeks, at East Boston. He was then running a well-smack, named the *Swampscott,* of 41 tons, old measurement. The season extended from the 1st of March until about the 4th of July, after which time the lobsters were supposed to be unfit for eating. The black lobsters, or shedders, were even considered poisonous. During this season of four months, Captain Oakes made ten trips, carrying in all 35,000 by count. He continued in this trade about six years, taking the combined catch of about five or six fishermen. At this period the smack *Hulda B. Hall,* 50 tons, of New London, Conn., Captain Chapell, was carrying lobsters from Cape Porpoise, Gloucester, Ipswich Bay, and occasionally Provincetown, to Boston, making 15 trips in the season of four months, and taking about 3,500 lobsters each trip. Captain Chapwell was supplied with

[1] Part of the lobsters used in the Eastport factories come from New Brunswick. It is impossible to separate the two lots.

lobsters by four men at Cape Porpoise, and by the same number at both Gloucester and Ipswich Bay. For four months following the close of the lobster season on the Maine coast, or from July 4 until November, Captain Chapell ran his smack with lobsters to New York, obtaining most of his supplies at Provincetown."

Captain S. S. Davis, of South Saint George, states that about 1864, when he first began buying lobsters at the Muscle Ridges, three men tending from 40 to 50 pots each, caught all the count lobsters he could carry to market in his smack. He could load 5,000 lobsters at a time, and averaged a trip in 7 to 9 days. The traffic continued from seven to nine years. In 1879, Captain Davis bought his supply of lobsters from 15 men in the same locality, and at times was obliged to buy also of others in order to make up a load.[1]

There are no accurate figures showing the catch of lobsters previous to 1880. In that year there were 2,773 persons in Maine employed in the lobster fishery, of whom 1,843 were fishermen. The total capital invested in the fishery was $151,154. There were 14,234,182 lobsters taken, valued at $268,739, which was at the rate of 1.9 cents a pound. About six-sevenths of the catch was east of the Wiscasset district, or in the Passamaquoddy, Machias, Frenchman's Bay, Castine, Belfast, and Waldoboro districts.

A review of the lobster industry for the United States for 1880 shows that there were 2,819 fishermen employed; the amount of capital invested in the industry was $266,353; the quantity of lobsters taken, 20,128,033; and their value was $483,891. New York and New Jersey were the only states outside of New England where the industry was carried on, the value of the lobster industry of these states being only a little in excess of $10,000.

[1] Goode, Sec. V, Vol. II, p. 700.

A more detailed view of the lobster industry of the country in 1880 is presented in the accompanying table:[1]

States.	Fishermen.	Capital Invested in Fishery.	Quantity of Lobsters Taken.	
			Pounds.	Value.
Maine	1,843	$151,154	14,234,182	$268,739
N. H.	44	2,810	250,000	7,500
Mass.	595	69,572	4,315,416	158,229
R. I.	129	12,677	423,250	15,871
Conn.	148	37,750	613,385	23,002
N. Y.	32	1,360	135,000	5,062
N. J.	28	1,030	156,800	5,488
	2,819	$266,353	20,128,033	$483,891

Since 1880, a more careful record of the lobster industry has been kept, and as the abundance of this crustacean has diminished, steps have been taken to prevent further diminution of the supply. Certain grounds have been almost totally exhausted from over-fishing. This was due largely to the canning industry, which frequently made use of half-pound lobsters until the passage of an act in the State of Maine fixed the minimum length of lobsters caught at ten and one-half inches. The effect of the act was to close the canneries in 1896; but it has been of great benefit to the fishermen, as young lobsters now have a chance to reach maturity.

For a decade after 1880, the catch of lobsters increased yearly, but since 1890 the catch has been on the decline. The value of the catch, however, has increased yearly until 1906 when it was six times as great as in 1880. The number of the fishermen engaged in the fishery and the number of lobster pots used has also increased. The aver-

[1] Goode, Sec. V, Vol. II, p. 793.

age price per pound has increased from 1.9 cents in 1880 to 10.9 cents in 1905.

The table given below shows important data respecting the lobster fishery of Maine for certain years:[1]

Year.	Fishermen.	Pots.	Catch.		Average Stock Per Man.	Average Price Per Pound.
			Pounds.	Value.		
1880	1,843	104,456	14,234,182	$268,739	$146	$.019
1887	1,906	113,299	22,916,642	512,044	269	.022
1888	1,967	112,632	21,694,731	515,880	267	.024
1889	2,080	121,140	25,001,351	574,165	276	.023
1892	2,628	153,043	17,642,677	663,043	252	.038
1898	3,099	155,978	11,183,294	992,855	320	.089
1905	2,562	169,350	9,018,759	939,799	386	.109

The value of the lobster fishery of Maine for the year 1906 was $1,640,646, the number of men engaged was 2,672, and the average stock per man was $614. The average weight of lobsters taken in 1905 was slightly over one and one-half pounds each. From 1897 to 1907, the total number of lobsters caught in Maine was 70,057,949.

Although the lobster fishery is prosecuted from Maine to Delaware, it has always been comparatively unimportant outside of New England. In 1900, in the states of New York, New Jersey and Delaware only 109 persons were engaged in the fishery, and the catch was only 200,600 pounds, valued at $27,960. In New England the lobster fishery reached its climax in 1889, when 30,449,693 pounds of lobsters were taken, valued at $833,736. The apportionment of the New England lobster fishery among the five

[1] Compiled from Cobb, The Lobster Fishery of Maine, p. 257; Maine Report for 1905 and 1906.

states in different years is seen in the following table:[1]

	1889		1898		1905	
States.	Pounds.	Value.	Pounds.	Value.	Pounds.	Value.
Maine	25,001,851	$574,165	11,435,789	$1,001,797	9,018,759	$ 989,799
New Hampshire ..	137,175	6,415	128,468	14,863	256,052	32,575
Massachusetts	8,853,787	148,492	1,695,688	175,095	1,288,071	176,234
Rhode Island	456,000	21,565	397,305	39,488	529,827	64,358
Connecticut	1,501,290	88,099	871,650	40,719	436,790	56,141
	30,449,608	$838,736	14,661,808	$1,276,967	$11,524,499	$1,319,107

From the above comparison of the state of the industry in different years, it is easily seen that the lobster fishery of the country centers very largely in the State of Maine. Massachusetts furnishes about as much as the total for the other three states; but the total catch and value of the lobster fishery outside of Maine become inconsiderable in comparison with the larger output and value of the fishery for Maine. During the last few years the statement has been made that there has been no diminution of the lobster supply, the financial condition of the lobster fishermen being taken as a proof of the statement. The trend of the New England lobster may be seen in the comparative statistics of the fishery, which are given below. The amount of the catch has decreased greatly since 1889, but the value of the fishery and the average price per pound of lobsters have increased far out of proportion to the diminution of the supply. For example, in comparison with the banner year, 1889, the quantity of the output for 1898 was 47 per cent less, while the value was 51

[1] Lobster and Clam Investigations, 1904; Fisheries of the New England States for 1898 and 1905.

per cent greater; the output for 1905 was 37 per cent less, while the value was 59 per cent greater.[1]

The artificial propagation of lobsters has been undertaken for many years by the National Government, through the Bureau of Fisheries, co-operating with the states to maintain the lobster supply. The work of gathering brood lobsters has been carried on extensively in recent years. Through arrangements with the several states the lobster fishermen are permitted to retain the egg-bearing or "berried" lobsters until they may be collected by an agent of the bureau, who pays the fishermen a little more for their fish than the market price. The "berried" lobsters are taken to the hatcheries, stripped of their eggs, and later returned to their native localities to be liberated.

From 1888 to 1903, the number of lobster fry that were planted by the Bureau of Fisheries was about 880,000,000. The results of this work have been beneficial to the fishery, but hardly more than to check the decline. The process of artificially hatching the eggs is a comparatively simple matter. The problem has been to rear the young through the prolonged period of larvæ helplessness until they were able to care for themselves. The many difficulties of the

[1] Comparative statistics of the New England lobster catch.

Years.	Pounds.	Value.	Average Price Per Pound.
1880	19,836,233	$ 473,341	$0.024
1887	28,627,600	784,238	.024
1888	27,640,282	808,842	.029
1889	30,449,603	833,736	.027
1892	23,409,927	1,035,501	.044
1898	14,661,808	1,276,967	.087
1900	15,567,081	1,362,619	.088
1902	14,028,845	1,271,962	.091
1905	11,524,499	1,319,107	.114

problem prevented success until 1900, when the honor of having won the first and only solution was claimed by the Rhode Island Commission at Wickford. Since the discovery of the principle, it has taken five years of slow and tedious experiments to develop the scheme to the point where it is practical and economical. Nearly 50 per cent of the larvæ have been reared from the first to the fourth stages of development in lots of 20,000 by the Rhode Island Commission. At the fourth stage the fry begins to burrow and are able to care for themselves. The value of the Rhode Island experiments is better appreciated when it is understood that the best result in Europe was 6.6 per cent, starting with 1,500 fry in the second stage; and in this country 21 per cent, from an estimated 3,000 fry in the first stage at the government hatchery at Woods Hole, Massachusetts.[1]

THE OYSTER.

The oyster industry of the country belongs properly to the region of the Delaware and Chesapeake bays. The industry in New England has always been, until recently, a small part of the oyster fishery of the country. It has already been shown in a previous chapter how the industry was carried on in New England previous to the Civil War; and how a trade developed in transporting oysters from the Chesapeake to be laid down in northern waters to await further growth or a more favorable season before being placed upon the market. This trade was interrupted and destroyed by the opening of active hostilities.

The business was resumed after the restoration of peaceful relations between the two sections of the country, and seems to have been carried on as usual until about the

[1] Annual Report of Commissioners of Inland Fisheries, Rhode Island, 1906, pp. 122 and 123.

year 1880. There had been quite a diminution in the quantity of southern oysters bedded down in New England by 1880. The probable cause of the diminution of this trade was the transplanting of native seed oysters. The cultivation of oysters transplanted when young from the natural reefs where they were spawned to inshore grounds has long been followed in the United States. As soon as this method began to be employed in the water of Long Island Sound the competition became too great for the transplanting of southern oysters and by 1885 the trade had greatly diminished.[1]

The annals of the New England oyster fishery are necessarily brief. Except in the state of Connecticut they have been, until recent years, rather unimportant. When the extensive report on the country's fishery was made in 1880, there were over 22,000,000 bushels of oysters produced, having a first value of more than $9,000,000. New England's contribution to these totals was 536,650 bushels, valued at $654,725, about seven per cent the total value. The value of the Connecticut yield was greater than the combined Massachusetts and Rhode Island output, it being $386,625. The oyster fishery in the Middle Atlantic States for 1901 had a value slightly in excess of $10,000,000. In New England, for 1902, the value was above $2,000,000, and for 1905 almost $4,000,000. So the present day comparisons are very greatly in favor of the New England industry as far as rate of growth is concerned.

The value of the industry in Massachusetts was $41,800 in 1880; $156,235 in 1899; $133,682 in 1902, and $221,990 in 1905. Barnstable County ranks first in the production of oysters, Osterville, Wellfleet and Cotuit being the principal towns from which they are shipped. The oyster beds at the head of Buzzards Bay and in the waters of Wareham township are important.

[1] Goode, Sec. V, Vol. II, p. 523.

The oyster fishery of Rhode Island, like that of Connecticut, represents the most important branch of the fisheries in the State. Its value has shown a very steady increase for the last twenty-five years. The first value of the fishery in 1880 was $225,500; in 1898, $505,378; in 1902, $588,052, and in 1905, $929,963. The yield in bushels for 1898, 1902 and 1905 was over 450,000 bushels, 600,000 bushels and 900,000 bushels, respectively. Both the yield and the value show very satisfactory improvement in the industry.

The private cultivation of the oyster in Rhode Island is carried on under the supervision of the State, represented by a Commission, from whom leases of oyster grounds are obtained. The rate of rental depends upon the depth of water. For a depth of twelve feet and more at mean low tide it is $5 an acre; under twelve feet, $10 an acre. In 1864, the total annual rental derived from the grounds was $61. The annual yield increased rapidly and steadily until 1885 when it was almost $12,000. From that date to 1899 the average annual rental was between $7,000 and $8,000; from 1899 to 1905, inclusive, the average rental was almost $33,000 annually; and for 1906 it was estimated at $59,305 derived from the rental of a little more than 8,000 acres of oyster ground.[1]

Formerly star-fish were destructive enemies on many of the beds of Long Island Sound. Record is made in 1884 of the destruction of oysters on a single bed to the value of $90,000 solely from star-fish. They do not appear to be so destructive as formerly, owing to the systematic and persistent manner in which they are destroyed by the oyster planters, who use tangles for catching them. In 1898, 27,362 bushels of star-fish were caught by the oyster-growers of Narragansett Bay.

The yield and value of the Connecticut oyster fishery

[1] Report of Commission of State Fisheries, R. I., 1906.

since 1880 has increased steadily until at present the value is more than seven times that of 1880, while the yield is ten times as great. In 1889, the value had increased to $1,055,807. The value for 1898 was $1,249,071, and for 1902, $1,471,582, the yield being over 2,000,000 bushels each year. A large part of the seed oysters used in the Rhode Island grounds are supplied from Connecticut. In 1898, there were more than 390,000 bushels supplied, valued at a quarter of a million dollars.

The expansion of the Connecticut oyster industry was very marked between 1902 and 1905. The number of persons engaged in the industry increased in the three years under consideration from 1,678 to 2,127; the capital invested, from $848,051 to $1,063,908, and the value of the oysters rose from $1,471,582 to $2,809,832, or nearly 100 per cent. The increase in the yield of seed oysters was from 848,065 bushels, valued at $872,634 to 1,135,699 bushels, the first value to the fishermen being $1,206,217.[1] Not only do recent statistics show a gain in the lobster and oyster industries of Rhode Island and Connecticut, but a perusal of the latest reports of the commissioners of the several fisheries of the states shows that the development of the fisheries is receiving an unparalleled degree of attention from the citizens of the states. The oyster fishery of New England, confined almost wholly to the southern coast, had a value in 1905 of $3,961,785, which is far in excess of any other fishery of New England. The sardine canning business of Maine has an output worth above $5,000,000, but that is an industry arising from the fisheries. The oyster fishery, as such, almost equals the combined value of the cod, the lobster, and the mackerel fisheries of New England. Little wonder that the people of the states where oysters are produced seek to develop the industry.

[1] Fisheries of the New England States for 1905.

THE CLAM.

The clam, that well-known denizen of all our sandy and muddy shores, never has been an inspiring subject for writers of natural history, economics, or the fisheries. Like the oyster, the clam is restricted in its movements, is not taken by the hook and line, nor by weirs, pots, nets, seines, tongs or dredges. The clam is the humblest dweller of the deep, selecting as his home a place in the soft sand or mud of the beach six or eight inches below the surface. At high tide he extends his blackened tube from the shell up through the "clam hole" to near the surface of the bottom, in order to get his supply of food. "Happy as a clam at high tide" has a greater force and significance to those who know that the high tide brings food to this mud-dwelling bivalve and also prevents further depredations to its "bed" from the hand of man. As the water recedes, the clam draws back its tube within its shells to remain in a comparative state of quiet for four or five hours.

The soft clam fishery of New England is carried on north of Cape Cod. The hard clam, or quahog, is found in greatest abundance from Cape Cod to the Long Island Sound region. Statistics for this industry are meager. In 1879, the soft clam industry of Maine yielded 318,383 bushels, valued at $90,056. The greater part of this product came from Portland and the region to the south. Especial mention is made of the fishery at Deer Island, not only because of the large quantities of clams dug in that region, but also as being the only locality in the United States where women made a practice of digging clams.[1] The yield of soft clams in Massachusetts for 1880 was 158,626 bushels, valued at $76,195. The output for Rhode

[1] Goode, Sec. V, Vol. II, p. 584.

Island was 53,960 bushels, valued at $48,564. They were sent chiefly to New York. The total product of Connecticut was about 75,000 bushels each year, worth $38,000.

The early productiveness of the quahog at Cape Cod is shown by the presence of numerous shell-mounds, built up by the Indians and consisting almost entirely of the shells of the quahog. As a business, the quahog fishery dates from the beginning of the last century. "It grew in extent until 1863, and from that time until 1868, the tide was at its highest, since when it has diminished year by year, owing to the lack of good market rather than failure of the supply. Between 1863 and 1869, the average catch each year was not less than 2,500 bushels. Of this amount a comparatively small part was consumed at Wellfleet, and the rest was shipped to Boston, Providence, Salem, Newport, Manchester, and a few other New England ports. From 1870 to 1876, the quantity of quahog taken per year decreased from 2,500 bushels to 1,800 bushels."[1]

The comparative statistics of the New England soft clam yield show a steady condition of the fishery. During twenty-five years this fishery has been subject to less fluctuation in supply and value than any other. The trend of the fishery is exhibited in the following table:

	1880		1889		1898	
State.	Bushels.	Value.	Bushels.	Value.	Bushels.	Value.
Maine	318,388	$112,706	842,369	$200,761	918,552	$328,458
N. H.	17,960	8,980	300	150	600	360
Mass.	153,626	76,195	251,823	37,711	147,095	102,594
R. I.	53,960	48,564	33,375	82,475	15,015	25,569
Conn.	75,000	38,000	26,360	24,900	19,980	19,028
Total	623,929	$284,445	1,154,227	$395,997	1,101,242	$465,015

[1] Fisheries of the New England States for 1902 and 1905.

Since 1898 there has been a falling off in the extent of the industry, so that national and state investigations have been made to determine how the industry may be revived, or even increased beyond its former best output. The outcome of the investigations was to recommend the increase of the productiveness of the flats by planting small clams on parts of the flats entirely unproductive, and to determine the time required for small clams to grow to a suitable size for market. From experiments already carried on in Maine and Massachusetts, the practicability of planting unproductive beds has been demonstrated. The rate of growth of the clam, also, has been determined approximately, although that depends somewhat on the character of the flats and the currents of water. The recorded growth of a clam five-sixteenths of an inch in length was to a length of one and one-half inches within a month; a clam two inches long increased to four inches in sixteen months; a certain number of medium-sized clams increased to more than three times their volume in a year in a bed with practically no current.[1]

In the year of 1902, the yield of soft clams for New England was 777,023 bushels, valued at $395,178. During the next three years the quantity of the output fell off to 637,585 bushels, but the value increased to $406,252. The output of quahogs for 1905 was 196,033 bushels, valued at $336,709. Razor clams were worth $1,620. The total value of the New England clam fishery for 1902 was $744,581, ranking sixth in the value of the different fisheries of these States.[2]

[1] Maine Report, Sea and Shore Fisheries, 1905 and 1906; Mass. Report, Fish and Game, 1906, pp. 68–96; Investigation of the Lobster and Soft-shell Clam, p. 210.

[2] Fisheries of the New England States for 1902 and 1905.

CHAPTER XIII

METHODS OF INSHORE FISHERIES

The methods of catching and curing groundfish—cod, hake, haddock, halibut, etc.—in the inshore fisheries is essentially the same that is employed in taking the fish in the offshore fisheries. Lines, trawls and bait are used inshore as well as offshore, and the fish are prepared for market in a similar manner. Also, the catching of mackerel by the purse-seine is illustrative of the method used in taking herring, menhaden and alewives in deep water. But the capture of herring in shallow water and the process of curing and packing the fish, and the means employed in taking certain kinds of shellfish call for separate treatment of the methods used in inshore fisheries. One of the most important inshore fisheries of New England is the industry of taking and packing young herring for the sardine trade.

The principal method of capturing young herring is by brush weirs. Herring may be taken by means of torching, that is, the use of lights and dip-nets at night, but this method is employed to a limited extent. The brush weir was known and used in Nova Scotia previous to the beginning of the nineteenth century. It was introduced into the United States about 1820 at North Lubec, Maine. By 1880, American fishermen were engaged in fishing with 66 weirs in the Passamaquoddy region.

The best location for a weir is at a point of land that extends well out into the water, or in a channel between two or more islands or ledges. Weirs are usually placed where the tides run with considerable force, as it is known that the herring remain in strong currents. They are constructed by driving a row of posts into the mud several feet apart which serve as the basis of the weir. Smaller posts, closer together, are placed between the larger ones and fastened at the top by a stringer piece that extends from the top of one larger post to the other. Fine brush is then woven horizontally on the smaller posts after the manner of basket-making. In shallow water where the current is not strong, it is enough to place small brush trees vertically between the larger posts by shoving one end into the mud and nailing the upper portion to the cross-piece. Where the weir is to be in deep water the small posts are generally arranged in sections and woven with brush on shore, to be placed in position between the larger posts at low tide.

Weirs vary in size and strength according to their position. The cost varies from $40 or $50 for the smallest ones to $800 or $900 for the largest. The value of the average size mud weir is from $200 to $250. The best form of weir is that which is arranged to receive and hold the herring both at flood and ebb tide. Usually one wing is used, with two "pounds" on the deep-water end facing in opposite directions. An extra "pound," or pocket, constructed just outside the weir, is used, in some localities, for retaining the herring until they are needed at the factory.[1]

The sardine canneries are located on wharves, in order that they may be reached easily by the boats that collect the fish from the weirs. They are usually two-story frame buildings, varying in value from $1,500 to $40,000, according to the location and the completeness of wharves, buildings, and fixtures. The average value of the canneries of Maine to-day is $30,000, a considerable increase over their value a few years ago. In 1898, there were 78

[1] Goode, Sec. V, Vol. I, pp. 499–508.

canneries, of which 61 were sardine canneries, valued at $566,305; in 1905, there were 46 canneries, of which 33 were sardine, valued at $1,279,525. The ground floor of the cannery generally has rooms for the machinery, for the salting vats, and for packing boxes. The upper floor contains rooms for steaming and drying the herring, for packing them in oil, and for sealing the cans.

The process of preparing herring for the market as sardines begins where the fish are taken from the receiving pounds of the weirs by the collecting boats and ends with placing the cases of packed sardines aboard the steamboat for shipment to the South or Middle West. Many companies own the boats used in collecting the fish from the weirs, and pay the captain and crew by the month. The boats are now steamers or auxiliary sloops; the prices paid for the fish at the weirs is 30 or 40 cents per bushel. The fish are bailed from the receiving pounds into the holds of the boats and thus taken to the canneries. Upon reaching the wharves the herring are taken out in basketfuls by a small engine, the baskets of fish are taken by overhead carriers into the cannery, where the fish are dumped out upon the tables, and the empty baskets sent out on the return-carrier.

At the cutting tables the fish are seized by boys and girls, and by one operation of the knife the heads and entrails are separated from the body, which is thrown in a box holding about a peck. Five cents a box is paid for heading the herring. The boxes of be-headed fish are placed in cars and taken to large vats. After being washed they are placed in brine and remain for an hour or more. From the vats the herring are dipped by nets into movable stands on which streams of water are playing. As the fish escape from the incline of the stand, they receive a thorough washing and are caught at the floor in baskets. The latter are hoisted to the second floor of the cannery where the cooking and packing processes take place.

There they are dumped on tables to be spread on flakes, made of galvanized iron wire, two by three feet in size. The cost of spreading is 40 cents per 100 flakes. After the spreading the flakes are placed in racks to the number of 20 to 24 to a rack. The racks are upright wooden frames set on rollers and have cross-pieces about three inches apart for holding the flakes. After a rack has been filled it is rolled into the oven—a room a little larger than the car. Here the fish remain from ten to fifteen minutes while they are cooked by steam, which is let into the enclosure from the boiler-room. This process of cooking has supplanted the former method of preparing the fish for the cans, which was to place them in large frying pans of heated oil for one or two minutes.

When the cars emerge from the cooking-room they are rolled in front of the dryer, an enclosed place heated from the furnace and containing a huge horizontal wheel with ten large arms each holding eight flakes. The wheel is on the principle of the Ferris wheel, the arms remaining horizontal as the wheel rotates. Two persons work at the front of the dryer, one removing from the arms the flakes of fish that have already been dried and the other putting in fish flakes as the others are removed. When an arm has been re-supplied the wheel is turned so that the next arm comes before the workman to be relieved of its flakes of dried fish. In this way the wheel rotates once in about twenty minutes, which is the time required for drying. As the flakes of dried fish are removed from the dryer they are again placed upon the racks, and these are rolled to the packing tables. The fish are now ready to be packed in the cans.

A more economical method for drying the fish than that described above is employed in some canneries. There the cars containing the flakes of fish are rolled direct from the steam-chamber into a chamber heated for drying, thus avoiding the extra work that comes from unloading the flakes to the heater and again loading them on the car to be taken to the packing tables.

Sardines are packed in one-quarter and three-quarter pound cans. These are for the most part manufactured in Boston, but one factory in North Lubec is entirely devoted to the manufacturing of cans. The oil used for sardines is the better quality of cotton seed oil, usually purchased in hogsheads. Formerly the oil was placed in the cans individually. Now a tray containing twenty-five cans is placed under the filler, a valve is turned, and the regulation amount of oil flows into each can at one time. This process of machine-oiling requires about one-fourth the time required under the old method. The packers work with a dexterity acquired through long practice, sorting the fish into lots as they are picked from the flakes, packing them away in the cans so closely and neatly as to give rise to the expression in common use, "like sardines in a box." The size of the herring used for sardines varies from three to eight inches in length. The smaller fish are packed in oil in the one-quarter pound cans, and are known as "oils"; the larger fish are packed in mustard in the three-quarter size cans and are known as "mustards." Packers receive ten cents a case, the average woman worker being able to pack two cases an hour. Each case contains 100 of the small cans or 48 or 50 of the larger ones.

The last process in the canning of sardines is that of sealing the covers on the cans. In this process probably the greatest improvement has been made. Formerly covers were sealed by means of solder, each can being subjected to a separate handling by the operator. Soldering has been done away with. The cans are sealed by machinery, and the covers, instead of being soldered are united to the upper edge of the cans by a rimming process which presses closely together the edges of the cover and the can. There are two kinds of sealers, the smaller ones being not unlike the machines for sewing lasts on shoes. The workman places a can on a support, the cover is put on the can, a lever is turned which drops a heavy head on the cover and sets the machinery in motion about the can. When the circuit of the can has been completed the head is automatically raised, the can is removed by the workman and another is placed on the support.

The larger machines are in general structure like the letter Y placed horizontally. An endless belt feeds the cans into the machine. An operator stands at the side and, as the cans pass in front of him, places covers on them. This requires quick work and both hands have to be used in placing the covers. After the covers are placed the cans pass alternately to the right and left part of the head of the machine where the covers are rimmed to the cans much as they are in the smaller machines. In this method of sealing by machinery there is a great saving of solder, labor, and time. A small machine will seal seven or eight cases an hour, as many as one workman could do in a day; a large machine will seal twenty-five to thirty cases in an hour. After being sealed, the cans are immersed in a hot bath to be cleaned. Then they are lowered through chutes to the packing-room where they are placed in the wooden cases for shipment. The cases when filled weigh 50 pounds and occupy about a cubic foot of space.

Next to the preparation of herring as sardines that of curing them for market, by smoking, stands prominent. The principal kinds of fish prepared by this process are herring, alewives, halibut, and haddock—called after being smoked, finnan haddie. Other kinds of fish are preserved by smoking, but they are not of enough importance to be considered here. The process is an old one, and consists in exposing the fish fresh or in a slightly salted condition to the action of smoke that comes from a slowly smoldering fire. In preparation for smoking, alewives and small sea herring to be cured as hard herring are usually not dressed at all; bloater herring are usually split down the belly and eviscerated; haddock and salmon are split so as to lie open like cured codfish, and halibut are cut into pieces before smoking.[1]

There are two products of the sea herring of New England differing only in the preparation they receive in the process of smoking—the hard or red herring and bloater herring. The hard herring are smoked for several weeks until quite dry, the bloater herring are smoked for a few hours only, consequently they will not keep so long as the hard herring. The smoking of hard herring in this country is confined largely to the eastern coast of the State of Maine, about 95 per cent of the business being carried on in Washington County in the vicinity of Lubec and Eastport. Formerly the business prospered all along the eastern coast when herring were secured in abundance from the Magdalen Islands. After that source of supply began to fail, the business was confined to those parts of the coast where herring could be easily secured. The bulk of herring smoked in eastern Maine to-day is secured from the weirs of Passamaquoddy Bay, both American and Canadian sides. Other herring are received from the coast farther west, from the Magdalen Islands and from Newfoundland.

The smoke-house used for smoking herring is a frame structure boarded and shingled, having no floor, little chance for light to enter, and at the top an opening along the ridge pole to allow the smoke to pass out slowly. The

[1] Preservation of Fishery Products by Smoking, C. H. Stevenson.

fires are placed on the ground within the building, and the fuel used may be ship-carpenters' chips, drift wood, or birch. The interior arrangement consists of a series of vertical rows of scantling two by four inches in thickness, being about 38 inches apart laterally and fourteen inches apart vertically. The only other equipment necessary is a large supply of herring sticks. These are about one-half an inch square and forty inches in length.

When herring are brought to the smoke-house in a fresh condition they are placed in butts containing brine in order that the salt may strike in before the fish are smoked. Newfoundland and Magdalen salt herring have to be placed in tanks of fresh water in order to allow some of the salt to soak out; sometimes it is necessary for the fish to pass into several tanks before they are in the proper condition.

When the fresh herring have been properly salted and the salt herring sufficiently freshened the process of stringing takes place. This consists of placing the herring upon the herring sticks preparatory to hanging them up on the scantlings in the smoke-house. Fish are strung by inserting the sharpened end of the herring stick under the left gill and passing it through the mouth of the fish. This operation is continued with other fish until the stick is filled. Then the stick of fish is washed by being dipped into a trough of clean water, is placed upon the herring horses, and removed out of doors or to another part of the building to dry for a few hours. After drying they are ready for the smoke-house.

It may require several weeks to fill a smoke-house, especially where the supply of fish comes from the weirs. The herring when first placed in the smoke-house are put on the lower rows of scantlings. Then after they have been smoked for fifteen hours or so they are removed to the upper parts of the house to remain until the house has been filled. This may go on for several weeks, depending upon the supply of fish. After the house has been filled —and to fill one holding 20,000 herring requires about two weeks—the herring have to be smoked for two or three weeks before they are ready for packing. Herring are packed in thin wooden boxes about 16 inches long, 8 inches wide and 4 inches deep.[1] The price of the Maine pack of herring in 1905 and 1906 averaged between nine and ten cents a box. The value of herring smoked in Maine in 1906 was $317,480.[2]

The bloater herring industry is carried on principally at Gloucester and Boston. The fish used in this business come generally from Newfoundland, being brought to these ports usually in American vessels in a salted condition in bulk. These have to be soaked, strung on sticks, and dried much after the manner described for hard herring. Further processes in the smoking and preparing bloaters are described as follows:

"In order to 'bloat,' the herring must be thoroughly moist, and after they have commenced to dry in the smoke-house the heat must be increased. If they are permitted to hang 10 or 12 hours without heating they will not bloat, but will become hard herring. The smoking is continued from 2½ to 6 days, when the fish are usually sufficiently cured. They are removed from the houses, allowed to cool for a few hours, and placed in boxes holding 50 or 100 fish each, the larger size being by far the most numerous. The average weight of 100 bloaters prepared from Newfoundland herring is about 40 pounds, whereas an equal quantity prepared from the Gulf of Maine fish weighs 25 to 35 pounds, according to their size and the extent of the

[1] Ansley Hall, The Herring Industry of the Passamaquoddy Region, Maine.

[2] 29th An. Report, Com'r of Sea and Shore Fisheries, Maine, 1905 and 1906.

smoking. The Eastport bloaters weigh about 25 pounds per 100 fish, being smoked two or three days longer than the Boston bloaters, as they are intended to keep a greater length of time and in warmer climates. They are placed in boxes 18½ inches long, 11½ inches wide, and 7½ inches deep, inside measurement. The thickness of the ends is generally ⅞ inch and of other parts ⅜ inch, and the cost of boxes approximates $12 per 100. The boxes at Boston, Gloucester, and Portland are usually considerably larger. One barrel of round fresh herring yields about 5 boxes of 100 bloaters each. Those smoked 2½ days will keep usually 3 or 4 months under favorable conditions, while those smoked 5 or 6 days will keep until warm weather. Very few bloaters are sold after the month of May.

"The market for bloaters is principally in Boston, New York, Canada, and the West, and the average wholesale price for those prepared from Newfoundland salted herring is about $1.20 per 100 fish. The Boston-cured bloaters sold in 1859 at $1.25 to $1.50, and in 1865 at $1.80 per 100. In 1880, the value of the Eastport bloaters was about 95 cents, in 1893, it was 77 cents, and in 1898, is was about 80 cents per 100."[1]

The use of ice in preserving fish in markets has been a practice in this country for three-quarters of a century. It was not until about 1840 that ice was carried in vessels to preserve the fish until market was reached. At first, the method employed was the use of cooling rooms in which the fish were stored. In 1846, it became customary to crush the ice and mix it with the fish. Within the last forty years the practice of carrying ice in vessels that secure cargoes of fresh fish has developed greatly. Ice is now carried by vessels engaged in taking cod, haddock, hali-

[1] Preservation of Fishery Products by Smoking, Chas. H. Stevenson.

but, bluefish and mackerel. The fresh haddocker takes from ten to twenty tons of ice in her houses, the fresh halibuter requiring about ten tons more. In each case the ice is chiseled fine before being used. Halibut are laid in the ice pens one upon each other, the ice being placed in the interior parts of the fish after the viscera have been removed. No ice is placed between the layers of halibut as it tends to injure the flesh of the fish by pressing into it. The other fish that are iced at sea are placed on layers of ice in the ice bins and other layers of ice are sprinkled on top. In case of the mackerel where the port is near the fish are packed in ice in barrels for convenience in handling.

The method of the Newfoundland herring industry is somewhat different. The herring are frozen in the open air usually, on beaches or on platforms made for the purpose. After the fish have frozen solid they are placed in the hold of the vessel. Across the forward part of the hold a partition is built to keep out the heat from the forecastle. The herring are dumped into the hold in bulk and remain frozen until the vessel reaches its destination, Gloucester, Boston, New York or Philadelphia. Occasionally the cabin walls are covered over with canvas or boards to protect them from injury, and the cabin is filled with the frozen fish.

The demand for frozen herring in the cod and other ground fisheries has led to the erection of several freezing houses along the New England coast. These are located at Gloucester, Provincetown, Boothbay Harbor and North Truro, and have a capacity of about 15,000 barrels of frozen fish. The shipment of fish from seaports inland is by the usual method of refrigerator cars, being little different from the methods of shipping other kinds of fresh foods that require low temperature for preservation.

For securing herring and squid for bait from Newfoundland, vessels have been fitted out with freezing plants aboard; but it is the general rule to let Nature wait upon the fisherman in the freezing of fish.

The method of catching lobsters is by means of pots. These are made of laths nailed to end hoops about twenty inches across. The laths are apart enough to allow the water to pass in and out readily. Across each end is stretched a funnel-shaped piece of netting pointing inside. The inner extremity of the netting is fastened to a small hoop about five inches in diameter. Stones or other weight are placed inside, a line is attached to the outside, and the fisherman lets the car sink to the bottom on the fishing grounds. A light buoy is attached to the rope to enable the fisherman to locate his pot. Within the pot and midway between the ends of the two pieces of netting the bait is secured. As the lobster moves about in search of food it is attracted by the bait, which is usually the heads of cod, haddock, or some other fish. The lobster enters by the end of the pot, passes along the netting and enters the pot through the hole that has been left for the purpose. Once inside the pot the chances are very much against his getting out again.

Lobster pots are usually pulled once a day. If any lobsters are within they are removed, the pot is rebaited, and again allowed to sink to bottom. If the lobsters are not sold immediately they are placed in lobster cars. These are floating, box-like enclosures, allowing free circulation of water and capable of holding a large number of lobsters. In some instances artificial enclosures for retaining lobsters have been made by building a dam across a narrow body of water. When the collecting boat of the fish dealer or the cannery comes to collect the lobsters, they are bailed out of the car or enlosure by means of dip nets.

The digging of clams and preparing them for market is laborious work, often carried on during the severest days of winter. The digger provides himself with a hoe and several "rollers," the latter being long oblong baskets usually made of laths placed far enough apart to allow the water to pass in and out during the process of washing the mud from the clams. A good clam digger can dig from 8 to 10 bushels at a tide; occasionally one is found who can dig under the best of conditions from 12 to 15 bushels. As the clams are dug they are picked up and placed in the rollers. The digger regulates his place of digging by the tide, usually following close to the water as it recedes and keeping just in advance of the incoming tide. Clams are found in greater numbers and larger size near the low water mark.

On the coast of Maine, where soft clams are more abundant, the digging is carried on in the late fall and winter. The fishermen are those who are engaged in the shore fisheries during the warmer months. Many others also engage in this work in winter, especially the farmers whose lands border on the shore. Small houses are constructed on the banks of the shore just out of reach of the highest tides. To this place the clams are removed after being dug, to be placed in barrels, if shipped in the shell, or to be shucked if shipped for bait. While the tide is up the fishermen attend to the shucking of the clams and getting them ready for market. Frequently the long winter evenings are employed in removing the clams from their shells. In that event, the kitchen of the fisherman becomes the seat of action, where more comfortable quarters and perhaps assistance from the members of the family make the work easier and less prosaic.

CHAPTER XIV

THE MACKEREL FISHERY IN RECENT YEARS

For the last forty years no branch of our fisheries has occupied so public a position as that held by the mackerel fishery. This prominence is due largely to legislation, both domestic and foreign, that has been enacted in reference to the fishery. The industry does not possess one tenth the value that it had twenty-five years ago. Yet it holds the attention still, not only for what it has been but also for the possibilities that it possesses. The mackerel is the mysterious fish of the sea. Its habits have been closely and scientifically studied; its yearly haunts have been noted and visited by fishermen for generations; experts have been detailed by the National Government to examine into its methods of propagation with the hope of increasing artificially the number of the fish in the sea; many devices for its capture have been invented, from the fish-pole, the gaff and the jig of earlier times to the wholesale methods of nets and seines; improvement in craft structure and motive power have gone along with improved methods of capture; ice-houses now take the place of the well-boats of ante-bellum days, while the speed of schooners has been increased by sharpening and deepening the hull, by spreading more sail, and by employing auxiliary motor power.

Because the mackerel is the mysterious fish of the sea is the reason for many of the changes in the methods employed in taking the fish. Fishermen must be prepared for any emergency. With them, the race is in most cases with the swift. The uncertainty of appearance of the schools of mackerel increases this need of extreme care in preparation for catching them. Mackerel have certain well-defined habits of appearance. For example, they first make their appearance off the coast of the Hatteras region the last of March or the first of April. The body of fish advances northward as far as Block Island, reaching that place about the first week in June. At about the same time another body of fish appears on the coast of Nova Scotia, advances along shore, follows up the Cape Breton shore to disappear finally in the Gulf of Saint Lawrence before the first of July.

In the meantime, the Block Island body of mackerel may have moved eastward to George's Bank, or turned north to appear along the New England coast; or they may disappear without leave or notice for the remainder of the summer. Similar conditions may exist with the body of fish that entered the Gulf of Saint Lawrence. With the coming of the fall months the mackerel feels the instinct of migration as does the summer song bird. Immense schools of the fish may appear in almost any part of the Gulf of Maine or of the Gulf of Saint Lawrence, preparatory to their annual fall migration. When the one school leaves Nantucket behind and the other disappears south of Scatari and Cape Sable the mackerel vanish for the winter,—for their haunts during the colder months of the year are still unknown to the fishermen.

With the opening of another spring they will return to the surface of the ocean near their usual haunts to repeat the great movement that their ancestors have followed for ages. This much the fisherman can depend upon, generally. But there are a thousand and one other tricks of the fish that are beyond knowing. They may appear in small, detached schools which, if caught, will repay the fisherman with a few barrels only. Or they may be in an immense school too large for a twelve-hundred-foot seine to encompass. Many instances are on record of the schooling of the mackerel in so great a body that it extended for miles in length and breadth over the surface of the ocean. Under such conditions, it may be a difficult task to catch them. Usually mackerel are sensitive to maneuvers to catch them; when they are wild it is almost impossible for the fisherman to take them, and a dozen "water-hauls" may be made during the day without any success on the part of the fisherman. On the other hand, a fortunate streak of good fishing may continue for several days and a cargo of fish be secured in the time; such conditions are unusual and are attended by great physical exertions on the part of the vessel's crew.

The annual catch of fish is a very unsafe guide by which to compute the quantity of mackerel in the sea. The quantity of mackerel that exist in the sea probably does not vary greatly from one year to another. During periods covering ten or twenty years there may be a difference. The fisherman realizes, as the landsman cannot, that the amount of mackerel in the ocean to be caught is immense —a quantity beyond all possibility of computing; the season's catch may be only ten thousand barrels, yet his knowledge of conditions causes the fisherman to believe that there are millions upon millions of them left untouched,—the possibilities of several seasons' work in the future.

Thus it is that the mackerel fishery offers great hope of reward to the skillful seiner. In a successful year his vessel may stock from $15,000 to $40,000. It is with the hope of making a good season's catch that he continues in the business from year to year. It is with the same hope that the mackerel fleet continues to exist. If an equal division were made each year of the proceeds of the total catch it is likely that the individual share would fall below the amount required to support a fisherman and his family for the year. But any year it may happen that conditions are just right for the fishermen—plenty of mackerel to be found, favorable weather, good markets, and ease in taking the fish. When such a season exists the captain and crew make good wages and the vessel owners may have a share large enough to pay for the first cost of the vessel. Although the state of the mackerel fishery is now in a depressed condition it does not signify that the fishery will cease; it may be only the ebb-tide of a prosperity that has had periodical fluctuations since the beginning of the fishery ninety years ago.

When considered from its economic point of view the mackerel fishery since the Civil War may be separated easily into two periods. The first twenty years were years of continued prosperity. The prosperous state of the mackerel fishery had its beginning in 1845, after a decade of depression. There was a short period of depression from 1855 to 1860, but not serious. The prosperity of the fishery culminated during the Civil War when the country's catch of mackerel exceeded $6,000,000 in value in a single year. Between 1845 and 1885 the number of barrels of salt mackerel annually marketed never fell below 100,000 barrels. During twelve of these years, at least, the number was in excess of 300,000 barrels. The lowest catch of the New England fleet from 1867 to 1885 was 117,096 barrels, in the year 1877. The greatest catch of salt mackerel was in 1884, there being 478,076 barrels, the largest quantity ever caught in a year in the history of the fishery. During the first twenty years after the war, the quantity of the catch varied from year to year, as is usual with this industry. But the average held up well, it being about 215,000 barrels yearly. The total catch from 1867 to 1885 inclusive was 4,071,705 barrels of salt mackerel.[1]

Beginning with the year 1886 and continuing to the present day the mackerel fishery shows a remarkable and in-

[1] Compiled from the Boston Fish Bureau Reports.

explicable decadence. This change did not come gradually through a number of years, but, instead, the fishery dropped in a year to only one-fourth its former volume and never has risen since to any respectable semblance of its former prosperous condition. The years since 1886 have found the annual catch only 43,000 barrels, an amount in striking contrast with the average of 215,000 barrels of the nineteen years that preceded the present period, or the 225,000 barrels of the forty-eight years ending with 1866. The total catch of the mackerel fleet from 1886 to 1908 was 978,357 barrels. The catch of salt mackerel for the three years, 1883–1885, exceeded by 20,000 barrels the total catch of salt mackerel for the last twenty-three years. During this period of decadence the industry has fallen off to about twenty per cent of the volume maintained from the active inception of the industry to the end of the period of prosperity in 1885. Not since the year 1814 has there been so few salt mackerel taken as in the year 1906, when the catch was only 10,136 barrels.

What reason can be offered in explanation of this remarkable decline of the mackerel industry during the last twenty-three years? No certain cause, no satisfactory reason, has yet been found. Allowance should be made for the increase of the fresh mackerel trade during the past few years when considering the falling off in the salt mackerel output; also, the decrease in the amount of tonnage employed in the fishery means a decrease in the average annual catch. Other reasons have been advanced, but not any one of them, nor all of them taken together, can offer a satisfactory solution to the puzzle.

Viewed historically, the story of the mackerel fishery for the last forty years centers about three events, the development of the southern spring fishery by the introduction of purse-seines for catching the fish, and the use of ice for preserving them; the extension of the privilege of inshore fishing to American fishermen for twelve years in Canadian waters by the treaty of Washington in 1873; and the prohibition of the southern spring mackerel fishery by the National Government between the years 1888 and 1892.

The development of the southern fresh fishery led, in a few years, to a fresh fish business at New York that was prosperous for a little more than a decade. The rise of a fresh-mackerel trade caused jealousy on the part of the salt-fish owners and dealers, who raised the cry that the catching of mackerel in the spring was detrimental to their business and threatened to destroy the mackerel industry as a whole. Legislators were appealed to that the mackerel fishery might be saved, with the result that the very prosperity of the southern mackerel fishing business brought about its prohibition for the space of five years.

The extension of the privilege of inshore fishing in Canadian waters to our fishermen was in the nature of a barter, as it was originally intended. In its outcome it proved to be a trade in which the British government got the better of our government by several million dollars. In brief, American fishermen were privileged to catch fish in Canadian waters within the three-mile limit. This concession was especially beneficial to the mackerel fish rather than to the ground-fish industry. The inhabitants of the provinces were allowed to import their fish and fish-oil into the United States free of all tariffs. Provision was made for a Commission to determine which country was thus given the greater privileges and to assess on that country a sum of money equivalent to the balance of privileges. The matter was determined by the Halifax Commission in the fall of 1877; and, by the finding of that body, the United States paid Great Britain the sum of $5,500,000 for twelve years of fishing in the Gulf of Saint Lawrence. The Commission made a most thorough examination of the condition of American and Canadian mackerel fisheries in

Canadian waters. Witnesses were summoned by the score, and the report of the Commission, which covers nearly seven thousand pages, is made up largely of the testimony of fishermen and experts in the fishing business.[1]

The contention of the American members of the Commission that the extent of American fisheries in Canadian waters was on the decline has since been verified by the records of vessels that have been to the Gulf of Saint Lawrence for the last twenty-five years. Whereas formerly American mackerel schooners frequented the Gulf in large numbers the fleet to-day visits those waters in comparatively small numbers. Our fleet visits the coast of Nova Scotia for mackerel but the fishery is carried on in no enclosed bay or gulf, and consequently is less under the dominance of Canadian authorities. The average mackerel fleet on the New England shore for the decade of the eighties was 276 schooners and in the Gulf of Saint Lawrence for the same time, 37 schooners. For the twelve years ending with 1908 the mackerel fleet on our own shores numbered 122 sails and the Bay fleet 7.5 sails. The percentage of the Bay fleet to the total American mackerel fleet fell off from 11 per cent during the decade of the eighties to about 6 per cent during the last twelve years. The catch of the American fleet in the Bay during the operation of the Treaty of Washington, 1873 to 1885, was about 300,000 barrels of mackerel, or 25,000 a year. The total catch of salt mackerel in the Gulf of Saint Lawrence for the seven years ending in 1908 was 3,672 barrels, an insignificant amount in comparison to the prosperity of former years.

No single phase of the mackerel fishery has occupied so much discussion and been so barren in its results as the southern spring mackerel fishery. The fundamental question that arose for settlement was whether the catching

[1] For a more detailed account of the Halifax Commission, see Ch. XIX, The Fisheries Question.

of the fish during the months of April and May, when the mackerel are migrating northward to spawn, did not lessen materially the quantity of mackerel in the sea and also threaten to annihilate the mackerel as a food-fish. The question did not arise until the invention of the purse-seine made it possible for fishermen to take mackerel in wholesale quantities.

Purse-seines for catching mackerel were in use previous to the Civil War, but the use of seines did not become general with the mackerel fleet until the early seventies. Soon after the war vessels from Gloucester began to land fares of fresh mackerel at New York during the spring season. By 1870, twenty-five or thirty fares of fresh mackerel were annually brought into the city, having been caught in the vicinity of Sandy Hook. These fish were caught by handline, or jigging. In 1872 the schooner *Dreadnaught*, of Portland, fitted out with a purse-seine for fresh mackerel fishing, being the first vessel that carried no salt on the southern trip. The vessel was successful in the venture and netted a large stock by the close of the spring fishing.

The success of the *Dreadnaught* influenced other vessels to fit out for the fresh mackerel fishery in the following year. The fresh mackerel fleet immediately became very large. By 1873 this fleet was in full force during the southern spring fishery. At first New York was the market; but so many cargoes were landed there that the market was soon overstocked. Philadelphia, Baltimore and other cities each became a center for a small fleet of mackerel schooners. At these places the mackerel, brought in on ice by the schooners, are handled by fishmongers at a commission of 12½ per cent.

To understand the reasons that led Congress to suspend the southern spring mackerel fishery for the five years between 1888 and 1892 it is necessary to examine the condition of the industry for a few years previous to the passage of the act of suspension. In 1885 a fleet of about 175 schooners set sail for the southern fishing grounds. The first fares were landed at New York by the schooners *Emma W. Brown* and *Nellie M. Rowe* on March 28, each vessel having 125 barrels of mackerel. During the season unusually large quantities of mackerel were landed at New York and Philadelphia, the quantity at the former market being 125,000 barrels. The price of the fish was from seventy-five cents to one dollar per hundred fish and the total value of all fish taken was about $273,000.

At one time 130 vessels were in with fares of fish. The market was inadequate to care for so large a quantity. Many fish sold as low as fifty cents per thousand at this time, and large quantities of the fish were thrown away. Some reports place the number of fish thrown away at 40,000 barrels; other reports stated that as high as 150,000 barrels of mackerel were wasted in this way. Undoubtedly the reports were greatly exaggerated. Personal observations made at the time by representatives of the United States Fish Commission, confirmed by reliable authorities, "showed that there was little foundation for these estimates, and indicated that only from ten to fifteen thousand barrels of fresh mackerel were thus destroyed for want of a market, and that the most liberal estimate should not place the quantity at over 20,000 or 25,000 barrels." [1]

The southern mackerel fishery for 1886 was almost a failure. This was due in part to a smaller fleet, there being about 150 vessels employed. The greater part of the fleet failed to pay expenses. During the height of the season twenty days of stormy weather caused a suspension of the industry, greatly to the loss of the fishermen. There were 117 fares of fresh mackerel landed at New York, which were sold out of the vessels for $78,507.

[1] Smith, Southern Spring Mackerel Fishery, p. 197.

From twenty-five to thirty fares were landed at Philadelphia and several fares at Providence and other New England ports. In addition, there were about 2,500 barrels of salt mackerel landed from the southern catch.

In the spring of 1887 about 106 vessels entered the fishery. A backward season made the fishery late in its beginning. A larger quantity of salt mackerel was landed by the vessels, the quantity being 4,732 barrels, valued at $33,403. The total catch of fresh mackerel was 8,384 barrels, valued at $53,402. The average stock per vessel was $504 from fresh fish and $315 from salt mackerel. Taking the three years as a whole it is found that the southern fresh mackerel fishery previous to its suspension was worth only $876 per year for each vessel that was employed in the industry—an amount that did not enrich vessel owners or fishermen greatly, however dangerous the fishery may have been to other forms of industry.

As early as 1878, the question of suspending the southern mackerel fishery was taken hold of by some Boston fish dealers, who felt concerned for the safety of the summer mackerel fishery. The Boston Fish Bureau, also, held that the fishery should be suspended on the ground that in the long run only detriment to the fishery could result. The agitation for actual suspension of the fishery was begun by mackerel dealers and fishermen of Portland, who, in December 1885, petitioned the United States Commission of Fish and Fisheries that the necessity and desirability of protection to mackerel during their spawning season be urged upon Congress. Other petitions of a similar character followed elsewhere throughout New England. The desire for a suspension of the southern mackerel fishery came principally from dealers in salt fish and from vessel owners who were engaged in the salt mackerel industry.

When, in May, 1886, a bill came before the House of Representatives to suspend the spring fishery it called forth one of the most interesting discussions ever heard in our national halls concerning the fishing industry. Every possible phase of the subject was discussed during the progress of the debates. The principal arguments of the promoters of the bill were that the southern spring mackerel fishery is extremely uncertain and has usually been carried on at a loss; that the continued catching of the fish on a large scale before the period of spawning would exhaust the supply of fish and destroy the mackerel fishery; that the continued harassing of mackerel during their migration breaks up the schools and drives the fish from the New England shores into the Gulf of Saint Lawrence; that the fish of the early season are inferior in quality; that this fishery interferes with the trade of salt mackerel and reduces the price of that kind of fish.

The opponents of the measure, those who wished the spring mackerel fishery to continue, claimed that for Congress to interfere with the ocean fisheries was establishing a dangerous precedent; that there was no indisputable evidence that the general abundance of mackerel was affected by the spring catch; that the use of purse-seines in taking the fish did not necessarily drive them from our shores to the Gulf of Saint Lawrence; that a sincere demand for protection would place the time limit as late as July, when the spawning season is over; that these fish furnish a cheap and wholesome food for thousands of people who cannot afford fish of a higher grade; that the proposed law would be severe and sectional as it would prevent the capture of mackerel when they appeared off the coast of the southern states and would allow their capture when the fish had reached the New England shores.

The act was approved by the President in February, 1887. Congress sought to give protection to the mackerel industry by passing a law that forbade the importation of the fish during certain seasons of the year. No action was taken to prevent the capture of the fish in the waters of the Atlantic, since nobody would be encouraged to catch fish for which no markets were provided. The question of the right to legislate over deep-sea fisheries was, in this way, carefully and adroitly avoided. The first and third sections of the act are as follows:

"That, for the period of five years from and after the passage of this act no mackerel, other than what is known as Spanish mackerel, caught between the first day of March and the first day of June, inclusive, of each year, shall be imported into the United States or landed upon its shores. . . .

"That the penalty for the violation or attempted violation of this act shall be forfeiture of license on the part of the vessel engaged in said violation, if a vessel of this country, and the forfeiture to the United States, according to law, of the mackerel imported or landed, or sought to be imported or landed."

The resumption of the mackerel fishery in southern waters in 1893 was watched with interest. The mackerel fleet numbered 60 vessels, 43 of them being from Gloucester, 11 from Portland, and the others from Boston, Rockport and Dennisport. The season was unfavorable for fishing, there being an unusual amount of stormy weather which interfered with seining. Only 16 of the fleet landed fares of fresh fish. The total quantity of fresh mackerel landed was 1,158 barrels, valued at $21,000. Had this sum been divided among the crews of all the fleet each man would have shared about ten dollars. Since 1893 the southern spring mackerel fishery has been small. In 1894 only 700 barrels of mackerel were taken; in 1897 and in 1900 the number went above 11,000 barrels, but the average for the eight years following the resumption of spring fishing was about 4,000 barrels a year. In that time the seining fleet numbered 48 schooners. The southern catch of fresh mackerel for the years 1907 and 1908 was 27,000 and 20,000 barrels respectively.

The experiment of prohibiting the taking of mackerel for five years from 1888 to 1892 can now be viewed from the perspective of eighteen years. The experiment was negative in its results; there is no indisputable proof that the seining of mackerel before or since the passage of the act has ever diminished the body of fish to be taken. Neither is there cause to infer that the five years of protection increased in any appreciable degree the quantity of mackerel in the ocean. The experiment is suggestive of a truth well known to every mackerel fisherman that the agency of man is the least of the causes to diminish the quantity of mackerel in the ocean—a quantity that in good seasons and in poor seasons is a thousand times greater than the small amount taken by the fisherman.

The fresh mackerel fishery which had its beginning in the southern catch of mackerel has been extended in recent years so that many vessels carry ice-houses during the summer months. The fishery is carried on in the Gulf of Maine, on George's and on the southeast coast of New England. The fresh mackerel industry has been on the increase during the last twenty years. The catch from 1889 to 1904 averaged about 20,000 barrels; the average for the past nine years has been near 55,000 barrels yearly, that for 1908 being 57,566 barrels. The growth of the fresh mackerel output should be kept in mind when accounting for the decline in the salt mackerel catch; during the seven years ending with 1908 the New England catch of salt mackerel was 211,000 barrels, and the catch of fresh mackerel was 384,860 barrels.[1] However, a comparative view of the total catch of mackerel for different periods of years does not account for the decadence of the mackerel industry as a whole.

Mention already has been made of the change that has taken place in the mackerel fishery in regard to the de-

[1] Compiled from the Boston Fish Bureau Reports.

crease in the number of vessels that go to the Gulf of Saint Lawrence. Another change to be noted is the concentration of the industry in the State of Massachusetts, or more properly at the port of Gloucester. During the decade of the eighties there were 80 of the vessels of the New England fleet that hailed from Maine; this number was near one-fourth the entire fleet. For the twelve years ending with 1908 the mackerel fleet from Maine had been about six per cent of the total New England fleet, which had been reduced to a total of 124 vessels. So it has come about that Massachusetts has been left in these later days, as it was in the early period of the fishery, the principal state of the Union to engage in the mackerel fishery.[1]

The indications are that the New England mackerel fishery will be carried on in its present field of action, that is, the largest quantity of the fish will continue to be taken during the southern spring fishery, along the Cape Shore, about Block Island, on George's and in the Gulf of Maine. In recent years attempts have been made to extend the fishery to new fishing grounds, but the attempts have not proved satisfactory. In April, 1878, the schooner *Notice*, Capt. Knud Markuson, set sail from Gloucester for the coast of Norway with the intention of using the purse-seine on that coast. The venture did not prove successful, for Captain Markuson found that the mackerel on the coast of Norway, unlike those in New England waters, did not school together in large bodies.

A second attempt to enlarge the field of the New England mackerel fisheries was made in October, 1889, when the schooner *Alice* sailed for the coast of Africa. The first season the vessel secured considerable fish, which were shipped home. The second season the fish were not in

[1] In 1908 the mackerel fishery was carried on by vessels from ports as follows: Portland, 1; Gloucester, 52; Boston, 19; Duxbury, 3; Chatham, 1.

abundance and adverse legislation that prohibited the use of purse-seines made the season unprofitable. The vessel returned to the United States in 1891. Another innovation in the mackerel industry was made in 1899. Capt. Solomon Jacobs of the schooner *Ethel B. Jacobs* sailed in July for the coast of Ireland, making the passage across the Atlantic in 14 days. The schooner secured 353 barrels of mackerel which were shipped home. In October the vessel went ashore on Abbey Island and was a total loss. It is doubtful if further attempts will be made to extend the New England mackerel fishery into foreign waters.

An important development of the industry was the introduction of gasoline engines as motor power in 1900, when the *Helen Miller Gould* was fitted out as an auxiliary schooner. Three other auxiliary schooners were fitted out in the following year and two in 1902. At the present time there are about a score of schooners equipped with auxiliary power. The extra power is of especial value in increasing the speed of the schooner to and from the fishing grounds and in assuring a means of propulsion on the fishing grounds during days of calm weather.

The improvement in motor power among mackerel vessels is along the line of greater efficiency and better means of living aboard the Gloucester schooner of to-day. The successful mackerel schooner must be equipped with the very best that dealers can furnish. Sails, cordage, seines and boats need to be new and strong to stand the test of many severe strains of work and weather. Probably the world furnishes no fleet of vessels comparable to the New England mackerel fleet in staunchness, equipment, sailing qualities and general appearance. On the best schooners the captain has his own stateroom, finished in oak and furnished with a berth, a leather-covered couch, a clothes closet, medicine-case, chest and locker. The crew were never before provided for so bountifully. Each man has a berth, attractive, roomy and well-ventilated. Closets are furnished for drying wet oil-clothes. The cook is the second best man aboard the schooner and has a culinary department and larder furnished and supplied in a manner corresponding to the general outfit of the vessel and the demands of hungry men.

CHAPTER XV

THE COD FISHERY IN RECENT YEARS

When the members-elect of the House of Representatives of Massachusetts met in January, 1895, to organize and to make plans for removal from the accustomed chamber to the new chamber in the State House extension, the question of taking with them the "representation of a codfish," which for more than 100 years had never missed a roll call, was brought up for consideration. A committee was appointed to look up the history of this famous emblem and its significance in the Hall of Representatives, and the committee later made an interesting report of its findings. The House ordered that the ancient "representation of a codfish" be removed from the old chamber to the new one in order that the House of 1895 might further the intent and purpose of the House of 1784, wherein it was voted to "hang the representation of a codfish in the room, where the House sit, as a memorial of the importance of the codfishery to the welfare of this Commonwealth, as had been usual formerly."[1]

A committee, appointed for the purpose of removal, repaired to the old chamber. The emblem was lowered from its abiding place, wrapped in the American flag, deposited upon a bier, and borne by members of the House. As the procession entered the new chamber, the members arose, the historic emblem was received with a vigorous round of applause, and deposited upon the table in front of the Speaker's chair, between the two sets of central columns,

[1] A History of the Emblem of the Codfish.

and under the names "Motley" and "Parkman." There it remains to-day, "a memorial of the Pilgrim, his privations and simplicity; an emblem significant of the hardiness, courage, and faith of those who dare and defy the seas, and daily telling of the great and surpassing glories of Massachusetts and her sons."

It would be difficult to find an industry of New England that surpasses in the remoteness of its origin, in the continuance of its existence, and in the hardy manhood developed in its followers, the industry of the fisheries. Further, no other industry of the country ante-dates this one of New England, and it is equally probable that no other industry of historic importance in the land finds so complete an epitome of life-history in a small section of the country as do the fisheries of the United States find in the story of the fishery industries of New England.

Of all the fish of the sea none is dearer to the heart of the New Englander than the codfish. History has claimed it for her own, and thrown a halo about its name. For years the cod held supreme sway over all others of its kind. This was due to no sentiment arising from historic associations. The life of the colonist was staked upon the economic importance of the codfish. The Revolution witnessed a struggle in diplomacy in which the codfish was a central figure. Our war for independence upon the sea was won by cod fishermen from the capes and banks. The cod tells "of commerce, diplomacy, war; of victories won in all three fields." While the cod occupies so completely the foremost place in our fisheries until the second war with Great Britain, there arises in the more recent history, consideration for other fish.

The mackerel fishery began to assume a place of importance before the waning glories of the whale fishery had sunk into obscurity. The oyster fishery is a leader to-day in economic importance, even though it is doomed never to possess the interest that attaches to the cod and whale fishery. Lobsters and salmon occur to the mind as important branches of the fishery industry of the twentieth century. Nevertheless the cod has not lost its place among these other fisheries. But we associate with the history of the cod of to-day the other fish of its kind,—the hake, the haddock, the cusk, the pollock, and the halibut. All of these fish are grouped together as ground fish. Further mention of the ground fish will embrace all the different fish here included in that term.

Since the Civil War the New England deep-sea fishermen have enlarged the field of their activities by embarking on the seas of Greenland and Iceland in search of halibut. The first vessel that ever left Gloucester for a flitching halibut voyage to the Grand Bank of Newfoundland was the schooner *A. J. Chapman* in 1864.[1] The schooner returned in August after a voyage of 85 days, with a stock of nearly $5,000. Reports of the abundance of halibut were first brought to Massachusetts by Provincetown whalers. The first trip to the coast of Greenland for halibut was made in 1866, by the schooner *John Atwood.* The stock of the schooner was $5,500. Capt. John McQuinn in 1870 brought from Greenland a flitched halibut trip worth $19,000. During the next two years half a dozen vessels made halibut trips to Greenland. The fishery did not prove remunerative, so that it was practically abandoned between 1874 and 1877. The fishery was revived in 1878.

Captain McQuinn was the first of the American fishermen to go to the halibut fishing grounds of Iceland. In 1873, he set sail in the schooner *Membrino Chief.* The vessel remained in Iceland waters for five weeks, visiting various harbors and cruising on the fishing grounds whenever the weather would permit, but catching very few fish.

[1] Goode, Sec. V, Vol. II, p. 91.

The trip proved a failure, and deterred others from going there for a time. Notwithstanding the uncertainty of the season, the lack of knowledge of the fishing grounds, and the long distance from home, there were 31 trips made from Gloucester to the grounds of Davis Strait for halibut from 1866 to 1881. The total catch of salt halibut was 3,283,765 pounds, or 113,233 pounds for each of the vessels returning.[1] Another venture by the New England fishermen to new fishing grounds was made in 1890 by the schooner *Gatherer*, of Gloucester, to the coast of Norway for codfish. She arrived just too late for the season. The expedition to the coast of Norway was a failure, but the vessel, returning by the way of Iceland, secured a fair trip of halibut.[2]

The method of catching ground fish by hand-line from the vessel continued to a limited extent down to 1860. By 1880 the method had practically given way to the use of dories. Hand-line fishing from dories first appeared between 1855 and 1856. The use of the trawl in these fisheries was as early as 1851 or 1852.[3] The process of trawl fishing from dories a quarter of a century ago has changed but little during the years. By this method, dories are sent out from the vessel which is anchored on some good fishing ground. Each dory has two men, and a trawl to be set. In this way, the area over which the vessel fishes is increased as many fold as there are dories.

An account of the pursuit of the offshore fisheries from Gloucester in 1876 sets forth the condition of fishing thirty years ago, which, in most ways, is essentially the same today. The writer says, "The Grand and Western Bank Fishery is pursued to a greater or less extent during every month in the year. Last year this business employed 175

[1] Ibid, p. 92.
[2] Boston Fish Bureau Report, 1892.
[3] Goode, History and Methods of the Fisheries, I, p. 158.

vessels, and 499 fares were landed. The business may be classed in two departments, a portion of the fleet making short trips and bringing in their fares fresh, to supply the fresh fish trade, and the rest of the fleet making longer trips and dressing and curing their fish as they are caught. This business employs the best class of fishing vessels known to the waters of the coast. A modern 'banker,' of average tonnage, costs about $8,800. Such a vessel, manned by a crew of twelve men and making nine trips to the Banks, being at sea 302 days, will require an expense of $1,023 for trawl gear, $1,824 for vessel's expense, $1,426 for provisioning, and $1,135 for general charges, such as ice, bait, salt, etc. The man who ventures on a trip in a 'trawler' finds little of the 'pleasing content' described by the early voyager. For him at least there is little of romance in 'the apostles' own calling.' Life on the banks he finds a constant round of drudgery, so long as he is able to make his daily rounds. He must rise early and work late in order to visit his trawls, remove his fish, rebait and reset the lines, and take care of the day's catch. Tossed on the waves in his frail dory, at greater or less distance from his vessel, he is subject to perils unknown to the fishermen of the olden time. His frail boat rides like a shell upon the surface of the sea, but in experienced hands no description of small sea craft is safer. Yet a moment of carelessness or inattention, or a slight miscalculation, may cost him his life. And a greater foe than carelessness lies in wait for its prey. The stealthy fog enwraps him in its folds, blinds his vision, cuts off all marks to guide his course, and leaves him afloat on a measureless void. Instances are on record of many a wearisome trip, of days and nights without food or water, spent in weary labor at the oars, at last to find succor from some chance vessel or by reaching a distant port; and imagination revolts from the contemplation of the hardships experienced, the hopes awakened and dispelled and the torturing fate of the many 'lost in the fog,' of whose trying experiences nothing is ever known.''[1]

The condition of the ground fishery of New England proved to be prosperous for several years subsequent to the war. There was no departure from the general condition of prosperity for about twenty years. Then there occurred a marked decline, similar in many ways to the decline in the mackerel fishery. This period of decline was preceded, as in the case of the mackerel fishery, by a few years of extraordinary activity and success. For the six years from 1880 to 1885 the number of vessels employed in the ground fishery was 686 yearly, carrying crews of 7,665 men, the average being 11 for a vessel. The catch of cured fish for that time was 880,969 quintals, or 1,284 quintals per vessel. The catch, during these years, was very evenly divided between the New England shore and George's fleet, and the Grand and Western Bank fleet. The tonnage of the Grand Bank fleet was in excess of the other, as vessels for this fishery generally average a little larger than those fishing near home. Only once, in 1888, has the number of vessels employed in the Grand Bank fishery exceeded the number engaged in the New England fishery. In that year the numbers were 339 and 284 respectively. Since 1891 the number of vessels going to the Grand Bank has been on the decline, there being about 60 vessels there yearly for the last decade.[2]

The catch of cured codfish and other ground fish for 1880 was 646,426 quintals. For three years thereafter a steady and even increase in the catch carried the amount of the catch to 1,061,698 quintals in 1883, which is a record-breaking quantity. In 1884, the catch was still in excess of a million quintals. The fishery was on the decline,

[1] The Fisheries of Gloucester, pp. 56–58.

[2] Compiled from Boston Fish Bureau Reports.

however, and the catch continued to fall off in amount yearly until 1890, when only 436,650 quintals were taken. While the average yearly catch for the decade of the eighties was 786,000 quintals, for the last nineteen years it has been only 457,000 quintals.[1]

During the last twelve years there have been annually in the New England shore and George's fleet 372 vessels, and in the Grand and Western Banks fleet 58 vessels, the whole number employed in both fisheries being 430 vessels, carrying crews of 7,142 men, and securing 460,000 quintals of fish yearly. The average for each vessel of the fleet is 1,070 quintals yearly. While the Grand Bank fleet is less than 14 per cent of the total, that fleet has caught in this period of time 36 per cent of the entire catch of fish.

The fresh-fish industry of the deep-sea fisheries has been carried on with more or less activity in New England for above three-quarters of a century. The method of preserving the fish fresh for the market was at first by the use of well-smacks. These boats were fitted out with a compartment that allowed the ocean water to circulate freely. As soon as the fish were captured they were thrown into the well. In the case of the halibut, which often weighs several hundred pounds, they were stunned by a blow on the head in order to take them more easily and safely from the

[1] The change that has taken place is indicated in the following table, of the catch of cured ground fish in quintals for different periods.

Period.	New England Shore and George's Banks.	Grand and Western Banks.	Total.
1880–1889	3,757,775	1,915,764	7,861,478
1890–1899	2,823,048	3,103,803	4,738,812
1900–1908	2,465,368	1,457,699	3,923,067

water. After they had been placed in the wells they invariably revived. Another early method of preserving the fish, especially halibut, from becoming injured by lying in a body in the hold of the vessel was to hang them up by the tail on large hooks that had been arranged for the purpose. Both these methods were cumbersome; neither was suited for the making of large cargoes of fish. With the introduction of ice-houses aboard the schooner the fresh-fish industry was placed on a more economical basis.

The first ice-house was constructed on board a vessel engaged in the fresh halibut business in 1846. The method was to place the fish in the house itself, much as meat is now preserved in cold storage rooms. The process of packing the fish in chiseled ice was not employed until many years later. The question of bait has always been more or less serious for the fisherman who frequents the off-shore banks for his fish. The practice used to be for the vessels to take small seines with them for the capture, near the land, of some variety of bait-fish. This was especially true of the fleet of cod-fishermen that went to the coast of Labrador for their trips. Toward evening a boat would go out from the vessel to the shallow inlets of the bays and set the seine for the capelin that school at that time. Often fresh-fish vessels would in some way catch their bait in the bays adjacent to the fishing grounds. But the fresh-fish industry was not fully launched in its career of greatest development and value until the trade in frozen herring with Newfoundland was begun. That important industry made it possible for the New England fishermen to secure a cheaper bait when other bait was out of the market, and, as a consequence, since the middle of the last century there has been a great change in the quantity of business carried on in fresh ground fish.

By 1879, the business had reached such a profitable stage that forty vessels from Gloucester were employed exclusively in the fresh halibut industry. In addition to these there were several others from Massachusetts and Long Island Sound that were employed in the fishery during a part of the season. The New England catch of fresh halibut for 1879 was 14,637,000 pounds. After that year there was decline in the fishery. In 1885, the catch was light, although the same number of vessels were in use. For the ten years between 1890 and 1899 the total number of arrivals of fresh ground fish in Boston was 38,299, and the total amount of the catch of fresh fish for that decade, landed at Boston, was 632,325,496 pounds. The average amount for each trip was 16,500 pounds.[1]

From 1900 to 1906, there were landed at Boston 595,000,000 pounds of fresh ground fish, a gain of ten millions of pounds yearly over the decade of the nineties. During the year 1907, American fishing vessels landed at Gloucester and Boston 177,892,976 pounds of fresh ground fish, valued at $4,709,022.[2] A great stimulus to the fresh fishery in recent years has been the almost universal introduction of gasoline engines aboard the smaller craft used in the inshore ground-fishery. The prospects are that fresh ground-fishery will continue to increase in importance and annual output for many years.

1 Compiled from Boston Fish Bureau Reports.

2 Bureau of Fisheries, Statistical Bulletin, No. 208.

CHAPTER XVI

THE DECADENCE OF THE DEEP-SEA FISHERIES

The last fifty years of the history of the New England fisheries afford a more complete record of their condition than is given by any former period. It is a record filled with deeds of daring and of suffering equal to the hardy experiences of fishermen during colonial times. The fisheries have changed during the centuries in the methods of pursuit and of capture; but charts of the coast cannot lessen the severity of winter seas, neither are the men of the later period less daring and courageous because of lighthouses and patrols along the shore. The history of fishermen must always be a record of a superior class of men who are ready to brave the severest storms and exposures of the ocean in order to gain a living, men who apparently have little regard for their own life and safety, due to long continued experiences of a hazardous nature in battling for the safety of their little craft.

It will be remembered that one of the first acts of the Federal Government was the passage of an act that gave fishing vessels the benefit of a bounty. This measure of national assistance continued in force until 1866, since which time the fisheries have received no help in the form of bounties. That the benefit of bounties was a wise and beneficial provision cannot be doubted. However, from 1845 to 1885 the fisheries of the country were in a prosperous condition, in general, and would have prospered had there been no system of bounties. Those were years of abundance of deep-sea fish, and for a decade before and after the beginning of the war there was also a very large body of fish of the herring family on the New England coast. Since 1885, however, the tonnage employed in the fisheries has fallen off to such a marked degree that a revival of the system of tonnage-bounty would be a desideratum.

During the last quarter century the fisheries of New England have declined in a remarkable manner. This decadence has been most marked with the offshore fisheries. They are the kind that require larger vessels, that are most hazardous, and that develop able seamen. Our fisheries have been the nursery of our navy in the past. The capture of Louisburg, the naval exploits of the Revolution, the fishermen-gunners of the privateers of the war of 1812, as well as the naval history of the Civil and Spanish wars afford most glorious proof of that fact. The shipbuilding industry and our merchant marine date back to the fisheries for their origin and development. So that if the fisheries had no reason of their own for protection and development, their important relation to our navy and merchant marine, in the past and in the present time, would afford ample grounds for their continuance, even at the cost of national assistance.

The amount of tonnage employed in an industry of the sea is an index to the general prosperity of that industry. The tonnage of the merchant marine employed in the cod and mackerel fisheries since 1866 shows a general decline when compared with the tonnage employed in these industries during the Civil War period. The tonnage reached its highest mark in the history of the fisheries in 1862, when 204,197 tons were employed in the deep-sea fisheries. The registry for the last three years of the war averaged about 146,000 tons annually. A decline began even in the last years of the war, and since that time the total tonnage has reached beyond the 100,000 tons mark only once. Beginning with 1870, there was a revival of the fisheries which reached a climax in 1873, when 109,519 tons were employed. The total tonnage fell off immediately to 78,290 tons in 1874, but kept above the 70,000 tons mark until 1890, averaging 87,000 tons yearly.[1]

The tonnage in 1890 was 68,367 tons. For the next seven years there was little change. In 1898, there was a decline to 52,327 tons, and in 1899 to 50,679 tons, the lowest tonnage employed in the cod and mackerel fisheries of the country since the close of our last war with Great Britain in 1816. Since 1899, there has been a yearly increase to the 61,439 tons for 1906.

The tonnage employed in the cod and mackerel fisheries is made up of enrolled vessels, and licensed vessels under twenty tons. The fluctuations in the total tonnage from year to year have been principally in the class of enrolled vessels, that is, those that engage in the deep-sea fisheries. The year of greatest tonnage for enrolled vessels was 1873, when 1,558 vessels had a total of 99,542 tons. In 1899, the 545 vessels of this class aggregated 42,901 tons, the lowest tonnage on record. From 1867 to 1887, inclusive, the number employed was always above a thousand vessels. Since that time the number never has been above 968. The number of enrolled vessels employed in the cod and mackerel fisheries in 1906 was 560 vessels. The average tonnage of vessels of this class is a little above eighty tons.

Taking the last forty-four years as a whole it is easily seen that there are two well-defined periods. The first, from 1866 to 1885, is one of general prosperity in the amount of tonnage employed in the cod and mackerel fisheries. Yet even this better period is considerably below that period of greatest prosperity which had its beginning in the year 1845 and extended through 1865. The

[1] Compiled from An. Reports, Commissioner of Navigation.

second period, from 1885 to the present, has been one of general decline in the amount of tonnage employed in these fisheries, in spite of the upward tendency of the fishing industry, as a whole, since the opening of the present century.

The decadence of the deep-sea fisheries of New England during the last quarter century has led some people to the erroneous conclusion that our fisheries and fishery industries are in a declining condition. In the mackerel and codfishery there has been a falling off in the amount of tonnage employed, and consequently in the output of these fisheries. Instances of decay might be cited in other branches of the fisheries, as the menhaden industry. But the development of several inshore industries and the rise of new enterprises in connection with the fisheries make the value of the fisheries of New England greater to-day than for the past twenty-five years.

It is not difficult to discover several of the causes that have led to the decadence of the deep-sea fisheries. In a word, it has been due to important economic changes that have entered into the industrial life of the country at large. Contributing causes to this changed condition of things are the competition of fishery products from other parts of our own country and Canada, the development of cheap but wholesome food products of other varieties, the increase in the pound-net fisheries in southern waters, the improved methods of transportation and the rise of the refrigerator car system, and the passage of legislation that has favored the ready admission of food fish from the neighboring British Provinces.

The New England States held a monopoly of the fishing industry of the United States down to 1850. So complete was the control of the fisheries possessed by these states that a history of the fisheries of the United States for the first two hundred and fifty years after the inception of the industry on our shores is very nearly comprehended in a history of the industry as it has been carried on in New England for that time. By the middle of the last century, the oyster trade of the Middle Atlantic States had assumed definite form throughout the Middle West. The importance of the fisheries of the Great Lakes was being discovered by the people of those sections, and its growth was by leaps and bounds once the people realized its possibilities. The new settlers about San Francisco found oysters in the sands of their bay; but these were inferior to the eastern product. With Yankee enterprise, oysters from the East were introduced into the waters of the Pacific, that the markets might be supplied with home grown products. The canned fish products and dried codfish of New England found their way to the Pacific coast. But they could hold no monopoly of the market there after it was learned with what endless abundance the waters of the northern Pacific supplied markets with salmon and halibut.

The building of railroads throughout the land at first led to the extension of the New England fish trade over wider sections of the land. For one or two decades the effect was wholly to the advantage of the eastern fisheries. With the growth and development of an important lake fishery, and with an over-abundant supply of salmon every year upon the Pacific coast, however, the time was sure to come when the fishery products of the West invaded the markets of the East, and competition was set up in the very stronghold of the industry. This event marked the beginning of the time when the deep-sea fisheries had to contend not only for the control of the markets but even for their existence. Improvements in the refrigerator car service made it possible for the fresh fish of New England to be delivered a thousand miles from Gloucester. But the same service brought the fresh salmon of the Columbia and the fresh halibut of Alaska in open competition, in Boston and New York markets, with Penobscot River salmon and with halibut fresh from the banks.

The sardine industry of Maine has an annual value into the millions of dollars to-day. This means that a cheap food fish has been prepared that is palatable in all climates for months after it has been put up. Previous to its use as sardines, the small herring had little commercial importance. The enhanced value that the fish receives at the canneries not only makes the industry so important that it will be continued as long as the fish are in abundance but, economically, it represents something saved for mankind,—a new fish-food added to the list that was a comparatively short one while the cod and mackerel held the monopoly. The canning industry has become an important element in breaking up the old monopoly. Cod and mackerel cannot compete against cheap fish-foods and thrive as formerly. The amount of capital required to carry on the deep-sea fisheries makes that impossible. So it is with other fish that are now found in the markets,—the shad, the scup, the bluefish, the squeteague, the swordfish, the scallop and other kinds,—so great a variety is now offered to the purchaser that he will pass by mackerel worth twenty-five cents and buy a cheaper kind of fish. The herring family inhabit the ocean in incomputable masses. From this one family, several cheap products appear in the markets,—the hard and bloater herring, the sardine, the alewives, and the shad,—so numerous that people who formerly never thought of resorting to them now find them a pleasant substitute for the standard deep-sea fish.

To these domestic forms of competition must be added the unnatural, and to some extent unnecessary, competition from the fisheries of the British Provinces. The terms of the Reciprocity Treaty, which was in force from 1854 until 1866, were favorable to the provincial fishermen.

Further, there was developed at this period some phases of these fisheries that increased greatly the competition of the Provinces with New England in the fisheries. The United States may have benefited somewhat by the terms of this treaty; but undoubtedly Canada benefited the more and in a manner that led to greater possibilities of future competition. Again, in 1873, the two countries entered into an agreement that lasted twelve years, in which Canada got the better of the bargain. The privilege of admitting her fish free into our markets was, as it proved, of greater importance to her fishermen than were the privileges granted to our fishermen in provincial waters. Especially injurious to American interests was the free importation of Canadian herring during this period. Yet our Government had to pay Great Britain $5,500,000 for the balance of trade in this reciprocal arrangement, and this sum of money was expended in building up Canadian fisheries. It may be an ungenerous policy for a government to legislate so as to inflict injury upon competing industries of a rival state, but such a policy is more justifiable than legislation that promotes those industries directly, to the injury of home industries of the same kind.

Another contributing cause to the decline of offshore fisheries is found in the social changes that have taken place along the New England coast within the last twenty-five years. A few decades ago a certain place was the site of a fishing station. To-day Bar Harbor occupies the spot, a summer city for dwellers of the large towns. What has happened at that place has occurred, on a lesser scale, at hundreds of other places along the coast of New England. The fisherman's hut has given place to the cottage of the summer visitor. The unsightly fish-buildings must be removed from the neighborhood. The natives turn from the deep-sea industries to find employment in supplying the needs and demands of this transient population.

One of the demands is a bountiful supply of fresh fish. The long offshore voyages have been abandoned and a new home-market business has sprung up, important enough to maintain a large percentage of the coast people engaged in inshore fisheries.

In place of numerous towns engaged in sending fleets of vessels to the banks forty years ago, we find deserted wharves, buildings in ruins, and fish-stands already past repair. Only one town in Maine now employs vessels in the bank fishery. The business has gradually centered about Boston and Gloucester. The decadence of the New England deep-sea fisheries is to be regretted. A possible remedy will be the increase of those industries that are immediately dependent on the products of the deep-sea for their existence. That is, instead of the industry depending upon its existence through the first value of the products, as in the past, it may have to be maintained through the profits that arise from the enhanced values these raw products receive at the hands of the manufacturers of fancy brands of cod, mackerel, hake, and other deep-sea fish.

The subject of the decadence of our deep-sea fisheries should not be dismissed without a consideration of the status of the New England fisheries as a whole, both in their relation to their former prosperous state and to the fisheries of the country at large. A comparative view of the New England fisheries for the years 1880 and 1905 shows that there is little difference to be found in the number of men employed in the fisheries and the amount of capital invested. In 1880 there were 37,043 men employed and in 1905 the number was 37,339. The total value of the fisheries for the first-named year was $11,753,062, and for 1905 the value was $14,184,205. The average value to each man employed in 1880 was $312, in 1905 the average was $379, an increase of twenty-two per cent.[1] During the period the number of men employed at sea has decreased twenty per cent; the number of men employed on shore increased ninety per cent, or from 7,205 in 1880 to 13,114 in 1905.

The value of products taken and sold by the fishermen of the United States in 1905 was approximately $56,250,000. The number of persons who made a living in the industry was 232,000, and the capital invested was above $82,000,000. The individual value that the fisheries of the United States gave in 1905 was $213 per person. The New England fisheries compare favorably with other sections of the country; their value is one-fourth the total value of the country's fisheries and the individuals employed receive seventy-five per cent more returns for their labor than do the fishermen of other parts of the country. Whatever may be the future of our country's fisheries, there can never be elsewhere in the United States an industry more honored upon the land and more glorious upon the sea than the New England deep-sea fisheries.

[1] The status of the New England fisheries, 1880 and 1905.

	Men.	Capital Invested.	Value of Fisheries.
Total for 1880..........	37,043	$19,937,607	$11,753,062
Total for 1905..........	37,339	22,530,720	14,184,205

CHAPTER XVII

METHODS OF THE DEEP-SEA FISHERIES

The present day method of capturing mackerel is so universally confined to the purse-seine fishing that other methods hardly merit consideration. The use of nets set in the water or dragged slowly along by the vessel is confined to a few of the larger vessels and to a fleet of small boats in the early mackerel season. The crew of a mackerel schooner consists of a captain, cook, and sixteen other men. With the exception of the first two the crew take turns day and night in standing watches, there being two men in each of the watches that last two hours each. The men of a watch take turns alternately at the wheel and on the lookout. By day, the one at lookout is at the foremast-head where he scans the sea for mackerel over an area of three hundred square miles, if the sky is clear.

So frequently does the work of the day extend far into the night—sometimes to the dawn of another day—that it is difficult to mark the beginning of the fisherman's day. By two in the morning the cook must be astir, preparing the breakfast of the crew. At three or three-thirty the first "gang" is called to eat their breakfast,—for the accommodations of the forecastle allow only half of the crew to be seated at the table at a time. By four o'clock breakfast is over for the whole crew and everything is in readiness for fishing, providing there are fish to be caught. The most favorable times for catching mackerel are at dawn and at sundown, a condition of affairs that necessitates the three meals of the day at three in the morning, at nine o'clock, and at three in the afternoon, respectively.

While on passage to the fishing grounds the seine boat [1] is taken aboard the schooner on the port side; this is also done during heavy weather when there is danger of the boat careening over in the rough seas and spilling out the seine, or of capsizing. On reaching the fishing grounds, whether off Virginia, on George's Bank, or in the Gulf of Saint Lawrence, the seine boat is put overboard and the seine carefully stowed in the after part ready for immediate use. When a body of mackerel is seen schooling the lookout at masthead cries out, "School O!" and causes the helmsman to change the course of the vessel to the direction of the fish. At the cry of the lookout the crew below decks rush forth with their oil clothes in hand ready for instant action. The seine boat is hauled along side and at a word from the captain all get aboard. The captain and lookout stay on the schooner till the last moment watching the fish to discover in what direction they are moving and at what rate. The two men who go in the dory hoist it overboard, throw a bight of the painter around a bitt-head, and get into their boat. They are towed along by the vessel until the seine boat leaves the other side of the vessel at the captain's command. Then they can release their own boat by letting go the end

[1] NOTE.—Seine boats that are used with mackerel schooners are from 35 to 38 feet in length, 7½ feet in width, and 3 feet deep. Their cost is from $350 to $400, besides fittings, such as oars, rowlocks, purse-weights, oar cranes, davits, etc. The life of a seine boat is from four to six years. Seines are made 1,000 to 1,200 meshes deep and 900 yards in length, which reduces nearly one-half in tarring and hanging. The mesh is 2¼ inches, and the average seine is 240 fathoms long when hung. Its cost varies from $400 to $900, depending upon the length and depth. A seine lasts about three years.

of the painter and be in the vicinity of the larger boat when it is making a set of the seine around a school of fish. It is their task to take into their dory the free end of the purse line after the seine has been cast into the water from the larger boat, and to get it in readiness to be taken aboard that boat after it has made the circuit of setting the seine.

When the seine boat leaves the schooner the cook takes charge, he usually being the only man left aboard the vessel. An efficient man can perform marvels of seamanship aboard one of these vessels when occasion requires. He keeps the vessel in the vicinity of the other boats that no time may be lost in picking them up, if the mackerel disappear from the surface before a set can be made. On auxiliary schooners the cook is assisted greatly by the motive power of the engine and by the aid of the engineer.

The arrangement aboard the seine boat is for the captain to steer the boat, which he does by standing aft with an oar twenty to twenty-two feet long. Two of the crew stand just forward of the seine, which is piled up across the boat making a heap of seine four feet wide and between five and six feet high. One of the men passes "bights" of the seine to the other whose business it is to cast the seine as far as possible over the rail into the sea. The latter person is called the seine-heaver, and is usually noted for his strength and endurance. The farther the "bights" of seine are cast from the boat the less the other part of the seine will draw on the rail and retard the progress of the boat. Another man stands aft of the seine and throws the corks overboard. The other members of the crew are seated at the oars. They row the boat under the captain's direction with great speed, if necessary, to a position near the fish,—a little to the left and slightly ahead of the fish if the latter are in motion.

If the fish are schooling in a circle, "cart-wheeling," the set may be made from any position, but it is always made around the fish in the direction of the movement of the hands of a watch when held face upwards.

At the right moment the seine-heaver receives word to cast the twine, the rowers bend to their task for all they are worth, and within a few minutes the circuit has been made about the fish, in many cases without the fish becoming aware that anything unusual is occurring. When the seine boat nears the first end of the seine that was cast off it is met by the dory, the free end of the purse line is passed aboard the larger boat by the dory's crew, and the crew of the seine boat, by pulling in on the purse line, draw together that part of the seine which is lowest in the water, thus closing it below the fish much as an inverted bag would be drawn together by pulling on a puckering string. The crew work rapidly until the seine is "pursed up," after which they can work more leisurely in hauling the surplus seine into the large boat.

If fish have been caught the seine is gathered in until enough only is left to serve as a commodious bag to keep the fish well in hand until the vessel comes along side. Then the process of bailing the mackerel aboard the vessel takes place. This is accomplished by means of large dip nets guided by one of the crew and hoisted aboard by means of tackle blocks. Two of the crew dump the net as it comes over the rail, the fish are emptied out on deck, and the net is thrust back into the water for another load. About a barrel of mackerel are hoisted aboard at a time.

When all the fish have been taken out the process of dressing the fish begins. If the mackerel are to be carried to market fresh the process is simple, and the labor is comparatively light; part of the crew chisel ice, another part hoist it on deck and get barrels in readiness for others who pack the ice and fish into the barrels in alternate layers. When such a cargo reaches market the fish are counted out into baskets to be hoisted or carried on to the wharf or to the near-by stand.

The process of salting mackerel aboard ship requires more work and more handling of the fish. The fish are first scooped by small dip nets from the deck into gib-keelers, square boxes about four feet on a side and eight inches deep, placed on top of empty barrels. The fish are seized by the "splitters" who lay the fish on a board, and with a quick motion draw the splitting knife down through the entire length of the fish from the head to the tail, laying it open from its back to the inner ventral wall. The "gibber" receives the split fish from the splitter, seizes it in the left hand by the head and with the right pulls away the gills and viscera, casting the fish into a barrel. When all the fish have been split and gibbed the crew begin "plowing" them. This operation is done by an instrument fitted with a sharp-cutting edge. The instrument is grasped by its curved handle and the edge drawn twice across the inner ventral wall of the mackerel lightly and quickly, the object being to cut enough of the membrane to show how fat the fish may be. The process of "plowing" mackerel is not always resorted to. The final work remains to put the fish loosely in barrels and fill the barrels with water. There the fish are allowed to remain for ten or twelve hours in order that the blood may be soaked out of them.

All the work of catching a school of mackerel that has been described thus far, from the setting of the seine to the last act, must be done without any rest on the part of the fishermen. If a large school of fish, perhaps several hundred barrels, is taken towards night the crew work without stopping until the fish are taken care of. It may be daylight again before their work has been finished.

In case of all-night work the cook usually prepares coffee about midnight and the crew have a "mug up." When daylight finds the work only just completed it may happen that other fish are discovered schooling; then the work goes on as before, with no chance for sleep or rest and little opportunity for refreshment.

When the dressed fish have soaked long enough in the water they are dumped out and then carefully packed into the barrels, after being thoroughly salted. Here they remain one or two days to allow them time to contract and settle to the normal size of a pickled mackerel. The barrel is then filled with the fish, strong salt brine is poured over them, after which the barrel is headed and stowed away in the hold. If the trip is a long one the fish may be inspected in the hold and if any barrels are found in which the brine has leaked out they are hoisted on deck and refilled.

The life of the crew aboard a mackerel seiner is all work or no work. When mackerel are abundant and schooling well, there is a great deal of hard work for the men. If the weather is foggy or stormy, or if no fish can be found, the amount of work required is very light. The standing of watches must go on as ever, but the task is easy when divided among sixteen men. There may be some mending of the seine during these quiet spells, or a new purse line may need to be put in, or some light work done in repairing the rigging. With leisure on their hands the crew do as other men in other occupations do when loafing, in so far as their surroundings will allow. Some lie in their berths and read; others sleep a great deal; a knot of them gather on the quarter and spin yarns; there is always the sea-lawyer present to give final decision on all questions that come up for consideration; and always from early morn until nine in the evening one may find part of the crew playing cards. There is no observance of

Sunday on fishing vessels except in the cessation of card playing. It is a day for work; but it is rarely that games are carried on by the crew on the Sabbath.[1]

The pursuit of the mackerel and the cod, with his allied kindred, the haddock, hake, halibut, pollock and cusk, constitute the deep-sea fishery of New England. The capture of the swordfish, the bluefish, and the herring is not commonly looked upon as possessing the same significance in the fisheries that is possessed by the aspirant for those fish that are found in the deeper and broader stretches of the ocean. The methods of pursuit and capture of mackerel may be representative of the class of vessels that fish for their cargoes by means of seines. There is another class of vessels engaged in the fisheries that may be described by using a trawler for an example of the methods pursued. Mackerel and ground fish are caught almost entirely by the two methods of seining and of trawling, respectively. Net fishing, drag nets, and hand lines are occasionally resorted to to-day in the mackerel fishery; but this is to a limited extent as compared with the amount of fish captured by the purse-seine. Handline fishing from the vessel and from the dory for ground fish is resorted to somewhat; but the wholesale method employed is that of trawling. Imagine a stout cordline a mile or more in

[1] NOTE.—Some of the expense items of a mackerel schooner are ice, the quantity taken being from 10 to 30 tons; 50 to 75 barrels of salt worth $1.75 per hogshead; four hundred barrels at about one dollar each; food supplies for a crew of eighteen men that cost from $150 to $200 per month. The cost of ice, barrels and salt for the season depends largely upon the amount of fish taken. One-half the gross stock of the vessel, after deducting the value of salt, barrels and gasoline used in making the stock, is divided equally among the vessel's crew; the other half goes to the owners of the schooner. The average share per man for a crew of eighteen varies from $22 to $28 for every $1,000 worth of stock, depending on the expenses and the length of time of the trip.

length having tied to it at distances of six feet other lines smaller in size and about three feet in length, each fitted with a hook at the end. Such an arrangement of lines and hooks is called a trawl.

The build of the trawler differs little from that of the seiner except that the sails may be stronger, and there may be an extra riding-sail aboard. The complement of crew is about the same; food must be provided in abundance especially when the trawler goes to the Grand Bank of Newfoundland for several weeks or months; the ice-houses of the hold are furnished with a supply of frozen herring to be used for bait; instead of many empty barrels in the hold as in the case of the mackerel seiner there is an abundant supply of salt; on deck are found eight or more dories, one stowed away in the other; and around in different places are tubs made from barrels and casks in which the trawls are neatly coiled. The trawler-fisherman pursues his calling in winter as in summer, sometimes amid the severest hardships of the sea.

When the vessel reaches the fishing ground the captain awaits a favorable opportunity to set the trawls. This does not mean calm weather or even an approach to it. Often sets are made when the sea is running high, before the full fury of a storm has abated. But the weather conditions must be clear, else there is great risk of losing some of the dories and their crews. The frozen herring are taken from the ice-house, cut into pieces and the hooks baited each with a piece. As the trawl is baited it is taken from one tub and coiled into another. With this work done the tubs are placed in the dories, these are hoisted overboard, and two men go with each dory. One seats himself at the oars and pulls steadily away while the other is busied in flinging the trawl over into the sea. The first end of the trawl is made fast to a buoy which is usually a small cask, in order that the trawl may be readily found when necessary.

A second dory is lowered from the vessel about a quarter of a mile from the first, and so on until all have been sent away. The men in the dories set their trawls, sometimes a second one near the first, and soon begin to underrun them. This process consists in hauling the trawl aboard the dory again. The signal for underruning the trawls usually is given by the captain from the vessel in order that the trawls may be set no longer than he thinks proper, as he is a better judge of weather conditions than are the men who are busied about their work. When a trawl is thrown into the water it has a ground-line attached so that it can sink into the water the required depth and the trawl be kept in part of its length, at least, from resting on the bottom of the sea.

In the bow of the dory is placed a "gurdy," or broad wheel, over which the trawl is placed when it is hauled aboard the boat. As the trawl comes aboard fish are found on the hooks. These are freed from the hook and landed in the bottom of the boat by the one who hauls the trawl by a dexterous yank and twist. The other man receives the hooks as they come aboard, baits them, and coils the trawl again into the tubs for use at another time. When the trawl has been hauled, or all of them if more than one is used, the men seat themselves at the oars and pull for the vessel. Hauling a mile and a half of trawl from ocean bottom on a cold day in winter takes more strength and pluck than the average laborer possesses. These men who go down into the sea in winter are strong, tough, plucky fellows, the like of whom it would be hard to find in any other calling.

Usually the vessel meets the dories more than half way in picking them up. By the time the last dory is found it is dark; there may be some trouble in finding this one, especially if the fog suddenly shuts in and the wind springs up. Then comes the terror of being lost in the fog, of rowing for hours and days at a time in order to reach land, the dread experience of having to pass sleepless nights and disappointing days at the oars in the ceaseless labor of rowing. Many strayed dories are picked up by other vessels; others reach the coast and are cared for; some are overwhelmed by storms that arise and their crew meet a quick death in the sea; while to a few that cruel fate of insanity and starvation aboard their frail craft is reserved. The hardships that befall the life of a fisherman can only be realized by following in his steps and being an active participant in the hazardous, laborious, and ill-paid calling of the followers of the sea.

The process of curing cod, haddock, hake, pollock, and cusk is essentially the same as was practiced a century ago. As soon as the dories return to the vessel the dressing of the fish is begun. The usual gang aboard a Grand Bank vessel consists of a "throater," a "gutter," and a "splitter." The first-named person cuts the fish across the throat below the gills, slits open the abdominal walls and cuts off the fish's head. The "gutter" removes the organs that are contained within the abdominal walls, the livers being thrown into a barrel to be saved for their oil, the other parts being cast overboard. The fish then passes to the splitter who, with a knife that is rounded at the end, cuts along each side of the backbone from the ventral side towards the back and removes the bone from the fish. Care is taken that the cut does not extend too deep and appear through the fish, or too far from the bone and thus leave much flesh on the bone that is removed. Sometimes the "sounds," or air bladders, are removed from the backbones and saved. The fish, after being split, are thrown into tubs of salt water and thoroughly washed.

The salting of fish may be accomplished in either of two

ways, one being called the kench cure, the other the pickle cure. In the former process, the fish are thoroughly salted and placed in regular piles on top of each other, called kenches. In this way pickle that is formed can drain off freely. Pickle-cured fish are salted and then placed in large tubs or butts, where the pickle is retained. The kench-cured fish make a drier product and are better suited for a warm market. Generally the fish are salted in kenches aboard vessels and in butts on shore. The fish remain in kenches until the vessel reaches port. If the trip is a long one—the "Grand Banker" may be gone three or four months on a trip—there may be need to re-kench that part of the cargo that was first secured.

After the fish are landed on the dock they are culled into different grades, the principal being large cod, which includes all over 22 inches in length when salted; medium or small cod, between 16 and 22 inches in length; and snappers, which are below 16 inches. If the cargo is a mixed one, the cod are separated from the others. The first process ashore is to wash the fish and again place them in butts, after another salting. When the time comes for them to be dried they are removed from the pickle of the butts and piled on each other in order that the pickle may drain off. These kenches of fish are about three feet high. Sometimes weights are placed on top of the kenches in order that the pickle may be pressed out quicker. This process of draining the pickle from the fish is called "water-horsing." From the water-horsing the fish go to the flakes where they are spread out in the sun to dry. Fish flakes are raised about two and one-half feet from the ground, and are made by nailing narrow strips of wood about three or four inches apart on top of long stringers. This arrangement allows the free circulation of the air to all parts of the fish.

The time required for drying depends upon the market

for which the fish is being prepared. Some markets demand fish with fifty per cent of the moisture removed; others as high as sixty or seventy per cent. Fish that are used in the preparation of boneless-fish require little drying, perhaps eight or ten hours of a good day, while the fish for export trade may require a week or ten days. Every evening the fish are gathered together in small piles on the flakes and covered by a box to prevent them absorbing moisture from fog or rain.

The loss in weight in dressing and curing fish for the market varies from fifty to sixty-five per cent, according to the species, the time of year and the previous amount of salting received. Haddock and cod lose the most, cusk and hake the least. The following table shows the average quantity of each kind of fish required to make a gross quintal (114 pounds) of dried fish for the New England markets:

Species.	Pounds Required to Make a Quintal Cured.		
	Round.	From the Knife.	From the Butt.
	Pounds.	Pounds.	Pounds.
Haddock	299	296	133
Cod	288	193	131
Pollock	280	184	130
Hake	258	190	131
Cusk	246	178	132

The cost of curing a quintal of dried fish varies with the conditions of weather, the season of the year and other circumstances; but it generally runs from 38 to 50 cents per quintal, of which 15 or 18 cents represents the cost of salt. The value of the finished product varies likewise with the grade of the fish and the season of the year. The cured cod sells at a higher price than hake, haddock, cusk and pollock. The winter fish are worth more than fish caught in the warmer months.[1] For the local and nearby markets the fish are packed in bundles of a quintal each and tied up with cords, or in wooden boxes holding from 100 to 450 pounds each. At Gloucester, Boston, Vinal Haven and Portland large quantities of dried ground fish are prepared as boneless cod and placed upon the markets in neater shape and fancier coverings than is the uncut fish.[2]

The preparation of boneless cod originated with the patenting of a process in 1868 by William D. Cutler, of Philadelphia. By this new process the skin, bones and refuse matter of the fish were separated from the fiber of the fish, the latter was run through a machine which thoroughly broke up the fibers into a disintegrated mass, and the new product was spread upon some hard surface, heated and pressed. The finished product was known as "dessicated fish." Within nine months, improvements on this process were devised at New York City, Brooklyn, and Provincetown. At first the inferior grades of fish were used in the production of the article, but as an increased demand arose for the food more attention was paid to the preparation of it for the market. So rapidly did the business increase that by 1875 over 500,000 pounds of boneless fish were prepared at Gloucester alone; in 1879, about 12,000,000 pounds were prepared at Gloucester, and about 6,000,000 pounds elsewhere in New England. By 1898, the New England production of boneless cod was about

[1] For the year 1898 the price per quintal of large cured cod was from $4.50 to $6.25, for medium cod $3.00 to $4.00. The values of other kinds of cured ground fish were for hake, $1.75 to $2.50; haddock, $2.00 to $2.50; cusk, $3.25 to $3.62; and pollock, $2.75 to $3.00.

[2] Preservation of fishery products by drying and dry-salting, p. 398.

25,000,000 pounds, most of the fish being prepared at Gloucester.

Practically all the fish thus prepared goes to the market under the designation of codfish. Of the total quantity of boneless fish, it has been stated that an average of 60 per cent is prepared from cod, 28 per cent from hake, 8 per cent from haddock, and 4 per cent from cusk. The cost of preparing and packing the smaller bricks of boneless fish is about two dollars per 100 pounds. The refuse skins and bones are sold to the companies that manufacture glue and fertilizers. This by-product of the fish amounts to a considerable sum when millions of pounds are used annually.

CHAPTER XVIII

THE EVOLUTION OF THE FISHING SCHOONER

A comparison between the appearance of citizens of the Massachusetts Bay Colony of 1630 and those of Boston to-day would afford no greater contrast than would appear if one of the fishing craft used by the first settlers of Gloucester should arrive at T Wharf with a cargo of fish and mingle with the fleet of modern fishing vessels that frequents that great fish-market of New England. Important differences would also be seen in the furnishings of the vessels, in the apparatus used in fishing, and in the methods employed in curing the catch. The condition of the fisherman has improved greatly during the period of three centuries that fishing has been carried on along our shores, but the change has not been gradual. For the first one hundred years only one important change is to be noted, one affecting the arrangement of the sails of vessels. Old-fashioned methods of catching and curing fish that had been in use since the days of Bradford and Winthrop were employed by fishermen well down towards the middle of the last century. The most numerous and many of the most important changes have been instituted within the lifetime of the last two generations of fishermen.

The shallop, the sloop, and the ketch were the kinds of boats most generally employed by the early colonists for their fishing and coast trade. The sea-going craft of the day were of the caravel type, not differing greatly from the vessels employed by Columbus. In these ships the bows and sterns were constructed high above the waist line, the hull was not deep, consequently so much superstructure erected on a moderate-sized hull rendered the vessel unseaworthy in windy or stormy weather, and of little use in making headway against adverse winds.

The first boat used by the settlers was the wooden canoe, made by the Indians. In Maine these were constructed by fastening strips of birch bark on to light wooden frames; in Massachusetts, the trunks of large trees were hollowed out for the purpose. The majority of the canoes were capable of holding four or five men, yet larger ones were built which held twenty, thirty, and even forty men. This boat was so light, cheap, and convenient that it became the universal oyster boat on the Atlantic coast, being still in use in some localities.

Ship's boats, called shallops and pinnaces, were employed in the shore fisheries until the colonists had boats constructed for the purpose. At Plymouth, boat-building began in 1624. In addition to the names shallop and pinnace, it is probable that the term lugger was used for the early small craft. They were small boats, often constructed by the settler in his barn or shed during the winter season and hauled to the shore for launching by yokes of oxen. The boats were usually provided with two unsupported masts, each carrying a large square or lug sail. The colonists have left us representations of much of their personal and household belongings, but there is no picture left by them of the rig and appearance of their earliest small fishing boats.

For deeper water the catch, or ketch, was employed. The Dutch called them pinkie, a term derived from the Mediterranean region and signifying round at both ends. The first ketches had one mast amidships with a square sail suspended to it cross-wise of the boat. Afterwards, a small mizzenmast was added away aft; the mainmast was then set up a little forward of amidships. The mainmast bore two square sails, the mizzenmast a lateen, or triangular shaped sail.

Ketches were from nine to ten feet deep in the hold and drew seven or eight feet of water. Their width was about two-fifths their length, they were decked throughout, and had cabins aft. The average size was about thirty tons burden, although a few were constructed of eighty tons. Salem was a famous center for the building of ketches. The cost for building was about £3 5s per ton. The *Sparrow Hawk,* which sailed from England in 1626 with forty passengers and was wrecked on Cape Cod, was a typical representative of this early class of vessel.[1]

The ketch was an improvement over the lugger in that the sails could be handled more easily, as there were two sails on a mast instead of one large one. This type was a popular form of fishing craft in the seventeenth century, but was unsuited to economy in sailing along the New England shores where the winds are variable. When the vessel was tacking the lateen sail had to be lowered and changed, an operation involving much labor and hardship among the seamen.[2] The lateen yard was hung on the mizzenmast diagonally, with the forward end a few feet above deck and the after end tipped up so that it came nearly on a level with the top of the mizzenmast. When changes had been made in the arrangement of sail and in the shape of the hull the ketch developed into the brigantine of to-day. The shallop was the forerunner of two types, the Chebacco boat, or pink, and the sloop.

The change in the arrangement of sail, brought about by the schooner type with sails rigged fore-and-aft, was a

[1] Henry Hall, Report on the Ship-building Industry of the U. S., 10th Census, Vol. VIII.

[2] Capt. J. W. Collins, the Evolution of the Fishing Schooner, New England Magazine, May, 1898.

welcome one to the fishermen. Probably a step in the evolution of the schooner-rig was the cutting away of that part of the lateen sail which extended forward of the mast, although the yard itself projected forward as formerly. The complete transformation was made in 1713. Capt. Andrew Robinson, of Gloucester, "built and rigged a ketch, as they were called, masted and rigged in a peculiar manner." His vessel had gaffs and booms to the sails instead of lateen yards, previously in use, and the luff or upright edge of the sail was fastened to hoops on the masts. She was also provided with a jib-sail forward. As the vessel entered the water at her launching, she was christened a "scooner" by her builder, a name ever since applied to vessels with two or more masts rigged with fore-and-aft sails. There is no evidence that Capt. Robinson attempted any change in the shape of the hull.[1]

The economy of the schooner was soon evident, as the sails could be handled easily on deck, changes in the vessel's course could be made without lowering and setting sail, and the arrangement of sails allowed the vessel to sail several points nearer the wind. Schooners did not come into general use at Gloucester immediately, but by 1741 there were about seventy of them owned there.[2] For nearly two centuries, the schooner rig of sails has been found to be well adapted to our ocean fisheries, and is the popular style for yachts and pilot boats. Among the deep-sea merchant marine of the world this rig of sail has well nigh superseded the square rig style.

For many years after 1713 the schooner represented a compromise between the square sails of the ketch and the fore-and-aft sails of the vessels of the Robinson class. The square sails served in place of the top sails

[1] Babson, History of Gloucester, p. 250.
[2] Ibid, p. 254.

of to-day. The vessels were called top-sail schooners, which was the rig of the privateers of the Revolution and of the War of 1812.[1]

Modifications were made in the shape of the hull and the arrangement of the decks. A Marblehead cod-fishing schooner of 1750 is represented with two stout masts supported by stays, a short maintopmast, and a bowsprit. The schooner had a wide, square stern, there was scarcely any curve to the deck, while the quarter-deck was raised several feet above the waist of the schooner. The bows were blunt, practically a semi-circle. This class of schooner was known as a "heel-tapper," from the fancied resemblance that the deck gave to an inverted shoe, the quarter-deck simulating the heel. These schooners generally were without bulwarks forward of the quarter-deck, or at most with a strip spiked to the top timbers to serve as a low rail. It was believed in those days, and the idea was held well into the next century, that it was unsafe to have any rail to prevent the free sweep of water across the main deck. Capt. Collins, writing in 1898, describes this type at length. He says, "The same idea found expression in building sea-going vessels for commercial purposes until after the beginning of the present century, and even as late as 1815, many had no bulwarks forward of the fore rigging. The quarter deck of one of these old "heel tappers," being so much higher than the main deck, was comparatively dry in a gale when the main deck would be all awash; therefore, when anchored on the banks in rough weather the crew stood there to fish, which they could do very well, since at that time only half the men engaged in fishing at the same time, as their rule was to fish, watch and watch, each half of the crew taking its turn of four hours in regular rotation.

"The cabin was aft, and entered through a small com-

1 W. L. Marvin, The American Merchant Marine, p. 23.

panionway, in the after end of which the binnacle was usually located. A rude fire-place, built of brick or stone, served for heating and cooking. The culinary operations were of the most primitive sort. Berths were arranged around the cabin, which were also equipped with a board table and locker seats. Sometimes this apartment was painted, but generally its dingy hues were attributed solely to smoke and grime. A rude log windlass worked by hand-spikes, heavy hemp cables and a long tiller were details of equipment that continued in use until near the middle of the present century, though in the meantime fishing vessels had undergone material changes in other respects. There was in these early schooners no attempt at ornamentation except in rare instances, when a broad white or yellow band stretched along the side, or when representations of windows were painted on the stern. Ordinarily they had only a plain gannon-knee head and were coated with tar, at least as high as their bends. They had long floors, with barrel-shaped bottoms, and excessively full ends. This form continued in vogue until early in the nineteenth century.

"A long high-steeved bowsprit, two rather short, clumsily made masts—the foremast stepped well forward—a main-topmast, and rather short mainboom, projecting just beyond the taffrail, constituted the more important features of the spar plan. The hemp sails had very little angle to their peaks; they were baggy and ill-formed, when considered from present standards, and one marvels how such vessels, with no light sails and with such a diminutive sail area, managed to make passages to and from the distant banks or, stranger still, to European ports. Only three sails were usually carried—jib, foresail and mainsail; but the desire for supplementary sail area in light winds led to a modification in the rig, to the extent of having in addition a single square foretopsail and sometimes a flying jib. The gafftopsail and maintopmast staysail, now so generally used by fishing vessels, had not been adopted.''[1]

The Revolution annihilated our fisheries and with them the fishing craft that had been employed. When peace had been made and the fisheries were again resumed the fishermen could not afford to build vessels of large size. The craft that came into general use subsequent to the war was the chebacco boat, so named because they were constructed at first at the parish of Chebacco, now a part of the town of Essex. These were from three to five tons at first, but were constructed of larger size as more of them came into use. They usually had sharp sterns, were partially decked, having standing-rooms in which the crew stood to fish, and had a little cabin, or cuddy, beneath the forward deck which served for sleeping quarters and cooking. The stem of the chebacco boat always stood high above the bow and served as a bitt-head for the boat's hawser when she was at moorings. Later the chebacco boats were constructed as large as twenty tons burden, and some had square sterns. These were called ''dog-bodies,'' to distinguish them from the other kind. Both kinds were cat-rigged, with two masts, the foremast standing as far in the bow as possible.

From the round sterned chebacco boat to the pinky was an easy and natural step in vessel construction. Boats were needed for offshore fishing, so the chebacco boat was enlarged, a bowsprit and jib were added, and thus a new type was brought into use. The pinky was very much in favor from the close of the war of 1812–1815 down to 1840 or later. The special feature of the pinky was the extension of the rail and bulwarks away aft until they met and were fastened to a V-shaped board, like the stern of a dory, with a hollow crotch on top to serve as a rest for

[1] Collins, pp. 337–338.

the main boom. This was the "pink," and was usually curved upward sharply. The cabin was forward, with a brick chimney and fireplace. It was dingy and dark, being lighted only by the companionway and two or three dead-lights of thick glass. The sails were of hemp, or what was called Raven's duck, as cotton duck was then unknown. With these sails went a "scout horn," to wet them down when the wind was moderate. As hemp sails were porous, and did not have the wind-holding properties of duck, the "scout horn" was an indispensable instrument. It consisted of a pole about fifteen feet long, with a leather pocket on the end holding about a pint of water. By means of the "scout horn," water was thrown on the sails to make them less porous.[1]

The square-sterned chebacco boat, or dogbody, was the forerunner of the square-sterned schooner, a vessel used extensively in the mackerel fisheries when the fish were caught with the jig; hence the name "jigger" was applied to this kind of a schooner. This vessel had several points of superiority over the old type with high quarter deck. Although her bow was as full as an apple and her sides as round as a barrel still her lines were more symmetrically proportioned. "Instead of an open waist, the bulwarks were filled in from deck to rail; the long low quarterdeck made such vessels better adapted to the mackerel hook and line fishery; and the introduction of jib-booms and additional light sails increased their speed and efficiency in summer weather. They were much better adapted to the needs of ocean fishery than the pinky, since the yawl-boat or dingy had to be stowed on the deck of the latter, thus limiting the deck space, while it could be hoisted to the stern davits of the schooner, leaving the deck clear for fishing or for temporarily stowing the catch. Then, too, the cabin was aft, and the accommodations for

[1] Fishermen's Own Book, pp. 38–40.

sleeping and cooking were more commodious, if not yet luxurious."[1]

After a few years of depression in the mackerel fishery, ending with the year 1845, the business became prosperous again and there arose a demand for larger and swifter schooners. In 1847 an attempt was made to produce a clipper schooner at Essex when the schooner *Romp* was built. This vessel had easier lines than her predecessors, her stern was narrower, and she was provided with a longer head. In her day the *Romp* was regarded as a very sharp craft; but measured by present day standards the new type represented merely a natural step towards the models of the present time.

For a decade after the appearance of the *Romp* there was a revolution in the building of schooners of this class. Rivalry was keen among the fishermen to procure models of schooners that would be both seaworthy and speedy. The tendency at the time was to build them on the "cod's head and mackerel tail" principle. The clipper schooner as evolved at the time had leaner bodies and sharper lines under water; their bows were longer and cleaner, with a long pointed cut-water; the angle of entrance forward was reduced from the old style of 85° to an angle of 45°; the spars were lengthened; the schooners were rather straight on top, they had little draught forward, with a greater draught aft. The hull was painted green under the water and black above, with one or two white stripes. Most schooners had gilded figure-heads forward, and carved and gilded moldings on the trail-boards along the sides.[2]

From 1860 to 1885 changes were made in the clipper schooners but they were undesirable or unimportant modifications in most cases. The schooners were constructed longer and wider, with little change made in their depth.

[1] Collins, pp. 341–342.

[2] Hall, Report on the Ship-building Industry of the U. S., p. 22.

The ballast consisted of stone carried in the hold; consequently the center of gravity was high and the righting power of the vessel reduced almost to a minimum. In the matter of tonnage there was a decided change from 1845 to 1885, which represents the first stage of the development of the clipper type of schooner. Previous to 1850 the vessels employed in the New England fisheries seldom exceeded 45 or 50 tons, old measurement, representing about 25 or 30 tons of the present system of rating. In the mackerel fishery there arose a demand for larger vessels, and this demand was increased when the trade in frozen herring with Newfoundland began to assume a place of importance between the fifties and the sixties, as vessels that were employed in the mackerel fishery in the summer would be engaged in the frozen herring trade in winter. Thus the average size of the offshore vessels increased gradually, so that by 1880 they were of 75 tons burden, new measurement, or 100 to 110 tons by the old measurement.

The final stage in the development of the clipper schooner was inaugurated with the building of the *Grampus*, a schooner made for deep-sea work of the United States Fish Commission, in 1886. The work of designing the new schooner was placed in the hands of Capt. J. W. Collins, a man of wide experience and thorough knowledge in all matters pertaining to the fisheries, who describes his new model as follows:

"The *Grampus*, as the schooner was named, was a radical departure from the prevailing form and was in reality a new type of American fishing vessel. She was about two feet deeper than the ordinary schooner of the same length; the after section was more V-shaped, with easier horizontal lines; the stern was not so wide and had a much stronger rake; while the stem was nearly perpendicular above water, though curved strongly below. Many other new features were introduced in her construction to insure greater strength or adaptability. Among these the most noticeable were in the rig. The foremast was made considerably shorter than the mainmast, and the foretopmast, instead of being of the same length as the maintopmast, as had previously been the custom, was not so long by several feet. The schooner was also rigged to carry a forestaysail —the forestay coming down to the stem head—and comparatively small jib, this arrangement of head sails being considered preferable to the large jib until then in almost universal use. Wire rigging was used instead of hemp."[1]

The appearance of the *Grampus* among the fishing vessels easily demonstrated the superiority of the new type over the old. Immediately other vessels were constructed along the new lines and rig, so that within a decade the appearance of the fleet had undergone a revolution. The deep-sea fishing fleet of New England to-day represents grace of outline combined most efficiently with the requirements of speed and the demands of stability and capacity. The modern schooner is immeasurably swifter than the clipper of the early eighties, besides being safer in heavy weather. Many instances are on record in recent years where the new type has been able to beat to windward and escape shipwreck when caught on a lee shore in the face of a raging gale, when schooners of the old type could hope for no escape.

All vessels engaged in foreign commerce or domestic business under the American flag are listed and documented by the Government. Those that are engaged in foreign commerce are "registered," while those employed for domestic purposes are "enrolled"; if under twenty tons measurement they are "licensed." The tonnage of a vessel may be of three kinds: displacement, gross register, and net register. Displacement tonnage is the weight of the vessel, and it may be found by multiplying by 62.4 the

[1] Collins, pp. 347–348.

number of cubic feet of water displaced by the vessel when afloat; it corresponds to the old carpenter's measurement. In 1854 the English method of tonnage measurement was adopted by the United States in order that our vessels would not be at a disadvantage in comparison with those of foreign measurement. This method of measurement corresponds to the gross register, which is found by dividing the number of cubic feet in the capacity of the vessel by 100. In 1882 the Government established a net tonnage which was 5% off the gross tonnage for sailing vessels. In 1895 there was a further reduction made by taking out all space for officers and crew, so that the net register now corresponds to the cargo capacity of the vessel. "The net register tonnage is obtained by dividing by 100 the capacity in cubic feet of the space available for cargo and passengers. From the entire capacity of the ship are deducted the spaces occupied by machinery, crew accommodations, and certain other housings, carefully designated by law; and then the number of cubic feet in the remaining capacity is divided by 100 to obtain the net register."[1]

The deep-sea fishing schooners of New England are so nearly uniform to-day in their build, rig and equipment that data concerning one of them may apply easily to most of the others, provided allowance is made as the different vessels vary in size. The data given has been furnished by Capt. J. W. McFarland, owner of the schooner *Victor*, of Gloucester. The *Victor* is an average sized mackerel schooner of the auxiliary type, with a gross tonnage of 122, or net tonnage of 75. She is 10 feet deep, 25 feet beam, and 112 feet over all. The mainmast is 85 feet in length, the mainboom, 70 feet, and the topmast 45 feet. The sail area is of 2,000 square yards, made up of a mainsail, foresail, forestaysail or "jumbo," jib, baloon jib,

[1] E. R. Johnson, Ocean and Inland Water Transportation, Ch. I.

maintopsail, foretopsail, and mainstaysail. There are 75 tons of ballast, mostly of pig iron, and the vessel has a capacity of 400 barrels outside of the store rooms. The present-time cost of the hull, sails and rigging is $12,000. The engine is of 85 horse power, valued at $6,000. The cost of gasoline averages about $170 per month for the season.

CHAPTER XIX

THE FISHERIES QUESTION

By the Fisheries Question is meant the international complications that have arisen between the governments of the United States and Great Britain over the interpretations placed upon the intent and meaning of the several treaties that have been made regulating the fishing interests of the two countries. The fisheries question has come up before the American people several times for settlement, on each occasion presenting new issues and meeting with unexpected difficulties in the way of permanent settlement. The question has to deal, largely, with issues that have arisen from the New England fishing interests. In the earlier stages of its history the contentions had to do with the enjoyment of fishing privileges in waters adjacent to British coasts in America; at the present time, the question has to deal mostly with the securing of Newfoundland herring for bait in the American fisheries.

The first grant of fishing privileges was secured in 1783 when the treaty of peace was concluded between the United States and Great Britain. The contentions of the American commissioners at the time for fishing rights and privileges were so strong that, for a time, there was grave danger that a treaty of peace could not be effected. The conditions of the Treaty of 1783 held until the opening of the war of 1812. When arrangements were made at the close of hostilities for a new treaty of peace, it was found that Great Britain maintained the proposition that the United States, by going to war with Great Britain, had abandoned her rights to the provisions of the former treaty that had to do with the fisheries.

Accordingly, as has been noted in Chapter VII, a special commission was appointed in 1818, which made different arrangements for the fisheries and which was a very serious limitation of the rights enjoyed under the first treaty. The provisions of the convention of 1818 have been the basis of our fishing rights in Canadian and Newfoundland waters much of the time since 1818, and, with certain modifications, are still in force.

It appears that serious misunderstandings arose soon after the ratification of the Convention of 1818 because of different interpretations of several of its provisions. The British Parliament, in June, 1819, passed an act following closely the fisheries article of the convention and providing for its enforcement.[1] Between 1818 and 1854 the Provincial Legislatures of Canada, Nova Scotia, and New Brunswick passed various statutes purporting to be based on the treaty. They were more stringent and even more specific in their application than the English Act.[2] Within half a dozen years after the ratification of the treaty, correspondence had arisen between the two governments over the seizure of American fishing vessels in the Bay of Fundy and their subsequent rescue from Canadian authorities by Eastport enthusiasts. But from 1824 to 1836 little trouble seems to have occurred.

In 1836 the Legislature of Nova Scotia passed what was commonly designated the "hovering act," in which the hovering of American vessels within three miles of the coast or harbors was sought to be prevented by imposing various regulations and penalties. Subsequently "claims were asserted to exclude fishermen from all bays and even from all waters within lines drawn from headland to

[1] Sabine, p. 22.
[2] Elliott, p. 61.

headland, to forbid them to navigate the Gut of Canso, and to deny them all privileges of traffic, including the purchase of bait and supplies in the British colonial ports. From 1839 down to 1854 there were numerous seizures, and in 1852 the home government sent over a force of war vessels to assist in patrolling the coast."[1] It was contended on the part of the United States that the three miles as stated in Article I of the Treaty of 1818 meant three miles from the shores of those bays, creeks, etc., following their sinuosities. Therefore all bays with an entrance more than three miles broad would be accessible to Americans for the purpose of fishing, except within three miles of the shores of the bay.

On the other hand, the British interpretation of this clause was that three miles from a bay meant three miles from the mouth of the bay, "in other words, three miles from a line drawn from headland to headland of the bay. Moreover this construction was applied not only to the great arms of the sea, as the bays of Fundy and Chaleurs, but also to all indentations of the coast, as the north coast of Prince Edward Island from North Cape to East Cape, and the northeast coast of Cape Breton, from North Cape to Cow Bay."[2]

Two notable cases of seizure arose from an attempt to enforce the principle of the "headland doctrine." On May 10, 1843, the American schooner *Washington,* while fishing in the Bay of Fundy ten miles from the shore, was seized by a revenue schooner on the charge of violating the treaty of 1818. She was carried to Yarmouth, Nova Scotia, where she was decreed in the Vice-admiralty Court to be forfeited to the Crown, and was ordered to be sold with her stores. Under the Claims Convention of 1853, the case of this vessel was referred to a joint commission

[1] Snow, American Diplomacy, p. 441.

[2] Moore, International Law Digest, I, p. 783.

which disagreed. Then the case was given to Mr. Joshua Bates, an American financier connected with the house of Baring Brothers in London, who allowed to the owners of the *Washington* the sum of $3,000 damages, on the ground "that the Bay of Fundy is not a British bay, nor a bay within the meaning of the word as used in the treaties of 1783 and 1818."[1] On the fourth of August, 1844, the American schooner *Argus* was seized while fishing off the coast of Cape Breton at a distance of twenty-eight miles from shore. The same umpire, Mr. Bates, found no case of violation of the treaty, and awarded to the owners of the *Argus* the sum of $2,000 for the loss of their vessel and its stores. The "headland" rule as applied to the Bay of Fundy was relaxed by the British Government in 1845.

As early as 1847 England proposed a reciprocity treaty governing the fisheries of the two countries, but no agreement was reached by the two governments and the relations between the two became more and more critical. When the excitement had reached a high pitch and war was talked of, the British Government, in 1854, sent Lord Elgin to Washington for a conference with Mr. Marcy, Secretary of State. The result of the conference was the Reciprocity Treaty of 1854, a treaty in relation to the fisheries, commerce and navigation. In the first Article of the Treaty it was

"agreed by the high contracting parties, that, in addition to the liberties secured to the United States fishermen by the above-mentioned Convention of October 20, 1818, of taking, curing, and drying fish on certain coasts of the British North American colonies therein defined, the inhabitants of the United States shall have, in common with the subjects of Her Britannic Majesty, the liberty to take fish of every kind, except shell fish, on the seacoasts and shores, and in the bays, harbors and creeks

[1] Moore, I, pp. 785–87.

of Canada, New Brunswick, Nova Scotia, Prince Edward Island, and of the several islands thereunto adjacent, without being restricted to any distance from the shore; with permission to land upon the coasts and shores of these colonies and the islands thereof, and also upon the Magdalen Islands, for the purpose of drying their nets and curing their fish; *provided,* that, in so doing, they do not interfere with the rights of private property or with British fishermen in the peaceable use of any part of the said coasts in their occupancy for the same purpose." [1]

By the Reciprocity Treaty of 1854 the American fishermen gained fishing liberties analogous to those enjoyed under the Treaty of 1783; but the liberty thus defined applied solely to the sea-fishery. The salmon and shad fisheries, and all fisheries in rivers and the mouths of rivers, were expressly reserved for British fishermen. It was provided by the second article of the treaty, that British subjects should have, in common with the citizens of the United States, "liberty to take fish of every kind, except shellfish, on the eastern seacoasts and shores of the United States north of the 36th parallel of north latitude, and on the shores of the several islands thereunto adjacent, and in the bays, harbors, and creeks of the sea coast and shores of the United States and of the said islands," on precisely the same conditions, including the reservation of salmon, shad, and all river fisheries, as were made in the first article with respect to the reciprocal liberty secured to American fishermen.[2]

For a number of years the treaty seems to have worked to the satisfaction of both parties. The number of American vessels frequenting the Canadian waters increased, as did the extent of their business. On the other hand, it was a period of unexampled prosperity for Canadian interests. "From the making of the reciprocity treaty until

[1] Elliott, p. 75.

[2] Moore, I, p. 792.

its abrogation, Nova Scotia increased in wealth and prosperity at a most extraordinary rate; from its abrogation to the present, (1869) we have retrograded at a most frightful rapidity."[1]

The causes that led to the abrogation of the treaty in 1866 were, first, the Americans realized in a few years that the loss of revenue from the remission of duty on Canadian imports far exceeded the value of the fishing liberties conceded to the American fishermen; second, the high taxes brought about by the Civil War so increased the cost of living and the price of labor in this country that the American farmer found it impossible to compete with the Canadian farmer in the production of the articles included in the free list of the reciprocity treaty; and finally, there is reason to believe that the hostile attitude assumed by Canada towards the North during the war hastened the abrogation of the treaty on the part of our statesmen. The resolution for the abrogation of the reciprocity treaty passed the House by a vote of 88 to 54; it passed the Senate by a vote of 33 to 8, and went into effect on the seventeenth of March, 1866.[2]

The stipulations of the Reciprocity Treaty in regard to the fisheries were "in addition to those of the convention of 1818"; therefore, when the treaty was abrogated, the fisheries arrangements between the two governments again rested upon the Convention of 1818. The ambiguous clauses of that arrangement were still unsettled, so that all the old disputes were revived later.

After the abrogation of the Reciprocity Treaty in 1866, the British authorities resorted to the system of licenses, which were issued to American vessels wishing to enjoy the privileges of inshore fishing in Canadian waters. The rate at first was fixed at fifty cents per ton but within two

[1] Elliott, p. 75, quoting Halifax Chronicle.

[2] Snow, p. 445.

years it was increased to $2 per ton. The license was so large that it became practically prohibitive, there being but twenty-five vessels that availed themselves of this privilege in 1869. In 1870 the system of issuing licenses was discarded. Preparations were made at once for enforcing the provisions of the treaty of 1818. During the summers of 1870 and 1871 many seizures of American fishing vessels were made, although the instructions given by the British Government in 1870 were very explicit that no seizures should be made unless it was clearly evident that there had been illegal fishing and the vessel itself was captured within three miles of land. In the case of two vessels, at least, the Canadian authorities had caused seizures to be made on the grounds of entering harbors to secure bait, and of securing bait.[1]

In 1871, the Joint High Commission met at Washington to settle the disputes between the two governments, chief of which were the Alabama claims. The fisheries dispute also received much of the attention of the Commission. The Treaty of Washington, signed by the Commission the 8th of May, 1871, contained nine articles regulating the fishing relations of the two countries.

By Article XVIII of that treaty it was provided that, in addition to the liberty secured by the treaty of 1818 to take, cure and dry fish on certain coasts of the British North American provinces, the inhabitants of the United States should have, in common with the citizens of Her Majesty's government, the liberty to take fish of every kind, except shellfish, on the coasts and shores, and in the bays, harbors and creeks of the provinces of Quebec, Nova Scotia, New Brunswick and Prince Edward's Island, with their adjacent islands, without being restricted to any distance from the shore. These liberties applied only to the sea-fisheries, the shad and salmon fisheries being

[1] Moore, Vol. I, pp. 796–797.

excluded from the treaty provisions. On the other hand, similar rights were granted to British fishermen on the eastern coasts of the United States north of the parallel of 39 degrees. Fish and fish-oil, the product of sea-fisheries, were to be admitted into each country, respectively, free of duty. The Newfoundland fisheries were placed on the same basis as the other provinces mentioned, as far as free fishing on the coast and the free admission of their fish into the United States were concerned.[1]

It was claimed by the British commissioners, but not admitted by the commissioners of the United States, that, by this treaty, greater benefits would accrue to the fishermen of the United States than to those of the British provinces, and it was provided by one of the articles of the treaty that commissioners should be appointed to determine the amount of compensation to be paid by the United States to the British Government for this excess of privileges granted to our citizens under the fishery articles of this treaty.

The fisheries provisions of the Treaty of Washington were to last ten years from the time that they went into force, and could not be abrogated by either party without a notice of such intention two years previous. On account of delay in passing the necessary laws, the fisheries articles of the treaty did not go into effect until the first of July, 1873. They continued in force for twelve years.

The Halifax Commission, provided for by the Treaty of Washington, met in June, 1877, to determine the amount of compensation, if any, that should be paid to Great Britain in return for the fisheries privileges accorded to the citizens of the United States under that treaty. The Commission consisted of Mr. Maurice Delfosse, the Belgian minister at Washington, Sir Edward Galt, on the part of

[1] Snow, p. 98.

Great Britain, and Ensign H. Kellogg, on the part of the United States. The agent of the United States was Judge Dwight B. Foster, assisted by Richard H. Dana and W. H. Trescott, as counsel. The British agent was Francis C. Ford, assisted by counsel from each of the provinces involved and from Newfoundland, six in all.

The aggregate amount of compensation claimed by Great Britain for the twelve years that the treaty was certainly to remain in force was $14,880,000, or a yearly amount of $1,240,000. Of this amount the sum of $2,880,000 was claimed on behalf of Newfoundland. On the other hand, the United States claimed that the privilege of free admission of Canadian and Newfoundland fish into our markets was sufficient compensation to offset the privileges of free fishing in British provincial waters. The Commission, by a vote of two to one, the American commissioner dissenting, awarded the sum of $5,500,000 on November 3rd, 1877. The United States paid this amount with some grumbling and with a protest[1] against

[1] This most unjust decision illustrates the folly of assuming that the status of the fisheries for one period of years may stand as a criterion for a future period. During the first year of the operation of the fisheries provisions of the Treaty of Washington, 1873, which was by far the most prosperous one for the Americans of the whole period of twelve years, there were 254 American vessels fishing for mackerel within the Gulf of Saint Lawrence; the number of barrels caught within the three-mile limit on a liberal estimate was 25,670 barrels, which were valued at $268,508 in the United States. (U. S. Fish Com., Report, 1881, p. 520.) The amount of the catch and the value of the same fell off rapidly and steadily during the following years of the treaty arrangement for inshore fishing, so that by 1877, the year in which the Commission made the award, only 60 American schooners availed themselves of the privilege of fishing within the three-mile limit. These schooners secured 2,439 barrels of mackerel, valued at $27,072. In 1880 three schooners went to the Bay and in 1881 only one American schooner fished there, securing 95 barrels of mackerel valued at $717. Yet

its being made a precedent for any future arrangement.

The fishery provisions of the Treaty of Washington came to an end the first of July, 1885, but, with the hope of negotiating a new treaty, the American fishermen were permitted, by mutual consent of the two countries, to fish in British waters during that summer. Again, the basis of the fisheries arrangements was thrown back upon the convention of 1818.

In his annual message to Congress in December, 1885, President Cleveland recommended that provision should be made for the appointment of a joint commission to arrange for a new fisheries treaty. This recommendation was voted upon adversely by the Senate on April 13, 1886, petitions against such action having been received from many New England fishermen who believed that their interests could be best served under the provisions of the convention of 1818, then in force.

Immediately, however, new trouble arose between the Canadian officials and the American fishermen, and seizures of American schooners were made. The seizures led to controversies between the two governments for a period of two years over new issues set forth by Canadian officials. The first case of the seizure of an American vessel in 1886 was that of the *David J. Adams*, in Digby Harbor, May

for this paltry cargo our Government paid Great Britain $458,333, that being the yearly average of the Halifax award for the twelve years. For the ten years from 1873 to 1882 the total catch of the American fleet within the three-mile limit was 78,827 barrels of mackerel valued at $598,429. For the twelve years that these privileges were in force the United States paid about eight times as much for the privilege as our fishermen secured from it, besides remitting the duty, at one cent per pound, on millions of pounds of Canadian fish imported into this country. Little wonder that the United States gave notice, as soon as the treaty provisions allowed, of the abrogation of the treaty in so far as it affected the fisheries interests.

7, the schooner being sent to Saint John, New Brunswick, for trial. The question involved in this trial was the right of an American fishing vessel to buy bait in the British waters not free to such a vessel for the purposes of fishing. Other seizures were made that year, which brought up the whole subject of the right to buy bait and supplies, other than wood and water mentioned in the convention of 1818. The substance of the controversy that followed is set forth by Freeman Snow in his "Topics of American Diplomacy":

"To follow the question back to the date of that convention, we find that the United States, by the said convention, 'renounced the liberty to fish' in certain defined waters; but American fishermen were permitted to enter the same for the *'purposes of shelter, of repairing damages therein, of purchasing wood and of obtaining water, and for no other purpose whatever.'* The act of parliament passed the next year, 1819, to carry this convention into effect, imposed no penalties for the purchase of bait or other supplies, and it would seem that American fishermen went on buying these articles as they had done before; and there is no record of any proceedings against them therefor until 1870. . . .

"The counsel for the owners of the *Adams* argued that American fishermen had a right to buy, in Canadian waters, bait and other supplies necessary for fishing, to be used in the deep-sea fisheries.

"The Canadian counsel insisted upon the strict interpretation of the treaty, that American vessels may enter their waters for purposes of shelter, repairing damages, the purchase of wood and obtaining water, and 'for no other purpose whatever.' As the British courts do not administer treaties, but only the statutes that are passed to carry the treaties into effect, the Canadian courts encountered a difficulty, from the fact that the Imperial Act of 1819, to carry into effect the convention of 1818, imposed no penalty for buying bait or supplies; the only penalty was for 'fishing or preparing to fish' within the prohibited waters.

In order to make the purchase of bait or other fishing supplies an offense under the treaty, it was therefore necessary to consider this act as 'preparing to fish' within the prohibited waters. . . .

"The cause of complaint on the part of American fishermen in the summer of 1886 was not so much on account of not being permitted to buy bait and other supplies, as it was on account of the harsh and unreasonable manner in which the Canadian authorities carried out the regulations under the convention of 1818 and their own customs laws. A vessel, according to this interpretation, could enter a port to repair damages, but was not permitted to buy a rope or anything that might be wanted for such repairs; she could procure water, but if her water casks were out of repair, she was not allowed to buy new ones, or even a hoop with which to mend the old ones; she could enter a harbor for shelter, but if she put a man on shore or took one on board, she was subject to detention and fine: and in any case she was often so pestered and insulted by revenue officials that it was better to run the chances of shipwreck than to enter a harbor where it was easy to find an excuse for her detention. . . . Such acts as these revealed the animus of the Canadians, which would seem to have been to take advantage of every technicality of law or treaty, to annoy and provoke the American fishermen, with the probable motive of ultimately forcing the United States to make a new reciprocity treaty." [1]

By an act of Congress, March 3, 1887, called the retaliatory act, the President, in his discretion, was empowered by proclamation to deny to vessels, their masters and crews, of the British dominions of North America any entrance into the waters of the United States, except cases of distress or necessity, and also to deny entry of fresh fish or salt fish or any other product of the dominions in question or other goods coming from such dominions into the United States. The President did not exercise his authority under this act. But negotiations were continued

[1] Snow, pp. 451–458, passim.

for an adjustment of the controversies, which resulted in the meeting of the representatives of the United States and Great Britain in conference at Washington, November 22, 1887. On February 15, 1888, a treaty was signed, which was subsequently rejected by the Senate, August 21, 1888.

The provisions of the unratified treaty of 1888 were not so favorable to United States fishermen as the earlier reciprocity treaties. By this treaty, delimitation of what were considered exclusively British waters under the convention of 1818 was made. The three marine miles mentioned under that convention were to be measured, under the new arrangement, seaward from low-water mark, and all bays, harbors and creeks, where it was not otherwise provided, were to be measured from a straight line drawn across the part nearest the entrance, at the first point where the width did not exceed ten marine miles. Specific stipulations were made for establishing the limits of exclusion in the Baie des Chaleurs, Bay of Miramachi, Egmont Bay, Saint Ann's Bay, Fortune Bay, Sir Charles Hamilton Sound, Barrington Bay, Chedabucto and Saint Peter's Bays, Mira Bay, Placentia Bay, and Saint Mary's Bay. More liberal regulations were provided for regarding the entrance of American vessels into Canadian harbors for a short time, or during stress of weather or accidents to vessel or crew. The penalties for unlawful fishing and for preparing to fish were made specific to a certain degree. Finally, provision was made possible for the free interchange of fish and fish products into each country, with the privilege of American vessels to obtain licenses to purchase certain supplies, to transship the catch, and to ship crews.

After the treaty was signed the British plenipotentiaries, desirous of preventing a repetition of the former friction and irritation which might arise before there had been time for the treaty to be ratified, presented a paper in which it was stated that "with the further object of affording evidence of their anxious desire to promote good feeling and to remove all possible subject of controversy, the British plenipotentiaries are ready to make the following temporary arrangement for a period not exceeding two years, in order to afford a '*modus vivendi*' pending the ratification of the treaty:

1. For a period not exceeding two years from the present date, the privilege of entering the bays and harbors of the Atlantic coasts of Canada and Newfoundland shall be granted to United States fishing vessels by annual licenses at a fee of $1.50 per ton for the following purposes:

> The purchase of bait, ice, seines, lines, and all other supplies and outfits.
>
> Transshipment of catch and shipping of crews.

2. If during the continuance of this arrangement the United States should remove the duties on fish, fish-oil, whale and seal-oil (and their coverings, packages, etc.), the said licenses shall be issued free of charge.

3. United States fishing vessels entering the bays and harbors of the Atlantic coasts of Canada or of Newfoundland for any of the four purposes mentioned in Article I of the convention of October 20, 1818, and not remaining therein more than twenty-four hours, shall not be required to enter or clear at the custom-house, providing that they do not communicate with the shore.

4. Forfeiture to be exacted only for the offenses of fishing or preparing to fish in territorial waters.

5. This arrangement to take effect as soon as the necessary measures can be completed by the colonial authorities.

Washington, February 15, 1888.

This arrangement was agreed to by the American commissioners, and since that time it has been, by renewal, the fisheries arrangement between the two countries. An attempt to settle again the disputes that arose between our fishermen and the citizens of Newfoundland within recent years was the proposed Hay-Bond Treaty of 1902 which failed of ratification. Its provisions looked to the free admission of salted and dry-cured fish and green fish into our markets from Newfoundland, with the privilege to the Americans of securing free bait in Newfoundland waters. The fishing interests of New England could not see wherein they could be bettered by such a change, as, by the terms of the *modus vivendi,* our fishermen can now obtain bait by the payment of a license fee that in the aggregate amounts to from $120 to $200 a vessel only.

As the question now stands it is a three-sided one in which Canada, Newfoundland, and the United States are each striving to hold fast to all the privileges they now possess and, if possible, make new arrangements whereby their commercial interests will be furthered. Canada is desirous of securing the free entry of her fish because her own markets are inadequate for the annual supply of fish, and because she believes that the removal of the present duty into the United States would cheapen the cost of fish for the consumer in the United States, and thus develop a greater market for the supply that she could furnish. She also occupies an advantageous position in regard to the mackerel fishery, and has excellent train and steamboat facilities for shipping her fresh food into American markets.

Newfoundland occupies a commanding position in the fact that the supply of bait used in the deep-sea fisheries centers very largely on her coasts. She contends that the bait supply is indispensable to the United States, and that the present arrangement of license fees, which gives her only about $15,000 annually, is altogether inadequate.[1] This colony, too, would like to have the free

[1] During the season of 1907 the American fishermen at Bay of

admission of her fish into the United States. The United States realizes the immense importance of the Newfoundland bait to her fishing interests and will do nothing that will tend to jeopardize these interests. She does not want the free admission of Canada's fish and fish products; such a step to-day would seriously cripple the New England fishing industry. The three-mile limit for catching mackerel in Canadian waters is of little importance to New England fishermen at present, so that Canada can hold out little to this country, in the fishing industry, that should tempt our government to allow the free entry of Canadian fish. The fisheries of Newfoundland are not held in so great fear by the Americans, however, because Newfoundland is more remote from our markets and the transportation facilities are not so good as Canada possesses.

Why, then, do not the United States and Newfoundland enter into an agreement that might be mutually beneficial? Such a proposition was tried in the Hay-Bond arrangement, but the measure received the veto of the Imperial Cabinet. During the progress of the negotiations, Canada entered a protest against such a treaty on the grounds that Newfoundland should not be permitted to make an arrangement with the United States and barter away her inshore fisheries, because these were the joint possession of all the North American provinces of Great Britain. The importance of the interests of Canada and the insignificance of Newfoundland as a colony led to the sus-

Islands, Newfoundland, agreed to forego temporarily their rights of fishing and become what they were in 1904, traders. By this arrangement there was no license fee charged by the Newfoundland officials and the Americans agreed to pay not less than $1.25 per barrel for herring. There were fifty-seven American vessels at that port which purchased cargoes of herring valued at $251,652. (Annual Report, Dept. of Marines and Fisheries, Newfoundland, 1907.)

pension of negotiations that would have been beneficial to Newfoundland.[1] The whole question of these treaty rights has been submitted for settlement to the Hague Tribunal. It is earnestly hoped that a satisfactory and permanent solution of the much-vexed fisheries disputes may be effected.[2]

[1] When, in 1908, the writer was at Bay of Islands, Newfoundland, to investigate the status of the Fisheries Question in that country he found that the most cordial feeling exists between the fishermen of Newfoundland and the United States. The Newfoundland fishermen, when asked the cause of trouble between them and the American fishermen, replied, "The trouble is not with us; it is between *our Government* and the American fishermen."

[2] The full text of the Award of the Tribunal of Arbitration, with Dr. Drago's Dissent, may be found on pages 373 to 449.

CHAPTER XX

A BIBLIOGRAPHY OF NEW ENGLAND FISHERIES

GENERAL SURVEY OF THE LITERATURE.

The principal works that may be regarded as histories of the American fisheries are two reports made at the instance of departmental officers, which embody a history of the fisheries of the United States down to the time of their publication. These works are the "Report on the Principal Fisheries of the American Seas," prepared for the Treasury Department in 1853 by Lorenzo Sabine, and the "Fisheries and Fishery Industries of the United States," prepared through the co-operation of the Commissioner of Fisheries and the Superintendent of the Tenth Census, by George Brown Goode and a staff of associates in 1880. Mention may also be made of a "Report on the Cod and Whale Fisheries," by Thomas Jefferson in 1791; but this is a summary statement and by no means a history of the fisheries.

Sabine's Report on the Principal Fisheries of the American Seas is the only attempt thus far made to place the history of our fisheries within the compass of a single volume. It was written at a time when the fisheries were confined principally to the whale, the cod and the mackerel, and before industries arising from these fisheries had sprung up. Previous to the negotiations that led to the ratification of the Reciprocity Treaty of 1854, it was highly important that our Government should have data on the condition of our fisheries. The Secretary of the Treasury requested of the Collector of Boston that a report of limited size be submitted from the Boston office. The result of the correspondence that ensued was that the Secretary authorized Mr. Lorenzo Sabine, of Framingham, to write a report somewhat more elaborate than what was at first intended.

Mr. Sabine was well qualified for such a mission. At the time of writing he was Congressman from Massachusetts, he had served three terms in the Maine legislature, had been collector of the Passamaquoddy district in Maine for several years, and had witnessed personally much of the trouble between our fishermen and the Canadian officials in the waters of eastern Maine. Furthermore, he was probably the best informed person in America on the subject of the fisheries. For twenty years he had been collecting papers and documents with the purpose of writing a comprehensive history of the fisheries of the United States. His valuable work, "The American Loyalists," appearing six years previous, had placed him in the ranks of historians of the time. So that the selection of Mr. Sabine to submit a report on the fisheries was most opportune.

An examination of the Report furnishes strong internal evidence that Mr. Sabine had already completed a portion of his history of the fisheries before he was commissioned to submit a report upon the subject. This view would be strengthened by the fact that less than ten months were used in writing the report. That part of the report that deals with the fisheries down to the Declaration of Independence—about one-half of the work—was of the least importance to the statesmen of 1854, for whose benefit the report was especially prepared. Yet in his consideration of the colonial fisheries, Mr. Sabine goes into details of time and place as no other writer or set of writers has done. That part of the report remains to-day the most valuable authority on the subject of colonial fisheries.

In his treatment of the fisheries under the Federal Government, Mr. Sabine pays particular attention to the diplomatic questions that arose out of the different constructions of the treaties of 1783 and 1818. This part of the work gives no good comprehensive treatment of the fisheries as an industry. The last third of the work is an historical view of the controversy as to the intent and meaning of the fisheries article of the convention of 1818, which remained the best contribution to the Fishery Question literature for about a generation. The historical treatment of the fisheries subsequent to the Revolution is disappointing, especially in view of the fact that the writer must have had material at hand for a more complete treatment. The whale fishery is practically ignored; probably, because that fishery bore no part in the forthcoming negotiations of reciprocity. On the other hand, the cod, mackerel, and herring, that would receive particular attention in the negotiations, receive separate, though imperfect, treatment at his hands. Sabine's Report is not complete regarding fisheries other than those of New England.

Viewed as a whole, Mr. Sabine's work may be regarded as a highly interesting and valuable report. It is to be regretted that footnote references to his sources of information are not more numerous; it would have saved future writers much labor in collecting material had he made such references, or even supplied a bibliography. There is no index, but the arrangement of the work was so logical that the seriousness of this defect is not so apparent. As the historian of the fisheries of the United States, Mr. Sabine possessed the means and ability to write a work that would have been a classic for all time. As a result of governmental pressure the contemplated history was hurriedly and incompletely finished. For the occasion, it was excellent; but the country is the loser because circumstances did not permit the completion of a maturer work. For thirty years the report remained the best treatise on American fisheries, and even to-day it is unsurpassed in the excellence of treatment given to the British colonial fisheries.

The principal work upon our fisheries is the voluminous compilation of seven large books prepared by George Brown Goode and a staff of eighteen associates, from 1880 to 1887. Mr. Goode was at the time assistant secretary of the Smithsonian Institution. His associates were men of recognized authority in the subjects assigned to them, among the number being Prof. David S. Jordan, A. Howard Clark, R. Edward Earll, Ernest Ingersoll, Richard Rathbun, and Capt. J. W. Collins, a Gloucester fisherman of wide experience in the fisheries. This work upon the fisheries may be termed a compilation of material, part of which was gathered from previous works upon different phases of the fisheries, part was from articles already published by the Fish Commission in its reports, and a part from investigations made at the time of the authorization of the work.

The article on the mackerel fishery is made up essentially of a report that had been published by the Commission a short time previous. So with the oyster fishery. The whale fishery adds little to what had already been written upon the subject, as quotations are freely made from the two principal sources in existence at that time. In many parts of the work Sabine's Report is quoted freely. All this would imply that with some writers and on some subjects little effort was used in getting back of material that already existed. Other parts of the work represent great and valuable additions to the historical material for use in this subject.

Especial mention should be made of the Geographical Review in Section II. That volume was largely written after extensive inquiries in the field. An index to this section contains valuable references to sources of information, and, while it is in no sense a bibliography, it remains the most comprehensive survey of the early fisheries that has been written. In another section, the different fisheries are treated separately and the work is well done.

The principal defect in the work is the lack of a comprehensive view of the fisheries of the country in their relation to the development of the people. No mention is made of the effects of territorial expansion, the increased means of communication and transportation, the wars subsequent to the Revolution, and other important matters. The subject of international disputes finds little place; little mention is made of the value of bounties to fishing vessels or the abolishment of the same; and the reader can gather no adequate idea of the development of the fisheries as a whole, of the improvement in apparatus used except as it relates to the separate industries, or of the gradual betterment of the condition of the fisherman through the necessity of constructing schooners of a better class. Lastly, the size of the work is a serious defect for the average reader. It has not been placed in a form convenient either for the reader or the student, and while it will always remain invaluable to the latter it never has been and can never be, in its present form, what it should have been made,—a history of the fisheries for the people. In all probability, not one fisherman in a hundred knows of the existence of the work.

The principal source of statistics and historical material of the fisheries during the last thirty-nine years has been the publications of the United States Fish Commission, which is now designated the Bureau of Fisheries. In 1871, the Fish Commission began the issue of Annual Reports of the Commissioner, and these have continued as bound volumes to the year 1902. By legislative enactment, the volume known as the Annual Report of the Bureau has been discontinued, and is now represented by a brief report of the Commissioner of Fisheries to the Secretary of Commerce and Labor. The papers formerly published as appendixes to the Commissioner's report, with it constituting the annual report volume, are now issued as independent pamphlets, designated "Bureau of Fisheries Documents." Beginning with the year 1881, the Bulletin of the Bureau of Fisheries has been issued annually in bound volumes down to 1904. At present, the Bulletin is issued in bound volumes in a limited quantity; for general distribution by the Bureau, the miscellaneous papers composing the volume are issued as separate pamphlets, the supply being regulated by the demand for the subject-matter. The Bureau prints from time to time a list of its publications that are available for distribution.

The literature of the Fishery Question is far more satisfactory and abundant than what refers to the history of the fisheries. Sabine's contribution to the subject has already been referred to. In 1877, two small volumes appeared, each independent of the other and treating the subject from different points of view, those of Isham and of Elliott. The former represents original investigation, as the author apparently was not informed of Sabine's work; the latter knew Sabine far too well in parts of his treatment. Yet, in his discussion of the questions, he is original and presents a clear statement of the principles involved. Many periodical articles by eminent men have been contributed to the subject.

By far the best treatises on the Fisheries Question are those of Freeman Snow and John Bassett Moore, the one in his "American Diplomacy," the other in volume one of "A Digest of International Law." Both these jurists treat the subject exhaustively and in a manner that is clearly stated. The latter has a longer account, quotes more freely from authorities, and gives abundant lists of references to sources. So ably and fully do the authors treat the Fishery Question that the general reader would be satisfied with a summary view of the treatment that either author has written. These two works are the most scholarly contributions to the literature of the American fisheries.

In the following bibliography of the fisheries, critical comment is offered on works of especial value. Each work has been examined with that end in view. No attempt has been made to list the publications of the Fish Commission and the Bureau of Fisheries, except to note those publications that bear most directly upon the fisheries of New England.

BIBLIOGRAPHY

Abbot, Willis J. American Merchant Ships and Sailors. New York, 1902. (372 pp. No index. Information concerning fisheries is very general.)

Adams, John. Works with a Life of the Author, Notes and Illustrations. (Edited by Chas. F. Adams.) 10 vols. Boston, 1850–1856.

Adams, John Quincy. Duplicate Letters, the Fisheries and the Mississippi. Documents relating to the transactions at the negotiations of Ghent, Washington, 1822.

Adams, Herbert B. Village Communities of Cape Ann and Salem. Johns Hopkins University Studies, I. Art. 9 and 10.

American Antiquarian Society, Transactions and Collections of the Society. 7 vols. Worcester, 1820–1885.

American Archives. (See doings of Congress.)

Anderson, Adam. Origin of Commerce. A history of the Great Commercial interests of the British Empire, etc. 4 vols. London, 1787.

Anderson and McPherson, Annals of Commerce from the earliest Times to the Year 1800, in 4 vols.

Annual Report, Dept. of Marines and Fisheries. Newfoundland, 1907.

Atwater, W. C. American Fisheries, Menhaden. Account of the Agricultural Uses of Fish. 529 pp. New York, 1880.

Auchmuty, Robert. The Importance of Cape Breton to the British Nation. 7 pp. London, 1745.

Babson, John J. History of Gloucester. 600 pp. Gloucester, 1860. (Pages 248 to 255 contain an account of ships and shipbuilding at Gloucester in the early days. Pages 378 to 385 give the best account to that date of the fisheries of the town, with statistics.)

Baird, Spencer F. The Sea Fisheries of Eastern North America. Prepared for the Consideration of the International Commission held at Halifax, 1877. In Report of the U. S. Fish Commissioner for 1886.

Bancroft, George. History of the United States. 6 vols. New York, 1883-85.

Bates, William W. American Marine. The Shipping Question in History and Politics. New York, 1893.

Belknap, Jeremy. The History of New Hampshire. 3 vols. Boston, 1742.

Benton, Thomas A. Thirty Years' View, etc. . . . from 1820 to 1850. 2 vols. New York, 1854-56. (Pages 194 to 198, Vol. II, contain a discussion against the continuance of the bounties granted to vessels engaged in the cod and mackerel fisheries.)

Bernard, Gov. Francis. Select Letters on the Trade and Government of America, etc. 1763-68. VII plus 130 pp. London, 1774.

Bertram, J. G. The Harvest of the Sea. 520 pp. Edin., 1865; New York, 1866.

Bibliotheca Piscatoria. A Catalogue of Books on Angling, the Fisheries, and Fish Culture, etc. By T. Westwood and T. Satchell. London, 1883. (397 pages. An exhaustive treatise on books relating to British fisheries.)

Biskerdyke, John. Angling in Salt Water. Ill. 112 pp. London, 1889.

Bishop, W. H. Fish and Men in the Maine Islands. New York, 1885. (In Harper's Handy Series.) Illustrated. 129 pages.

Blackford, Eugene G. American Fish Foods. (Pages 390–395, in Depew's One Hundred Years of American Commerce, Chapter LVI.)

Bliss, Alexander. A Review of the Halifax Fishery Award. How it strikes a private citizen. Pamphlet, 24 pp. Washington, 1878.

Bollan, William. The Ancient Right of the English Nation to the American Fishery; and its Various Diminutions examined and stated. With maps, etc. 105 pp. London, 1764.

Boston Fish Bureau. Annual Reports, 1875 to 1909. (The reports contain valuable information concerning the quantity and value of the fisheries yearly, as well as tables showing the catch and prices for many years previous to the beginning of the reports. They are of unquestioned authority.)

Bourne, Edward E. The History of Wells and Kennebunk from the earliest settlement to 1820. Portland, 1875.

Bradford, William. History of Plymouth Plantation. Boston, 1901.

Brief Remarks concerning the Cod Fishery, the Beaver Trade, and the Limits of New England and New France. 61 pp. In Boston Public Library. (A MS. ascribed by J. Sparks to the Baron de La Houtaine, and written after 1709.)

Brooks, William K. The Oyster—A Popular Summary of a Scientific Study. 225 pp. Baltimore, 1905.

Butler, Benjamin F. The Fishery Award. Speech in the House of Representatives, June 19, 1878. Pamphlet. Washington, 1878.

Calculation of the State of the Cod and Whale Fishery belonging in Massachusetts, in 1763. Copied from a paper published in 1764. In Massachusetts Historical Society Collections, VIII, pp. 202–203.

Calendar of State Papers, Colonial, 1661–1674.

Calendar of State Papers, Venetian, 1202–1509. London, 1889.

Catalogue of the Collection to Illustrate the Animal Resources and the Fisheries of the United States. Prepared under the direction of G. Brown Goode. 351 pp. Bull. of the U. S. National Museum, No. 14. Washington, 1879. (Pages 277–282 contain Statistics of Fisheries, compiled from the report of the Commissioner of the Bureau of Statistics, 1877.)

Catalogue of collection illustrating the fishing vessels and boats and their equipment; the economic condition of the fishermen; anglers' outfits, etc. Washington. Government Printing Office, 1884.

Catalogue of apparatus for capture of fish exhibited by the U. S. Nat'l Museum. By Robert E. Earll. Washington. Gov't Printing Office, 1884.

Catalogue of fishery products, and of the apparatus used in their preparation. By Alonzo Howard Clark. Washington. Gov't Printing Office, 1884.

Cleveland, Grover. Rights of American Fishermen in British North American Waters. Message to Congress, 210 pp. 49th Congress, 2nd Session, House Executive Docs. No. 19.

Compendium of the Census, 1840.

Connecticut, Biennial Reports of the State Commissioners of Fisheries and Game, 1896–1906.

Deane, Samuel. A History of Scituate. Boston, 1831.

Department of Commerce and Labor, Organization and Law, Washington, 1904.

Descriptive Catalogue of the Collection sent from the United States to the International Fisheries Exhibition. London, 1883. By G. Brown Goode. Bulletin of U. S. Nat'l Museum, No. 27. Washington, 1884.

Descriptive Catalogue of Government Publications. Subject: Fish and the Fisheries. (Contains a long list of references.)

Documents relating to the Colonial History of New York. Albany, 1853-1887. 15 vols.

Documents and Records relating to the Province of New Hampshire. 2 vols. 1686–1722. Manchester, 1868.

Doings of Congress, have appeared as follows:

American Archives, by Peter Force (9 vols.), 4th series, 1774–1776, six volumes; 5th series, 1776, three volumes.

Annals of Congress, 1789–1824.

Congressional Debates, 1824–1837.

Congressional Globe, 1833–1873.

Congressional Record, 1873—.

Niles' Weekly Register, 1811–1837, continued as Niles' National Register, 1837–1849. 75 volumes.

Doran, J. I. Our Fishery Rights in the North Atlantic. Philadelphia, 1888. 67 pp.

Douglass, William. A Summary, Historical and Political, of the State of British Settlements in North America. 2 vols. Boston, 1755.

Elliott, Charles B. The United States and the Northeastern Fisheries. A History of the Fishery Question. Map. 144 pp. plus appendix of 7 pp. Minneapolis, 1887. (Part I, p. 1–100, is historical, giving a hasty review of the rise of the fisheries, based largely on Sabine's Report, and a statement of the important diplomatic questions arising under the several treaties down to 1887. Part II, pp. 100–133, discusses the terms of the Treaty of 1818. In the treatment of Part I after 1850 and in the treatment of Part II the author made a valuable contribution to the discussion of the Fishery Question. The economic and commercial aspects of the fisheries are not touched upon to any extent beyond the range of Sabine's contributions.)

Exhibit of the Fisheries and Fish Culture of the United States. Catalogue of the United States Exhibit at London Fisheries Exhibit, 1883. Bulletin of the U. S. Nat'l Museum, No. 27. Prepared under direction of G. Brown Goode. 1280 pp. Washington, 1884.

Felt, Joseph. Annals of Salem. 2 vols. Salem, 1845.

Fisheries Exhibition, Literature. London, 1884. (In Vol. I, Handbooks, Part I, pp. 461–544, is "A Popular History of Fisheries and Fishermen of all Countries from the Earliest Times." It is of little value to American fisheries.)

Fishing Gazette, The. A History of the World's Fisheries. New York, 1885—. (A weekly publication devoted exclusively to the fisheries and fishing industries of the country, giving reliable weekly reports of the details of different phases of the fishery world, on the sea, in the several industries, and concerning the trade.)

Fisheries of New England. Published reports of the fisheries of the New England States are as follows:

The Fisheries and Fishery Industries of the United States. Prepared through the co-operation of the Commissioner of Fisheries and the Superintendent of the Tenth Census, by George Brown Goode, assistant director of the U. S. National Museum, and a Staff of Associates. Washington, Gov't Printing Office, 1884–1887. 7 vols. in 5 sections.

Section I. Vol. I. Natural History of Useful Aquatic Animals, 895 pp. Vol. II. An Atlas of 277 plates, to accompany Vol. I.

Section II. A Geographical Review of the Fisheries Industries and Fishing Communities for the Year 1880. 787 pp.

(Sections I–V, pp. 1–340, give a review of the New England States. Also the appendix, pp. 675–737, contains "Historical References to the Fisheries of New England," by A. Howard Clark. This part of the work is the most complete and the most valuable list of references now in print relating to the New England fisheries; it also has many pages of quotations from original sources. Especial emphasis has been laid upon Massachusetts towns.)

Section III. The Fishing Grounds of North America, with forty-nine charts. Edited by Richard Rathbun. 238 pp.

(This section gives all that may be desired in regard to the position of the fishing grounds, their shape and depth, the nature of the bottom, the kind of fish to be found at the different places, and the fishing that is carried on, with some historical material showing how long the grounds have been fished upon, and with what success.)

Section IV. The Fishermen of the United States. By G. Brown Goode and Joseph W. Collins. 178 pp. Sections III and IV in one vol.

(This section is arranged under the subjects, Nationality and General Characteristics, the Sailor-Fishermen of New England, Officers of Vessels, etc., Dangers of the Fisheries, Management of the Vessels, and Appendix.

Section V. History and Methods of the Fisheries. 3 vols. Vols. I and II. Text, pp. 908 plus 881.

(These two volumes treat of the fisheries by subjects. Those that are of special interest to New England are the Halibut Fisheries, pp. 3-119; the cod, haddock and hake fisheries, pp. 123-243; the Mackerel Fishery, 247-313; the Swordfish Fishery, pp. 315-326; the Menhaden Fishery, pp. 327-415; and the Herring Fishery and the Sardine Industry, pp. 419-524, in Vol. I, and the Oyster, Scallop and Clam Industries, pp. 507-615; and the Lobster Fishery, pp. 658-794, in Vol. II.)

Vol. II. An Atlas of 255 Plates, accompany Secs. I and II. (These cuts illustrate the apparatus used in the different fisheries, the vessels and their equipment, and methods of taking the fish.)

Report on the Conditions of the Sea Fisheries of the South Coast of New England in 1871 and 1872. By Spencer F. Baird, Report U. S. Fish Commission, 1871-72, pp. 1-41.

Sea Fisheries of Eastern North America, The. By Spencer F. Baird. Report U. S. Fish Commission, 1886, pp. 3-224.

Statistical Review of the Coast Fisheries of the United States. By J. W. Collins. Report U. S. Fish Commission, 1888, pp. 271-378.

Notes on the Oyster Fishery of Connecticut. By J. W. Collins. Bulletin U. S. Fish Commission, Vol. IX, 1889, pp. 461-497.

The Herring Industry of the Passamaquoddy Region, Maine. By Ansley Hall, Report U. S. Fish Commission, 1896, pp. 443–489. (The subject is treated in a thorough manner.)

Report on the Fisheries of the New England States. By J. W. Collins and Hugh M. Smith. Bulletin U. S. Fish Commission, Vol. X, 1890, pp. 73–176.

The Lobster Fishery of Maine. By John N. Cobb. Bulletin U. S. Fish Commission, Vol. XIX, 1899, pp. 241–265. (A good statement of the condition and methods of the fishery.)

Statistics of the Fisheries of the New England States. Report U. S. Fish Commission, 1900, pp. 311–386.

Statistics of the Fisheries of the New England States, 1902. Report Bureau of Fisheries, 1904, pp. 245–325.

Statistics of the Fisheries of the New England States for 1905. Bureau of Fisheries, Document No. 620. 93 pp. 1907.

Fish Commission, Annual Reports and Bulletins of the United States, 1871–1909. (The bulletins begin with 1880. By an act of Congress approved Feb. 14, 1903, the Bureau was made a part of the Department of Commerce and Labor, and the old name of Fish Commission was changed by order of the Secretary to Bureau of Fisheries.)

Fiske, John. Beginnings of New England. Boston.

Fiske, John. New France and New England. Boston, 1902.

Folsom, George. History of Saco and Biddeford. Saco, 1830.

Foster, Dwight. Arguments of, on behalf of the United States, before the Halifax Commission. 64 pp.

Foster, John W. A Century of American Diplomacy. (1776-1876.) Boston, 1900.

Freeman, Frederick. A History of Cape Cod. 2 vols. Boston, 1862.

Francis, Francis. The Practical Management of Fisheries. A Book for Proprietors and Keepers. Illustrated. 114 pp. London, 1883.

Franklin, Benjamin. Complete Works. (Edited by John Bigelow.) 10 vols. New York, 1887–88.

Forester, Frank. Fish and Fishing. Illustrated. 18—. New York, 359 pp. with an appendix of 60 pp., by William H. Herbert, 1851.

Gloucester and the New England Fisheries. Papers and Statistics in Regard to Gloucester and New England Fisheries. Dec. 4, 1902. 27 pp. 57th Congress, 2nd Session, Senate Doc. No. 14.

Goode, George Brown. The Fisheries of the United States. (See New England Fisheries.)

Goode, George Brown, and others. Materials for a History of the Mackerel Fishery. Report U. S. Fish Commission, 1881. pp. 89–531.

(This is an important compilation of extracts from original sources, of statistics of the catch, of notes from fishermen's records, together with the natural history of the fish, methods of fishing, and of preparing for the market. It is the basis of the article in Sec. V of the Fisheries of the U. S., some parts being there quoted entire.)

Goode, George Brown. A Review of the Fishing Industry of the United States and the Work of the U. S. Fish Commission. The Fisheries Exhibition Literature, V, Part II. London, 1884.

Hakluyt, Richard. Principall Navigations. London, 1810.

Halifax Commission. Award of the Fishery Commission; documents and proceedings. 3 vols. Washington, 1878. House Ex. Docs. Vols. XVIII–XX, 45th Congress, 2nd Session.

(This report contains a mass of testimony relative to the American and Canadian fisheries, and the arguments of the counsel.)

Hall, Henry. (Special agent.) Report on the Ship-building Industry of the United States. 10th Census, Vol. VIII. 276 pp. 47th Congress, 2nd Session, H. Misc. Docs., Vol. XIII, No. 42, Part 8.

Harris, W. C. Fishes of North America that are captured with Hook and Line. Colored plates. New York, 1898.

Henderson, J. B. Northeast Coast Fisheries. (In his American Diplomatic Questions. 1901. pp. 449–529. A well-balanced and clear account, with historical sketch.)

Higginson, Francis. New England's Plantation. (In Young's Chronicles.)

History of the Emblem of the Codfish in the Hall of the House of Representatives. Compiled by a Committee of the House. Ill. 62 pp. Boston, 1895.

Holder, Charles F. The Big Game Fishes of the United States. Illustrated. 435 pp. New York, 1903.

Holmes, Abiel. American Annals, or a Chronological History of America from its discovery in 1492 to 1806. 2 vols. Cambridge, 1805.

Holroyd, John B. (Lord Sheffield.) Observations on the Commerce of the American States, etc. Appendix with tables, etc. 2nd edition. London, 1783.

Hubbard, Rev. William. General History of New England. Boston, 1845. (In Mass. Historical Society's Collections, 2nd series, V–VI.)

Hudson, Frank. Sea Fishing for Amateurs, etc. Illus. 79 pp. London, 1888.

Huske, ——. The Present State of North America. London, 1755.

Ingersoll, Ernest. The Oyster Industry. (In Goode's Fishery Industries of the United States.)

Isham, Charles. The Fishery Question. Its origin, history, and present situation. With a map of the Anglo-American fishing ground, and a short bibliography. 89 pp. New York, 1887. (In "Questions of the Day.")

Jackson, Andrew, Message of. United States Fisheries. 279 pp. Washington, 1831. 22nd Congress, 1st Session, House Doc. 99.

Jefferson, Thomas. (Secretary.) Report on the Cod and Whale Fisheries, 1791. Report to House of Representatives, Feb. 1, 1791. Printed with Sabine's Report in House Ex. Docs., No. 23, 32nd Congress, 2nd Session, and in House Misc. Docs., No. 32, 42nd Cong., 2nd Session.

Johnson, Edward. Wonder Working Providence, etc. (Edited by W. F. Poole.)

Johnson, Emory R. Ocean and Inland Water Transportation. 395 pp. New York, 1906.

Johnson, Emory R. Elements of Transportation. 364 pp. New York, 1909.

Johnstone, James. British Fisheries. A short account of the Origin and Growth of British Sea-Fishery Authorities and Regulations. 350 pp. London, 1905.

Joncas, L. Z., M. P. The Fisheries of Canada. The World's Fisheries Congress, Chicago, 1893. Washington, 1894. (Pages 341-348, Bulletin U. S. Fish Commission for 1893.)

Jordan, David S., and Evermann, Barton W. American Food and Game Fishes. A popular account of the species found in America north of the Equator, with Keys for ready identification, life history, and methods of capture. Illus. 573 pp. New York, 1902. (The best popular work on this subject, as far as natural history is concerned.)

Jordan, David S. A Guide to the Study of Fishes. 2 vols., illustrated. 427 plus 624 pp. New York, 1905.

Knapton, John and Paul. The Importance and Advantage of Cape Breton, truly stated and impartially considered, with proper maps. 8vo., pp. vi plus 156, 2 maps. London, 1757. (Not to be confounded with "The Importance of Cape Breton considered; in a Letter to a Member of Parliament, from an Inhabitant of New England," 8vo. 73 pp. London, 1746.)

Lechford, Thomas. Plain Dealings: or News from New England. 80 pp. London, 1642. (In Mass. Historical Society's Collection, 3rd series, III.)

Lloyd's Report on the State of the Fisheries. 19th Congress, 1st Session, Jan. 30, 1826.

Longworthy, Charles F. Fish as Food. 30 pp. Washington, 1898. (U. S. Agricultural Department, Farmer's Bulletin, 85.)

McDonald, William. Select Charters and Documents. 1606–1775. New York, 1899.

McMaster, John B. History of the United States, 7 vols. New York, 1874–1910.

MacPherson, David. Annals of Commerce, navigation and manufacture, etc. Vol. II. 4 vols. London, 1805.

Macy, Obed. History of Nantucket. Boston, 1835.

Maine, Report of Industrial and Labor Statistics, 1895. Augusta, 1896.

Maine, Board of Agriculture. Second Annual Report, 1862, on the Natural History and Geology of the State. (Pages 13–22 contain a sketch of the fishery industry of Maine.)

Maine, Annual Reports of the Commissioner of Sea and Shore Fisheries, from 1867 to 1884; biennial reports since that date. Augusta.

Maine, Annual Reports of the Inland Fisheries and Game Commission, 1895—.

Marvin, Winthrop L. American Merchant Marine, 1620–1902. 436 pp. New York, 1902. (Chapter XIII, The Deep Sea Fisheries, pp. 285–318, gives a short sketch of American fisheries.)

Massachusetts, Records of the Governors and Colony of, in New England. 1628–1686. 5 vols., Vol. IV in two parts. Boston, 1853.

Massachusetts, Report of the Commissioners on Fisheries and Game, 1906.

Massachusetts Historical Society, Collections, 2nd series, Vol. III.

Mather, Cotton. The Fisher-mans Calling. Small 8vo. Boston, 1716.

Menhaden Fishery of Maine, The. Published by the Association of the Menhaden Oil and Guano Manufacturers. 96 pp. Pamphlet. Portland, 1878.

Monson, Sir William. Naval Tracts, in Churchill's Collection of Voyages and Travels. 6 vols. 3rd London Edition, 1745. (Book VI, pp. 473–508, has material about British fisheries.)

Moore, John B. A Digest of International Law. 8 vols. Washington, 1906.
(Vol. I, pp. 767–875, deals with the Northeastern Fisheries. This is the most thorough discussion of the Fishery Question, having a full discussion, copious quotations, and numerous references.)

Morton, Levi P. Fisheries and Fish Culture. Their importance to the industry and wealth of the country. Speech in the House of Representatives, Feb. 4, 1880. 19 pp. Washington, 1880.

Niles' Weekly Register. (See Doings of Congress.)

New Hampshire, Documents and Records relating to the Province of, from 1686–1722. 2 vols. (Compiled and edited by N. Bouton.) Manchester, 1868.

Oldmixon, John. The British Empire in America, etc. 2 vols. Maps. London, 1708.

Parkman, Francis. Pioneers of France in the New World. Boston, 1889.

Pew, John J. The Fishing Industry. (In Century Celebration Publication, held at City Hall, Gloucester, Jan. 1, 1901.) pp. 23–31.

Pitkin, Timothy. Statistical View of the Commerce of the United States, New York, 1817.

Plymouth Colonial Records (1620–1692). 12 vols. Boston, 1855–61.

Procter, George H. Fishermen's Memorial and Record Book. 172 pp. 1873.

Procter, George H. Fisheries of Gloucester, from 1623 to 1876. 88 pp. 1876.

Procter, George H. Fishermen's Own Book. Illustrated. 274 pp. 1882.

(These three works came from the press of Procter Brothers, of Gloucester, and contain information concerning the fisheries of Gloucester and the methods of carrying on the fisheries in the earlier days. More valuable as a reference work than as a history of the fisheries.)

Prince, Thomas. A Chronological History of New England, in the form of Annals, etc. Boston, 1736. New Edition, 1826.

Prouty, Lorenzo. Fish: their Habits and Haunts, etc. 115 pp. Boston, 1883.

Rhode Island Colonial Records, Vol. VI.

Rhode Island, Annual Reports of the Commissioners of Shell Fisheries.

Rhode Island, Annual Reports of the Commissioners of Inland Fisheries, 1871–1909.

Ricketson, Daniel. History of New Bedford (and other towns). 408 pages. New Bedford, 1858.

Rich, Shebnah. The Mackerel Fishery of North America. Its

Perils and its Rescue. A Lecture. 40 pp. Boston, 1879. (The author advocates the discontinuance of seine-fishing and of the southern spring mackerel fishing. Of little value.)

Roberts, Lewes. The Merchant's Mappe of Commerce, etc. Folio, 431 pp., with index. London, 1636; 2nd edition, 1671, enlarged.

Roads, Samuel. History and Traditions of Marblehead. Boston, 1880.

Roosevelt, Robert B. Superior Fishing, etc. Illustrated. 349 pp. New York, 1884.

Russell, Edward T., and Co. The Harvest of the Sea. Boston, 1891.

Sabine, Lorenzo. Report on the Principal Fisheries of the American Seas: prepared for the Treasury Department of the United States. Written from Framingham, Mass., Dated Dec. 6, 1852. 317 pp. Washington, 1853.

Part I. France, Spain, and Portugal.

Part II. Newfoundland, Nova Scotia, Cape Breton, Prince Edward Island, Magdalen Islands, Bay of Chaleur, Labrador, New Brunswick.

Part III. United States. (pp. 93-317.)

(Published at Washington, in 1853; also in Exec. Docs., Vol. III, No. 23, 32nd Congress, 2nd Session, and in House Misc. Docs. No. 32, 42nd Congress, 2nd Session. This is the first attempt at writing a history of American fisheries, and the only one until the appearance of the voluminous compilation authorized in 1880. In many ways it still remains a standard work, especially of the fisheries as carried on in American waters previous to the settlement at Plymouth. References are not frequent and usually indefinite. The writer has placed more emphasis upon the political and diplomatic phase of the subject than upon its commercial importance. About twenty years were spent in collecting papers and documents before the work was written. In addition to his admirable collection of material, the author possessed other qualifications that should mark his volume as one of the standard sources of the history of the American fisheries, for it may safely be considered both a history and a source.)

Schuyler, Eugene. American Diplomacy. New York, 1886. (The writer gives in a dozen pages a clear statement of the Fisheries Question as affected by the treaties of 1783, 1818, 1854, and 1871.)

Scott, Genio G. Fishing in American Waters. 484 pp. New York, 1869. (Pages 315–431 contain chapters on Commercial Fisheries and Ancient and Modern Fish Culture.)

Seybert, Adam. Statistical Annals, etc. of the United States of America, from 1781 to 1818. Philadelphia, 1818.

Smith, John. Description of New England. (In Mass. Hist. Soc. Coll.) 3rd Series, Vol. VI, pp. 95–140.

Smith, Joseph. Gleanings from the Sea. 399 pp. Andover, Mass., 1887.

Snow, Freeman. Treaties and Topics in American Diplomacy. Boston, 1894. (The Fisheries Question is treated in pp. 427–470 in a thorough and clear manner.)

Speeches of the Governors of Massachusetts, 1765–1775, etc. 1818.

Statistical Abstract of the United States. Department of Commerce and Labor. Washington. Pub'd annually.

Storer, David H. A History of the Fishes of Massachusetts. Boston, 1867. (Pages 343–355 contain an article on the Cod, by Benj. W. Hale, Esq., taken from the Newburyport Herald of Feb. 12, 1840.)

Tariff Acts passed by the Congress of the United States, from 1789 to 1897, etc. By Robert C. Proctor. Washington. Gov't Printing Office, 1898.

Treaties and Conventions, concluded between the United States of America and other Powers since July 4, 1776, etc. Department of State, 1889.

United States Commission of Fish and the Fisheries. Annual Reports and Bulletins, 1871—.

United States Commission of Fish and the Fisheries. Laws and Regulations. 29 pp. Index. Washington, 1882.

Van Doren, Louis O. The Fishes of the East Atlantic Coast

that are caught with hook and line; including, The Fishes of the East Coast of Florida, by Samuel C. Clarke. Illus. 97 plus 80 pp. New York, 1884. (Description of fish, their haunts, natural history, etc., but meager.)

Warden, D. B. A Statistical, Political and Historical Account of the United States of North America, etc. Edinburgh, 1819. 3 vols.

Weeden, William B. Economic and Social History of New England. 1620–1789. 2 vols. Boston, 1890.

Wells, Henry P. American Salmon Fishing. Illus. 166 pp. New York, 1886.

Whymper, F. The Fisheries of the World. Illus. 376 pp. London, New York. (An illustrated and descriptive record of the International Fisheries Exhibition, at London, 1883.)

Wilcocks, J. C. The Sea-Fisherman; comprising the chief methods of hook and line fishing in British and other seas, etc. Illustrated. 303 pp. 2nd edition, London, 1868.

Winsor, Justin. Narrative and Critical History of the United States, Vols. II–VI, 1889.

Winsor, Justin. From Cartier to Frontenac.

Winsor, Justin. History of the Town of Duxbury.

Winterbotham, W. An Historical, Geographical, Commercial and Philosophical View of the United States. 4 vols. London, 1795. New York, 1796.

(In Vol. II, pp. 116-118, is a short article on the cod fishery.)

Winthrop, John. History of New England, from 1630 to 1649. 2 vols. Boston, 1825.

Woodbury, Charles L. The Relation of the fisheries to the discovery and settlement of North America. An address. 26 pp. Boston, 1880.

Periodical References.

Boston Evening Post, February 18, 1754.

Boston News Letter, January 29, 1730; Dec. 18, 1735.

Boyd, W. The North American Fishery Question. Fortnightly Review, Vol. 45, 1886.

Carryl, Guy W. New England Fisher-Folk. Harper's Magazine. Vol. 105, p. 909.

Codfish and Currency. The Nation, Vol. 63, p. 434. Dec. 10, 1896.

Collins, Capt. Joseph W. The Decadence of the New England Deep-sea Fisheries. Harper's Magazine, Vol. 94, p. 608. March, 1897.

(A statement of the causes of decadence.)

Collins, Capt. J. W. The Outlook of the Fisheries. Century, Vol. 32, p. 959. Oct., 1886.

Copeland, Guild A. The Fishery Industry of the United States. Chatauquan, Vol. 26, p. 387, Jan., 1898.

Dimock, A. W. With the Gloucester Fishermen. Outlook, Vol. 75, p. 43, 1903.

Edmunds, George B. The Fishery Award. North American Review, Vol. 128, p. 1.

Eggleston, Edward. Commerce in the Colonies. Century Magazine, Vol. 28, p. 234. June, 1884. (Pages 234–236 relate to the fisheries.)

Fishing Adventures on the Newfoundland Banks. Harper's, Vol. 22, p. 456, 1861.

Fishery Question, The; its imperial importance. Westminster Review, Vol. 125, p. 393. January and April, 1886.

(A plea in behalf of Canadian fisheries, with statistics.)

Ford, Worthington C. The Fisheries Dispute. Forum, Vol. 2, p. 174. 1886–87.

Geffcken, F. Heinr. North American Fisheries Disputes. Fortnightly Review, Vol. 53, p. 741. May, 1890.

(An anti-American article.)

Gourand, A. H. Our Fish Supply and its Deficiencies. North American Review, Vol. 168, p. 254. February, 1899.

Gourand, A. H. The Herrings' Mysterious Migration. New Scientific Review, Vol. 2, p. 162. October, 1895.

Grenfell, Wilfred T. Among the Deep Sea Fishermen. Outlook, Vol. 74, p. 695. 1903.

Henderson, William J. The Catching of the Cod. Century, Vol. 72, p. 485.

Hunt's, Freeman, Merchants' Magazine and Commercial Review.

(This magazine begins with 1824; 63 vols. to May, 1871, when it was changed from a monthly to a weekly periodical and incorporated with the Commercial and Financial Chronicle. Valuable articles on the history of the fisheries occur in Vol. 26, pp. 20-30, 159-172, 287-299, 416-424. Vol. 29, pp. 147-156, contains a review of Sabine's Report, with criticisms and additional material for the fisheries.)

Huxley, Prof. T. H. The Herring. Nature, Vol. 23, p. 607. April, 1881.

Kenealy, A. J. Fishermen of the Deep Sea. Outing, Vol. 40, p. 733. 1902.

Leggo, William. The Fishery Question—A Canadian View. Scottish Review, Vol. 8, p. 309. October, 1886.

(This article gives some account of Canadian fisheries, with statistics.)

Lodge, Henry C. The Fisheries Question. N. A. Review, Vol. 375, p. 121, Feb. 1888.

Lorne, Marquis of. The Canadian Fishery Dispute. Fortnightly Review, March, 1887. Eclectic, May, 1887.

Marsh, A. H. Canadian Fisheries Question. Am. Law Review, May–June, 1887, Vol. 21, pp. 298 and 369.

Mather, Fred. Progress in Fish Culture. Century, Vol. 5, p. 900. April, 1884.

(An illustrated article showing the earlier methods of fish culture.)

McFarland, Raymond. The Newfoundland View, Boston Transcript, June 23, 1910.

McGrath, P. T. New England's Deep Sea Fishing Interests. Am. Review of Reviews, Vol. 33, p. 561. 1906.

McGrath, P. T. The Atlantic Fisheries Question. Atlantic Monthly, Vol. 90, p. 741. Nov., 1902.

(A Newfoundland point of view of the fisheries question by a St. John's editor and publisher.)

McGrath, P. T. The Atlantic Fisheries Question. N. A. Review, Vol. 167, p. 729. Dec., 1898. (Shows what the effect of destroying the existing relations would have upon the fisheries of New England, Canada, and Newfoundland.)

Milner, James W. The Shad and the Alewife. Harper's, Vol. 60, p. 845. May, 1880.

Moore, Frederick. Work of the Fisheries Commission. Scientific American, Vol. 89, p. 290. October 24, 1903. Illustrated.

North, Franklin H. Gloucester Fisheries. Century, Vol. 32, p. 815. October, 1886.

Oxley, J. Macdonald. History of the Fisheries Question. Mag. of Am. History, Vol. 16, p. 50. July, 1886.

Palmer, Dexter F. (Mechanic of Massachusetts.) Dominion of the Seas and the Fisheries. Hunt, Vol. 29, pp. 276–291, 420–436.

Pomeroy, Prof. Job N. The Northeast Fishery. Am. Law Review, Vol. 5, p. 389. April, 1871.
(A judicial criticism of the fishery question.)

Prouse, D. W. The Atlantic Fishery and Protection. Nation, Vol. 75, p. 111. 1902. (Written from Newfoundland.)

Prouse, D. W. Reciprocity and the British North American Fisheries. Nation, Vol. 72, p. 29. 1901.

Rich, George A. The Fisheries in American History. Chautauquan, Vol. 17, p. 161. May, 1893.

Sheffield, Hon. W. P. Canadian Fisheries. New England Magazine, Vol. 6, p. 329. August, 1888. (A discussion of the Commission and the proposed treaty, from the British point of view.)

Snow, Freeman. A Review of the Fisheries Question. Forum, Vol. 4, p. 343. December, 1887.

Stevenson, Charles H. The United States Fish Commission. N. A. Review, Vol. 176, p. 593. (Short, historical sketch, showing the methods of the Commission.) 1903.

Stevenson, Charles H. Fertilizers from Fish Waste or Refuse. Scientific Am. Supplement, Vol. 58, pp. 23, 875. July 23, 1904. (Historical and Statistical; from U. S. Fish Commission Report, 1902.)

Stevenson, Charles H. The Menhaden Industry. Scientific Am. Supplement, Vol. 57, pp. 23, 687. April 30, 1904. (Illustrated; a good treatment of the subject. From U. S. Fish Commission Report, 1902.)

Walsh, Robert F. Conservation of the Mackerel Supply. Popular Science Monthly, Vol. 42, p. 821. April, 1893.

Walsh, Robert F. Economic Uses of Non-edible Fish. Popular Science Monthly, Vol. 45, p. 61. May, 1894. (An interesting account of the Menhaden industry. Illustrated.)

Woodbury, Charles L. The Canadian Fisheries Dispute—An Open Letter to Senator Morgan. Am. Law Review, Vol. 21, p. 431. May, 1887.

Woolsey, Theodore S. The Fishery Question. N. A. Review, Vol. 142, p. 219. March, 1886.

APPENDIX

A.

"A CALCULATION OF THE STATE OF THE COD AND WHALE FISHERY, BELONGING TO MASSACHUSETTS IN 1763: COPIED FROM A PAPER PUBLISHED IN 1764.

300 vessels in the cod fishery caught 102,265 quintals of merchantable fish, at 12s..........................	£ 61,359.00
and 137,794 quintals of West India fish, at 9s........	62,007.06
90 mackerel vessels, at 200 barrels each, are 18,000 barrels at 18s..............................	16,200.00
Shad alewives and other pickled fish, 10,000 bbls. at 10s.	5,000.00
12 bbls. of oil to each cod fishing vessel are 1,600, at 30s.	5,400.00
15,000 hhds. for packing West India fish, at 6s........	4,500.00
West India fish from Nova Scotia, and Newfoundland in return for provisions, rum, sugar and molasses....	10,000.00
180 sail of whale fishing vessels, the exportations to Great Britain amounting, in oil and bone, to......	75,000.00
To the West Indies and Continent in do..............	3,500.00
Total	£242,966.06"

Mass. Hist. Soc. Coll., VIII, pp. 202–203.

B.

STATISTICS OF THE COD-FISHERY OF MASSACHUSETTS FROM THE YEAR 1765 TO 1775, AND FROM 1786 TO 1790.[1]

Towns.	From 1765 to 1775.			From 1786 to 1790.		
	Vessels an. employed	Tonnage	No. of men	Vessels an. employed	Tonnage	No. of men
Marblehead	150	7,500	1,200	90	5,400	720
Gloucester	146	5,530	888	160	3,600	680
Manchester	25	1,500	200	15	900	120
Beverly	15	750	120	19	1,235	157
Salem	30	1,500	240	20	1,300	160
Newburyport	10	400	60	10	460	80
Ipswich	50	900	190	56	860	248
Plymouth	60	2,400	420	36	1,440	252
Cohasset	6	240	42	5	200	35
Hingham	6	240	42	4	180	32
Scituate	10	400	70	2	90	16
Duxbury	4	160	28	9	360	72
Kingston	6	240	42	4	160	28
Yarmouth	30	900	180	30	900	180
Wellfleet	3	90	21	...	...	...
Truro	10	400	80	...	...	...
Provincetown	4	160	32	11	550	88
Chatham	30	900	240	30	900	240
Nantucket	8	320	64	5	200	40
Weymouth	2	100	16	3	150	24
In Maine	60	1,000	230	30	300	120
Total	665	25,630	4,405	539	19,185	3,292

[1] Sabine, Report of the Principal Fisheries of the Am. Seas, p. 174.

C.

STATISTICS OF THE BANK, BAY, AND LABRADOR COD-FISHERIES OF NEW ENGLAND, 1790–1810.[1]

Vessels employed in the Bank, Bay, and Labrador fisheries.	1,232
Tonnage	85,140
Number of men	10,459
Number of hogsheads of salt consumed	178,370
Number of quintals of fish taken	1,158,700
Number of barrels of oil made	37,520

RECAPITULATION OF THE COD AND MACKEREL FISHERIES OF NEW ENGLAND, 1790–1800.[1]

Vessels	2,332
Tonnage	115,940
Men	15,059
Salt, hogsheads	265,370
Fish, quintals	1,353,700
Oil, barrels	50,520
Mackerel, barrels	50,000

[1] Adams, Duplicate Letters, quoting a letter from a gentleman of Boston.

D.

QUANTITY AND VALUE OF THE PRINCIPAL PRODUCTS TAKEN IN THE FISHERIES OF THE NEW ENGLAND STATES IN 1905.[1]

PRODUCTS.	Pounds.	Value.
Oyster, market and seed	33,220,194	$3,961,785
Cod, salted and fresh	75,065,224	1,994,572
Lobster	11,524,499	1,319,107
Haddock, salted and fresh	76,617,136	1,240,812
Mackerel, salted and fresh	15,981,034	1,084,781
Clam, hard and razor, salted and fresh	7,984,614	744,581
Herring, salted, fresh and smoked	83,390,554	692,854
Hake, salted and fresh	35,470,667	407,307
Pollock, salted and fresh	28,949,359	303,687
Halibut, salted and fresh	3,715,776	237,876
Swordfish	3,296,369	204,637
Squeteague	8,609,180	200,890
Scup	6,581,351	166,082
Flounders, salted and fresh	5,761,473	151,756
Scallop	19,875,112	151,032
Cusk, salted and fresh	9,079,866	139,954
Alewives, salted, fresh and smoked	2,743,234	92,957
Eels	1,266,287	77,151
Smelts	618,705	67,851
Whiting, salted and fresh	4,812,840	25,563
Sea bass	438,761	24,239
Bluefish	7,323,592	23,667
Butterfish	451,322	23,523
Salmon	86,332	20,143
Other products	....	218,065
Total	....	$13,744,603

[1] Not including whale products, $439,602. 439,602

Total......... $14,184,205

E.

NEW ENGLAND CATCH OF SALT MACKEREL FOR THIRTY-TWO YEARS, COMPARING THE NORTH BAY WITH THE SHORE CATCH.

AS COMPILED BY THE BOSTON FISH BUREAU.

	SHORE. Bbls.	BAY. Bbls.	TOTAL. Bbls.
1878	134,545	61,923	196,468
1879	209,803	10,796	220,599
1880	342,373	7,301	349,674
1881	291,187	470	291,657
1882	378,863	....	378,863
1883	198,019	28,666	226,685
1884	458,439	19,637	478,076
1885	302,271	27,672	329,943
1886	16,915	63,083	79,998
1887	70,814	17,568	88,382
1888	37,742	*10,463	48,205
1889	15,889	6,029	21,918
1890	15,934	3,108	19,042
1891	46,357	1,459	47,816
1892	49,189	2,179	51,368
1893	52,942	2,695	55,637
1894	40,483	5,838	46,321
1895	19,008	5,931	24,939
1896	74,154	3,310	77,464
1897	9,177	*3,977	13,154
1898	9,758	*4,528	14,286
1899	22,078	*1,390	23,468
1900	85,466	*2,501	87,967
1901	65,583	†1,808	67,391
1902	39,160	†6,374	45,534
1903	32,977	*11,415	44,392
1904	19,920	*9,053	28,973
1905	13,653	*15,648	29,301
1906	5,762	*4,376	10,138
1907	19,612	*11,784	31,396
1908	4,686	*16,581	21,267
1909	6,002	*11,540	17,542
Total bbls	3,088,761	379,103	3,467,864

* Includes catch on Cape Shore.

† All caught on Cape Shore.

F.

THE WORLD'S CATCH OF SALT MACKEREL, BARRELS, 1878 TO 1909, INCLUSIVE.

AS COMPILED BY THE BOSTON FISH BUREAU.

Year.	United States.	Canada.	Great Britain.	Norway and Sweden.	Total.
1878	196,468	183,919	....	....	380,387
1879	220,599	191,448	....	....	412,047
1880	349,674	233,669	....	....	583,343
1881	291,657	105,722	....	....	397,379
1882	378,863	110,352	....	....	489,215
1883	226,685	124,093	....	....	350,778
1884	478,076	180,170	....	....	658,246
1885	329,943	148,429	....	....	478,372
1886	79,998	147,962	....	....	227,960
1887	88,382	129,610	....	10,000	227,992
1888	48,205	62,756	15,000	10,000	135,961
1889	21,918	62,237	22,993	10,000	117,148
1890	19,042	96,246	28,390	10,000	153,678
1891	47,816	139,261	8,762	10,000	205,839
1892	51,368	95,044	400	14,000	178,812
1893	55,637	67,912	51,252	20,000	194,801
1894	46,321	53,087	45,133	14,050	158,591
1895	24,939	35,554	39,610	5,726	105,829
1896	77,464	37,765	75,375	10,257	200,861
1897	13,154	19,220	48,352	9,784	90,510
1898	14,286	24,913	54,261	8,785	102,245
1899	23,468	21,145	84,751	16,310	145,674
1900	87,967	70,436	16,421	18,857	193,681
1901	67,391	68,649	25,240	26,664	188,244
1902	45,534	34,742	35,713	12,889	128,908
1903	44,392	64,799	64,646	19,612	193,459
1904	28,973	27,320	67,781	28,717	152,791
1905	29,301	40,409	81,367	34,017	185,094
1906	10,138	52,075	42,604	28,999	133,816
1907	31,396	34,962	38,643	25,445	130,446
1908	21,267	66,314	*63,862	42,999	194,442
1909	17,542	*20,000	*50,000	39,651	127,193

* Estimated.

G.

NEW ENGLAND CATCH OF CODFISH AND OTHER GROUND FISH, COMPARING THE GEORGE'S AND NEW ENGLAND SHORE CATCH WITH THE GRAND AND WESTERN BANK FOR THIRTY YEARS.

COMPILED BY THE BOSTON FISH BUREAU.

	N. E. Shore and George's. Qtls.	Grand and Western Banks. Qtls.	Total Qtls.
1880	335,236	300,990	636,226
1881	419,387	355,640	775,027
1882	424,826	474,078	898,904
1883	482,963	578,735	1,061,698
1884	507,570	493,733	1,001,303
1885	528,027	374,428	902,455
1886	346,302	482,270	828,572
1887	306,427	370,296	676,723
1888	214,443	371,138	585,581
1889	192,594	306,395	498,989
1890	229,750	206,900	436,650
1891	344,548	223,165	567,713
1892	227,289	243,784	471,073
1893	232,803	185,337	418,140
1894	339,670	172,145	511,815
1895	274,510	210,469	484,979
1896	188,040	154,720	342,760
1897	239,476	119,003	358,479
1898	350,599	169,733	520,332
1899	396,363	226,517	622,880
1900	271,599	184,909	456,508
1901	255,799	166,700	422,497
1902	277,940	195,100	473,040
1903	243,137	141,231	384,368
1904	265,660	174,485	440,145
1905	299,435	140,040	439,475
1906	283,170	142,465	425,635
1907	315,487	153,366	468,853
1908	253,143	159,403	412,546
1909	195,608	232,100	427,708
Grand Totals	8,958,629	7,566,801	16,525,430

H.

STATEMENT, BY MONTHS, OF QUANTITIES AND VALUES, OF CERTAIN FISHERY PRODUCTS LANDED AT BOSTON AND GLOUCESTER BY AMERICAN FISHING VESSELS, FOR THE YEAR 1907.

MONTHS.	No. of Trips.	TOTAL. Fresh. Pounds.	TOTAL. Fresh. Value.	TOTAL. Salted. Pounds.	TOTAL. Salted. Value.	GRAND TOTAL. Pounds.	GRAND TOTAL. Value.
January	377	6,647,500	$196,601	....	....	6,647,500	$196,601
February	263	6,075,800	208,667	....	....	6,075,800	208,667
March	422	6,893,700	218,878	....	....	6,893,700	218,878
April	389	6,490,000	149,729	....	....	6,490,000	149,729
May	353	6,316,370	158,491	....	....	6,316,370	158,491
June	253	6,089,200	186,216	93,000	$5,415	6,182,200	191,631
July	404	8,728,121	326,871	197,600	13,027	8,925,721	339,898
August	408	9,594,475	306,300	68,400	5,495	9,662,875	311,795
September	364	8,549,540	242,063	35,000	3,350	8,584,540	245,413
October	488	9,798,730	281,033	....	....	9,798,730	281,033
November	302	7,112,900	184,450	....	....	7,112,900	184,450
December	360	5,421,000	184,727	....	....	5,421,000	184,727
Total landed at Boston	4,383	87,717,836	2,644,026	394,000	27,287	88,111.336	2,671,313
January	89	5,221,936	141,359	6,779,313	129,629	12,001,249	270,988
February	51	1,110,442	48,916	122,294	5,728	1,232,736	54,644
March	119	2,832,013	77,151	112,930	5,135	2,944,943	82,286
April	195	3,554,920	83,598	474,550	19,965	4,029,470	103,563
May	322	5,343,906	95,624	1,323,268	47,626	6,667,174	143,250
June	330	7,595,085	142,012	2,729,476	109,162	10,324,561	251,174
July	314	9,948,786	153,965	5,460,076	267,652	15,408,862	421,617
August	237	6,546,506	119,609	3,660,345	224,026	10,206,851	343,635
September	243	6,517,126	128,558	3,958,897	196,654	10,476,023	325,212
October	343	6,158,231	81,186	4,170,955	153,423	10,329,186	234,609
November	397	8,297,635	92,367	4,444,459	130,495	12,742,094	222,862
December	62	930,845	34,289	6,166,422	103,661	7,097,267	137,950
Total landed at Gloucester	2,702	64,057,431	1,198,634	39,402,985	1,393,156	103,460,416	2,591,790
GRAND TOTAL	7,085	151,774,767	3,842,660	39,796,985	1,420,443	191,571,752	5,263,103
Grounds E. of 66° W. Long.	535	25,973,613	728,726	28,889,530	855,089	54,863,143	1,583,815
Grounds W. of 66° W. Long.	6,550	125,801,154	3,113,934	10,907,455	565,354	136,708,609	3,679,288
Landed at Boston in 1906	4,505	89,610,170	2,107,665	83,200	9,659	89,693,370	2,117,324
Landed at Gloucester in 1906	2,401	46,907,324	887,957	33,800,516	1,067,081	80,707,840	1,955,038

(Statistical Bulletin, No. 208, Bureau of Fisheries.)

I

AWARD OF THE TRIBUNAL OF ARBITRATION, SEPTEMBER 7, 1910, UPON THE NORTH ATLANTIC COAST FISHERIES QUESTION.

PERMANENT COURT OF ARBITRATION

at the Hague

THE NORTH ATLANTIC COAST FISHERIES

PREAMBLE

Whereas a Special Agreement between the United States of America and Great Britain, signed at Washington the 27th January, 1909, and confirmed by interchange of Notes dated the 4th March, 1909, was concluded in conformity with the provisions of the General Arbitration Treaty between the United States of America and Great Britain, signed the 4th April, 1908, and ratified the 4th June, 1908;

And whereas the said Special Agreement for the submission of questions relating to fisheries on the North Atlantic Coast under the general treaty of arbitration concluded between the United States and Great Britain on the 4th day of April, 1908, is as follows:

Article I.

Whereas by Article I of the Convention signed at London on the 20th day of October, 1818, between Great Britain and the United States, it was agreed as follows:—

Whereas differences have arisen respecting the liberty claimed by the United States for the Inhabitants thereof, to take, dry and cure Fish on Certain Coasts, Bays, Harbours and Creeks of His Britannic Majesty's Dominions in America, it is agreed between the High Contracting Parties, that the Inhabitants of the said United States shall have forever, in common with the Subjects of His Britannic Majesty, the Liberty to take Fish of every kind on that part of the Southern Coast of Newfoundland which extends from Cape Ray to the Rameau Islands, on the Western and Northern Coast of Newfoundland, from the said Cape Ray to the Quirpon Islands, on the shores of the Magdalen Islands, and also on the Coasts, Bays, Harbours, and Creeks from Mount Joly on the Southern Coast of Labrador, to and through the Straits of Belleisle and thence Northwardly indefinitely along the Coast, without prejudice, however, to any of the exclusive Rights of the Hudson Bay Company; and that the American Fishermen shall also have liberty forever, to dry and cure Fish in any of the unsettled Bays, Harbours and Creeks of the Southern part of the Coast of Newfoundland hereabove described, and of the Coast of Labrador; but so soon as the same, or any Portion thereof, shall be settled, it shall not be lawful for the said Fishermen to dry or cure Fish at such Portion so settled, without previous agreement for such purpose with the Inhabitants, Proprietors, or Possessors of the ground.—And the United States hereby renounce forever, any Liberty heretofore enjoyed or claimed by the Inhabitants thereof, to take, dry, or cure Fish on, or within three marine Miles of any of the Coasts, Bays, Creeks, or Harbours of His Britannic Majesty's Dominions in America not included within the above-mentioned limits; provided, however, that the American Fishermen shall be admitted to enter such Bays or Harbours for the purpose of Shelter and of repairing Damages therein, of purchasing Wood, and of obtaining Water, and for no other purpose whatever. But they shall be under such Restrictions as may be necessary to prevent their taking, drying or curing Fish therein, or in any other manner whatever abusing the Privileges hereby reserved to them.

And, whereas, differences have arisen as to the scope and meaning of the said Article, and of the liberties therein referred to, and otherwise in respect of the rights and liberties which the inhabitants of the United States have or claim to have in the waters or on the shores therein referred to:

It is agreed that the following questions shall be submitted for decision to a tribunal of arbitration constituted as hereinafter provided:—

Question 1.—To what extent are the following contentions or either of them justified?

It is contended on the part of Great Britain that the exercise of the liberty to take fish referred to in the said Article, which the inhabitants of the United States have forever in common with the subjects of His Britannic Majesty, is subject, without the consent of the United States, to reasonable regulation by Great Britain, Canada, or Newfoundland in the form of municipal laws, ordinances, or rules, as, for example, to regulations in respect of (1) the hours, days, or seasons when fish may be taken on the treaty coasts; (2) the method, means, and implements to be used in the taking of fish or in the carrying on of fishing operations on such coasts; (3) any other matters of a similar character relating to fishing; such regulations being reasonable, as being, for instance—

(*a*) Appropriate or necessary for the protection and preservation of such fisheries and the exercise of the rights of British subjects therein and of the liberty which by the said Article I the inhabitants of the United States have therein in common with British subjects;

(*b*) Desirable on grounds of public order and morals;

(*c*) Equitable and fair as between local fishermen and the inhabitants of the United States exercising the said treaty liberty and not so framed as to give unfairly an advantage to the former over the latter class.

It is contended on the part of the United States that the exercise of such liberty is not subject to limitations or restraints by Great Britain, Canada, or Newfoundland in the form of municipal laws, ordinances, or regulations in respect of (1) the hours, days, or seasons when the inhabitants of the United States may take fish on the treaty coasts, or (2) the method, means, and implements used by them in taking fish or in carrying on fishing operations on such coasts, or (3) any other limitations or restraints of similar character—

(*a*) Unless they are appropriate and necessary for the protection and preservation of the common rights in such fisheries and the exercise thereof; and

(*b*) Unless they are reasonable in themselves and fair as between local fishermen and fishermen coming from the United States, and not so framed as to give an advantage to the former over the latter class; and

(*c*) Unless their appropriateness, necessity, reasonableness, and fairness be determined by the United States and Great Britain by common accord and the United States concurs in their enforcement.

Question 2. Have the inhabitants of the United States, while exercising the liberties referred to in said Article, a right to employ as members of the fishing crews of their vessels persons not inhabitants of the United States?

Question 3. Can the exercise by the inhabitants of the United States of the liberties referred to in the said Article be subjected, without the consent of the United States, to the requirements of entry or report at custom-houses or the payment of light or harbour or other dues, or to any other similar requirement or condition or exaction?

Question 4. Under the provision of the said Article that the American fishermen shall be admitted to enter certain bays or harbours for shelter, repairs, wood, or water, and for no other purpose whatever, but that they shall be under such restrictions as may be necessary to prevent their taking, drying, or curing fish therein or in any other manner whatever abusing the privileges thereby reserved to them, is it permissible to impose restrictions making the exercise of such privileges conditional upon the payment of light or harbour or other dues, or entering or reporting at custom-houses or any similar conditions?

Question 5. From where must be measured the "three marine miles of any of the coasts, bays, creeks, or harbours" referred to in the said Article?

Question 6. Have the inhabitants of the United States the liberty under the said Article or otherwise to take fish in the bays, harbours, and creeks on that part of the southern coast of Newfoundland which extends from Cape Ray to Rameau Islands, or on the western and northern coasts of Newfoundland from Cape Ray to Quirpon Islands, or on the Magdalen Islands?

Question 7.—Are the inhabitants of the United States whose vessels resort to the treaty coasts for the purpose of exercising the liberties referred to in Article I of the treaty of 1818 entitled to have for those vessels, when duly authorized by the United States in that behalf, the commercial privileges on the treaty coasts accorded by agreement or otherwise to United States trading-vessels generally?

Article II.

Either party may call the attention of the Tribunal to any legislative or executive act of the other Party, specified within three months of the exchange of notes enforcing this agreement, and which is claimed to be inconsistent with the true interpretation of the Treaty of 1818; and may call upon the Tribunal to express in its award its opinion upon such acts, and to point out in what respects, if any, they are inconsistent with the principles laid down in the award in reply to the preceding questions; and each Party agrees to conform to such opinion.

Article III.

If any question arises in the arbitration regarding the reasonableness of any regulation or otherwise which requires an examination of the practical effect of any provisions in relation to the conditions surrounding the exercise of the liberty of fishery enjoyed by the inhabitants of the United States, or which requires expert information about the fisheries themselves, the Tribunal may, in that case, refer such question to a Commission of three expert specialists in such matters; one to be designated by each of the Parties hereto, and the third, who shall not be a national of either Party, to be designated by the Tribunal. This Commission shall examine into and report their conclusions on any question or questions so referred to it by the Tribunal and such report shall be considered by the Tribunal and shall, if incorporated by them in the award, be accepted as a part thereof.

Pending the report of the Commission upon the question or questions so referred and without awaiting such report, the Tribunal may make a separate award upon all or any other questions before it, and such separate award, if made, shall become immediately effective, provided that the report aforesaid shall not be incorporated in the award until it has been considered by the Tribunal. The expenses of such Commission shall be borne in equal moieties by the Parties hereto.

Article IV.

The Tribunal shall recommend for the consideration of the High Contracting Parties rules and a method of procedure under which all questions which may arise in the future regarding the exercise of the liberties above referred to may be determined in accordance with the principles laid down in the award. If the High Contracting Parties shall not adopt the rules and method of procedure so recommended, or if they shall not, subsequently to the delivery of the award, agree upon such rules and methods, then any differences which may arise in the future between the High Contracting Parties relating to the interpretation of the Treaty of 1818 or to the effect and application of the award of the Tribunal shall be referred informally to the Permanent Court at The Hague for decision by the summary procedure provided in Chapter IV of The Hague Convention of the 18th October, 1907.

Article V.

The Tribunal of Arbitration provided for herein shall be chosen from the general list of members of the Permanent Court at The Hague, in accordance with the provisions of Article XLV of the Convention for the Settlement of International Disputes, concluded at the Second Peace Conference at The Hague on the 18th of October, 1907. The provisions of said Convention, so far as applicable and not inconsistent herewith, and excepting Articles LIII and LIV, shall govern the proceedings under the submission herein provided for.

The time allowed for the direct agreement of His Britannic Majesty and the President of the United States on the composition of such Tribunal shall be three months.

Article VI.

The pleadings shall be communicated in the order and within the time following:—

As soon as may be and within a period not exceeding seven months from the date of the exchange of notes making this agreement binding the printed case of each of the Parties hereto, accompanied by printed copies of the documents, the official correspondence, and all other evidence on which each Party relies, shall be delivered in duplicate (with such additional copies as may be agreed upon) to the agent of the other Party. It shall be sufficient for this purpose if such case is delivered at the British Embassy at Washington or at the American Embassy at London, as the case may be, for transmission to the agent for its Government.

Within fifteen days thereafter such printed case and accompanying evidence of each of the Parties shall be delivered in duplicate to each member of the Tribunal, and such delivery may be made by depositing within the stated period the necessary number of copies with the International Bureau at The Hague for transmission to the Arbitrators.

After the delivery on both sides of such printed case, either Party may, in like manner, and within four months after the expiration of the period above fixed for the delivery to the agents of the case, deliver to the agent of the other Party (with such additional copies as may be agreed upon), a printed counter-case accompanied by printed copies of additional documents, correspondence, and other evidence in reply to the case, documents, correspondence, and other evidence so presented by the other Party, and within fifteen days thereafter such Party shall, in like manner as above provided, deliver in duplicate such counter-case and accompanying evidence to each of the Arbitrators.

The foregoing provisions shall not prevent the Tribunal from permitting either Party to rely at the hearing upon documentary or other evidence which is shown to have become open to its investigation or examination or available for use too late to be submitted within the period hereinabove fixed for the delivery of copies of evidence, but in case any such evidence is to be presented, printed copies of it, as soon as possible after it is secured, must be delivered, in like manner as provided for the delivery of copies of other evidence, to each of the Arbitrators and to the agent of the other Party. The admission of any such additional evidence, however, shall be subject to such conditions as the Tribunal may impose, and the other Party shall have a reasonable opportunity to offer additional evidence in rebuttal.

The Tribunal shall take into consideration all evidence which is offered by either Party.

Article VII.

If in the case or counter-case (exclusive of the accompanying evidence) either Party shall have specified or referred to any documents, correspondence, or other evidence in its own exclusive possession without annexing a copy, such Party shall be bound, if the other Party shall demand it within thirty days after the delivery of the case or counter-case respectively, to furnish to the Party applying for it a copy thereof; and either Party may, within the like time, demand that the other shall furnish certified copies or produce for inspection the originals of any documentary evidence adduced by the Party upon whom the demand is made. It shall be the duty of the Party upon whom any such demand is made to comply with it as soon as may be, and within a period not exceeding fifteen days after the demand has been received. The production for inspection or the furnishing to the other Party of official governmental publications, publishing, as authentic, copies of the documentary evidence referred to, shall be a sufficient compliance with such demand, if such governmental publications shall have been published prior to the 1st day of January, 1908. If the demand is not complied with, the reasons for the failure to comply must be stated to the Tribunal.

Article VIII.

The Tribunal shall meet within six months after the expiration of the period above fixed for the delivery to the agents of the case, and upon the assembling of the Tribunal at its first session each Party, through its agent or counsel, shall deliver in duplicate to each of the Arbitrators and to the agent and counsel of the other Party (with such additional copies as may be agreed upon) a printed argument showing the points and referring to the evidence upon which it relies.

The time fixed by this Agreement for the delivery of the case, counter-case, or argument, and for the meeting of the Tribunal, may be extended by mutual consent of the Parties.

Article IX.

The decision of the Tribunal shall, if possible, be made within two months from the close of the arguments on both sides, unless on the request of the Tribunal the Parties shall agree to extend the period.

It shall be made in writing, and dated and signed by

each member of the Tribunal, and shall be accompanied by a statement of reasons.

A member who may dissent from the decision may record his dissent when signing.

The language to be used throughout the proceedings shall be English.

Article X.

Each Party reserves to itself the right to demand a revision of the award. Such demand shall contain a statement of the grounds on which it is made and shall be made within five days of the promulgation of the award, and shall be heard by the Tribunal within ten days thereafter. The Party making the demand shall serve a copy of the same on the opposite Party, and both Parties shall be heard in argument by the Tribunal on said demand. The demand can only be made on the discovery of some new fact or circumstance calculated to exercise a decisive influence upon the award and which was unknown to the Tribunal and to the Party demanding the revision at the time the discussion was closed, or upon the ground that the said award does not fully and sufficiently, within the meaning of this Agreement, determine any question or questions submitted. If the Tribunal shall allow the demand for a revision, it shall afford such opportunity for further hearings and arguments as it shall deem necessary.

Article XI.

The present Agreement shall be deemed to be binding only when confirmed by the two Governments by an exchange of notes.

In witness whereof this Agreement has been signed and sealed by His Britannic Majesty's Ambassador at Washington, the Right Honourable James Bryce, O.M., on behalf of Great Britain, and by the Secretary of State of the United States, ELIHU ROOT, on behalf of the United States.

Done at Washington on the 27th day of January, one thousand nine hundred and nine.

JAMES BRYCE. [seal.]
ELIHU ROOT. [seal.]

And whereas, the parties to the said Agreement have by common accord, in accordance with Article V, constituted as a Tribunal of Arbitration the following Members of the Permanent Court at The Hague: Mr. H. LAMMASCH, Doctor of Law, Professor of the University of Vienna, Aulic Councillor, Member of the Upper House of the Austrian Parliament; His Excellency Jonkheer A. F. DE SAVORNIN LOHMAN, Doctor of Law, Minister of State, Former Minister of the Interior, Member of the Second Chamber of the Netherlands; the Honourable GEORGE GRAY, Doctor of Laws, Judge of the United States Circuit Court of Appeals, former United States Senator; the Right Honourable Sir CHARLES FITZPATRICK, Member of the Privy Council, Doctor of Laws, Chief Justice of Canada; the Honourable LUIS MARIA DRAGO, Doctor of Law, former Minister of Foreign Affairs of the Argentine Republic, Member of the Law Academy of Buenos-Aires;

And whereas, the Agents of the Parties to the said Agreement have duly and in accordance with the terms of the Agreement communicated to this Tribunal their cases, counter-cases, printed arguments and other documents;

And whereas, counsel for the Parties have fully presented to this Tribunal their oral arguments in the sittings held between the first assembling of the Tribunal on 1st June, 1910, to the close of the hearings on 12th August, 1910;

Now, therefore, this Tribunal having carefully considered the said Agreement, cases, counter-cases, printed and oral arguments, and the documents presented by either side, after due deliberation makes the following decisions and awards:

QUESTION I.

To what extent are the following contentions or either of them justified?

It is contended on the part of Great Britain that the exercise of the liberty to take fish referred to in the said Article, which the inhabitants of the United States have forever in common with the subjects of His Britannic Majesty, is subject, without the consent of the United States to reasonable regulation by Great Britain, Canada, or Newfoundland in the form of municipal laws, ordinances, or rules, as, for example, to regulations in respect of (1) the hours, days, or seasons when fish may be taken on the treaty coasts; (2) the method, means, and implements to be used in the taking of fish or in the carrying on of fishing operations on such coasts; (3) any other matters of a similar character relating to fishing; such regulations being reasonable, as being, for instance—

(*a*) Appropriate or necessary for the protection and preservation of such fisheries and the exercise of the rights of British subjects therein and of the liberty which by the said Article I the inhabitants of the United States have therein in common with British subjects;

(*b*) Desirable on grounds of public order and morals;

(*c*) Equitable and fair as between local fishermen and the inhabitants of the United States exercising the said treaty liberty, and not so framed as to give unfairly an advantage to the former over the latter class.

It is contended on the part of the United States that the exercise of such liberty is not subject to limitations or restraints by Great Britain, Canada, or Newfoundland in the form of municipal laws, ordinances, or regulations in respect of (1) the hours, days, or seasons when the inhabitants of the United States may take fish on the treaty coasts, or (2) the method, means, and implements used by them in taking fish or in carrying on fishing operations on such coasts, or (3) any other limitations or restraints of similar character—

(*a*) Unless they are appropriate and necessary for the protection and preservation of the common rights in such fisheries and the exercise thereof; and

(*b*) Unless they are reasonable in themselves and fair as between local fishermen and fishermen coming from the United States, and not so framed as to give an advantage to the former over the latter class; and

(*c*) Unless their appropriateness, necessity, reasonableness, and fairness be determined by the United States and Great Britain by common accord and the United States concurs in their enforcement.

Question I, thus submitted to the Tribunal, resolves itself into two main contentions:

1st. Whether the right of regulating reasonably the liberties conferred by the Treaty of 1818 resides in Great Britain;

2nd. And, if such right does so exist, whether such reasonable exercise of the right is permitted to Great Britain without the accord and concurrence of the United States.

The Treaty of 1818 contains no explicit disposition in regard to the right of regulation, reasonable or otherwise; it neither reserves that right in express terms, nor refers to it in any way. It is therefore incumbent on this Tribunal to answer the two questions above indicated by interpreting the general terms of Article I of the Treaty, and more especially the words "the inhabitants of the United States shall have, forever, in common with the subjects of His Britannic Majesty, the liberty to take fish of every kind." This interpretation must be conformable to the general import of the instrument, the general intention of the parties to it, the subject matter of the contract, the expressions actually used and the evidence submitted.

Now in regard to the preliminary question as to whether the right of reasonable regulation resides in Great Britain.

Considering that the right to regulate the liberties conferred by the Treaty of 1818 is an attribute of sovereignty, and as such must be held to reside in the territorial sovereign, unless the contrary be provided; and considering that one of the essential elements of sovereignty is that it is to be exercised within territorial limits, and that, failing proof to the contrary, the territory is coterminous with the Sovereignty, it follows that the burden of the assertion involved in the contention of the United States (viz. that the right to regulate does not reside independently in Great Britain, the territorial Sovereign) must fall on the United States. And for the purpose of sustaining this burden, the United States have put forward the following series of propositions, each one of which must be singly considered.

It is contended by the United States:

(1) That the French right of fishery under the treaty of 1713 designated also as a liberty, was never subjected to regulation by Great Britain, and therefore the inference is warranted that the American liberties of fishery are similarly exempted.

The Tribunal is unable to agree with this contention:

(*a*) Because although the French right designated in 1713 merely "an allowance," (a term of even less force than that used in regard to the American fishery) was nevertheless converted, in practice, into an exclusive right, this concession on the part of Great Britain was presumably made because France, before 1713, claimed to be the sovereign of Newfoundland, and, in ceding the Island, had, as the American argument says, "reserved for the benefit of its subjects the right to fish and to use the strand";

(*b*) Because the distinction between the French and American right is indicated by the different wording of the Statutes for the observance of Treaty obligations towards France and the United States, and by the British Declaration of 1783;

(*c*) And, also, because this distinction is maintained in the Treaty with France of 1904, concluded at a date when the American claim was approaching its present stage, and by which certain common rights of regulation are recognized to France.

For the further purpose of such proof it is contended by the United States:

(2) That the liberties of fishery, being accorded to the inhabitants of the United States "forever," acquire, by being in perpetuity and unilateral, a character exempting them from local legislation.

The Tribunal is unable to agree with this contention:

(*a*) Because there is no necessary connection between the duration of a grant and its essential status in its relation to local regulation; a right granted in perpetuity may yet be subject to regulation, or, granted temporarily, may yet be exempted therefrom; or being reciprocal may yet be unregulated, or being unilateral may yet be regulated: as is evidenced by the claim of the United States that the liberties of fishery accorded by the Reciprocity Treaty of 1854 and the Treaty of 1871 were exempt from regulation, though they were neither permanent nor unilateral;

(*b*) Because no peculiar character need be claimed for these liberties in order to secure their enjoyment in perpetuity, as is evidenced by the American negotiators in 1818 asking for the insertion of the words "for ever." International law in its modern development recognizes that a great number of Treaty obligations are not annulled by war, but at most suspended by it;

(*c*) Because the liberty to dry and cure is, pursuant to the terms of the Treaty, provisional and not permanent, and is nevertheless, in respect of the liability to regulation, identical in its nature with, and never distinguished from, the liberty to fish.

For the further purpose of such proof, the United States allege:

(3) That the liberties of fishery granted to the United States constitute an International servitude in their favour over the territory of Great Britain, thereby involving a derogation from the sovereignty of Great Britain, the servient State, and that therefore Great Britain is deprived, by reason of the grant, of its independent right to regulate the fishery.

The Tribunal is unable to agree with this contention:

(*a*) Because there is no evidence that the doctrine of International servitudes was one with which either American or British Statesmen were conversant in 1818, no

English publicists employing the term before 1818, and the mention of it in Mr. GALLATIN's report being insufficient;

(*b*) Because a servitude in the French law, referred to by Mr. GALLATIN, can, since the Code, be only real and cannot be personal (Code Civil, art. 686);

(*c*) Because a servitude in International law predicates an express grant of a sovereign right and involves an analogy to the relation of a *praedium dominans* and a *praedium serviens;* whereas by the Treaty of 1818 one State grants a liberty to fish, which is not a sovereign right, but a purely economic right, to the inhabitants of another State;

(*d*) Because the doctrine of international servitude in the sense which is now sought to be attributed to it originated in the peculiar and now obsolete conditions prevailing in the Holy Roman Empire of which the *domini terrae* were not fully sovereigns; they holding territory under the Roman Empire, subject at least theoretically, and in some respects also practically, to the Courts of that Empire; their right being, moreover, rather of a civil than of a public nature, partaking more of the character of *dominium* than of *imperium*, and therefore certainly not a complete sovereignty. And because in contradistinction to this quasi-sovereignty with its incoherent attributes acquired at various times, by various means, and not impaired in its character by being incomplete in any one respect or by being limited in favour of another territory and its possessor, the modern State, and particularly Great Britain, has never admitted partition of sovereignty, owing to the constitution of a modern State requiring essential sovereignty and independence;

(*e*) Because this doctrine being but little suited to the principle of sovereignty which prevails in States under a system of constitutional government such as Great Britain and the United States, and to the present International relations of Sovereign States, has found little, if any, support from modern publicists. It could therefore in the general interest of the Community of Nations, and of the Parties to this Treaty, be affirmed by this Tribunal only on the express evidence of an International contract;

(*f*) Because even if these liberties of fishery constituted an International servitude, the servitude would derogate from the sovereignty of the servient State only in so far as the exercise of the rights of sovereignty by the servient State would be contrary to the exercise of the servitude right by the dominant State. Whereas it is evident that, though every regulation of the fishery is to some extent a limitation, as it puts limits to the exercise of the fishery at will, yet such regulations as are reasonable and made for the purpose of securing and preserving the fishery and its exercise for the common benefit, are clearly to be distinguished from those restrictions and "molestations," the annulment of which was the purpose of the American demands formulated by Mr. Adams in 1782, and such regulations consequently cannot be held to be inconsistent with a servitude;

(*g*) Because the fishery to which the inhabitants of the United States were admitted in 1783, and again in 1818, was a regulated fishery, as is evidenced by the following regulations:

Act 15 Charles II, Cap. 16, s. 7 (1663) forbidding "to lay any seine or other net in or near any harbour in Newfoundland, whereby to take the spawn or young fry of the Poor-John, or for any other use or uses, except for the taking of bait only," which had not been superseded either by the order in council of March 10th, 1670, or by the statute 10 and XI Wm. III, Cap. 25, 1699. The order in council provides expressly for the obligation "to submit unto and to observe all rules and orders as are now, or hereafter shall be established,'' an obligation which cannot be read as referring only to the rules established by this very act, and having no reference to anteceding rules ''as are now established.'' In a similar way, the statute of 1699 preserves in force prior legislation, conferring the freedom of fishery only ''as fully and freely as at any time heretofore.'' The order in council, 1670, provides that the Admirals, who always were fishermen, arriving from an English or Welsh port, ''see that His Majesty's rules and orders concerning the regulation of the fisheries are duly put in execution'' (sec. 13). Likewise the Act 10 and XI, Wm. III, Cap. 25, (1699) provides that the Admirals do settle differences between the fishermen arising in respect of the places to be assigned to the different vessels. As to Nova Scotia, the proclamation of 1665 ordains that no one shall fish without license; that the licensed fishermen are obliged ''to observe all laws and orders which now are made and published, or shall hereafter be made and published in this jurisdiction,'' and that they shall not fish on the Lord's day and shall not take fish at the time they come to spawn. The judgment of the Chief Justice of Newfoundland, October 26th, 1820, is not held by the Tribunal sufficient to set aside the proclamations referred to. After 1783, the statute 26 Geo. III, Cap. 26, 1786, forbids ''the use, on the shores of Newfoundland, of seines or nets for catching cod by hauling on shore or taking into boat, with meshes less than 4 inches''; a prohibition which cannot be considered as limited to the bank fishery. The act for regulating the fisheries of New Brunswick, 1793, which forbids ''the placing of nets or seines across any cove or creek in the Province so as to obstruct the natural course of fish,'' and which makes specific provision for fishing in the Harbour of St. John, as to the manner and time of fishing, cannot be read as being limited to fishing from the shore. The act for regulating the fishing on the coast of Northumberland (1799) contains very elaborate dispositions concerning the fisheries in the bay of Miramichi which were continued in 1823, 1829 and 1834. The statutes of Lower Canada, 1788 and 1807, forbid the throwing overboard of offal. The fact that these acts extend the prohibition over a greater distance than the first marine league from the shore may make them nonoperative against foreigners without the territorial limits of Great Britain, but is certainly no reason to deny their obligatory character for foreigners within these limits;

(*h*) Because the fact that Great Britain rarely exercised the right of regulation in the period immediately succeeding 1818 is to be explained by various circumstances and is not evidence of the non-existence of the right;

(*i*) Because the words "in common with British subjects" tend to confirm the opinion that the inhabitants of the United States were admitted to a regulated fishery;

(*j*) Because the statute of Great Britain, 1819, which gives legislative sanction to the Treaty of 1818, provides for the making of "regulations with relation to the taking, drying and curing of fish by inhabitants of the United States in 'common.'"

For the purpose of such proof, it is further contended by the United States, in this latter connection:

(4) That the words "in common with British subjects" used in the Treaty should not be held as importing a common subjection to regulation, but as intending to negative a possible pretention on the part of the inhabitants of the United States to liberties of fishery exclusive of the right of British subjects to fish.

The Tribunal is unable to agree with this contention:

(*a*) Because such an interpretation is inconsistent with the historical basis of the American fishing liberty. The ground on which Mr. ADAMS founded the American right in 1782 was that the people then constituting the United States had always, when still under British rule, a part in these fisheries and that they must continue to enjoy their past right in the future. He proposed "that the subjects of His Britannic Majesty and the people of the United States shall continue to enjoy unmolested the right to take fish . . . where the inhabitants of both countries used, at any time heretofore, to fish." The theory of the partition of the fisheries, which by the American negotiators had been advanced with so much force, negatives the assumption that the United States could ever pretend to an exclusive right to fish on the British shores; and to insert a special disposition to that end would have been wholly superfluous;

(*b*) Because the words "in common" occur in the same connexion in the Treaty of 1818 as in the Treaties of 1854 and 1871. It will certainly not be suggested that in these Treaties of 1854 and 1871 the American negotiators meant by inserting the words "in common" to imply that without these words American citizens would be precluded from the right to fish on their own coasts and that, on American shores, British subjects should have an exclusive privilege. It would have been the very opposite of the concept of territorial waters to suppose that, without a special treaty-provision, British subjects could be excluded from fishing in British waters. Therefore that cannot have been the scope and the sense of the words "in common";

(*c*) Because the words "in common" exclude the supposition that American inhabitants were at liberty to act at will for the purpose of taking fish, without any regard to the co-existing rights of other persons entitled to do the same thing; and because these words admit them only as members of a social community, subject to the ordinary duties binding upon the citizens of that community, as to the regulations made for the common benefit; thus avoiding the "bellum omnium contra omnes" which would otherwise arise in the exercise of this industry;

(*d*) Because these words are such as would naturally suggest themselves to the negotiators of 1818 if their intention had been to express a common subjection to regulations as well as a common right.

In the course of the Argument it has also been alleged by the United States:

(5) That the Treaty of 1818 should be held to have entailed a transfer or partition of sovereignty, in that it must in respect to the liberties of fishery be interpreted in its relation to the Treaty of 1783; and that this latter Treaty was an act of partition of sovereignty and of separation, and as such was not annulled by the war of 1812.

Although the Tribunal is not called upon to decide the issue whether the treaty of 1783 was a treaty of partition or not, the questions involved therein having been set at rest by the subsequent Treaty of 1818, nevertheless the Tribunal could not forbear to consider the contention on account of the important bearing the controversy has upon the true interpretation of the Treaty of 1818. In that respect the Tribunal is of opinion:

(*a*) That the right to take fish was accorded as a condition of peace to a foreign people; wherefore the British negotiators refused to place the right of British subjects on the same footing with those of American inhabitants;

and further, refused to insert the words also proposed by Mr. ADAMS—"continued to enjoy"—in the second branch of Art. III of the Treaty of 1783;

(*b*) That the Treaty of 1818 was in different terms, and very different in extent, from that 1783, and was made for different considerations. It was, in other words, a new grant.

For the purpose of such proof it is further contended by the United States:

(6) That as contemporary Commercial Treaties contain express provisions for submitting foreigners to local legislation, and the Treaty of 1818 contains no such provision, it should be held, *a contrario*, that inhabitants of the United States exercising these liberties are exempt from regulation.

The Tribunal is unable to agree with this contention:

(*a*) Because the Commercial Treaties contemplated did not admit foreigners to all and equal rights, seeing that local legislation excluded them from many rights of importance, e.g. that of holding land; and the purport of the provisions in question consequently was to preserve these discriminations. But no such discriminations existing in the common enjoyment of the fishery by American and British fishermen, no such provision was required;

(*b*) Because no proof is furnished of similar exemptions of foreigners from local legislation in default of Treaty stipulations subjecting them thereto;

(*c*) Because no such express provision for subjection of the nationals of either Party to local law was made either in this Treaty, in respect to their reciprocal admission to certain territories as agreed in Art. III, or in Art. III of the Treaty of 1794; although such subjection was clearly contemplated by the Parties.

For the purpose of such proof it is further contended by the United States:

(7) That, as the liberty to dry and cure on the Treaty coasts and to enter bays and harbours on the non-treaty coasts are both subjected to conditions, and the latter to specific restrictions, it should therefore be held that the liberty to fish should be subjected to no restrictions, as none are provided for in the Treaty.

The Tribunal is unable to apply the principle of "*expressio unius exclusio alterius*" to this case:

(*a*) Because the conditions and restrictions as to the liberty to dry and cure on the shore and to enter the harbours are limitations of the rights themselves, and not restrictions of their exercise. Thus the right to dry and cure is limited in duration, and the right to enter bays and harbours is limited to particular purposes;

(*b*) Because these restrictions of the right to enter bays and harbours applying solely to American fishermen must have been expressed in the Treaty, whereas regulations of the fishery, applying equally to American and British, are made by right of territorial sovereignty.

For the purpose of such proof it has been contended by the United States:

(8) That Lord Bathurst in 1815 mentioned the American right under the treaty of 1783 as a right to be exercised "at the discretion of the United States"; and that this should be held as to be derogatory to the claim of exclusive regulation by Great Britain.

But the Tribunal is unable to agree with this contention:

(*a*) Because these words implied only the necessity of an express stipulation for any liberty to use foreign territory at the pleasure of the grantee, without touching any question as to regulation;

(*b*) Because in this same letter Lord BATHURST characterized this right as a policy "temporary and experimental, depending on the use that might be made of it, on the condition of the islands and places where it was to be exercised, and the more general conveniences or inconveniences from a military, naval and commercial point of view"; so that it cannot have been his intention to acknowledge the exclusion of British interference with this right;

(*c*) Because Lord BATHURST in his note to Governor Sir C. HAMILTON in 1819 orders the Governor to take care that the American fishery on the coast of Labrador be carried on *in the same manner* as previous to the late war; showing that he did not interpret the Treaty just signed as a grant conveying absolute immunity from interference with the American fishery right.

For the purpose of such proof it is further contended by the United States:

(9) That on various other occasions following the conclusion of the Treaty, as evidenced by official correspondence, Great Britain made use of expressions inconsistent with the claim to a right of regulation.

The Tribunal, unwilling to invest such expressions with an importance entitling them to affect the general question, considers that such conflicting or inconsistent expressions as have been exposed on either side are sufficiently explained by their relations to ephemeral phases of a controversy of almost secular duration, and should be held to be without direct effect on the principal and present issues.

Now with regard to the second contention involved in Question I, as to whether the right of regulation can be reasonably exercised by Great Britain without the consent of the United States:

Considering that the recognition of a concurrent right of consent in the United States would affect the independence of Great Britain, which would become dependent on the Government of the United States for the exercise of its sovereign right of regulation, and considering that such a co-dominium would be contrary to the constitution of both sovereign States; the burden of proof is imposed on the United States to show that the independence of Great Britain was thus impaired by international contract in 1818 and that a co-dominium was created.

For the purpose of such proof it is contended by the United States:

(10) That a concurrent right to coöperate in the making and enforcement of regulations is the only possible and proper security to their inhabitants for the enjoyment of their liberties as fishery, and that such a right must be held to be implied in the grant of those liberties by the Treaty under interpretation.

The Tribunal is unable to accede to this claim on the ground of a right so implied:

(*a*) Because every State has to execute the obligations incurred by Treaty *bona fide*, and is urged thereto by the ordinary sanctions of International Law in regard to observance of Treaty obligations. Such sanctions are, for instance, appeal to public opinion, publication of correspondence, censure by Parliamentary vote, demand for arbitration with the odium attendant on a refusal to arbitrate, rupture of relations, reprisal, etc. But no reason has been shown why this Treaty, in this respect, should be considered as different from every other Treaty under which the right of a State to regulate the action of foreigners admitted by it on its territory is recognized;

(*b*) Because the exercise of such a right of consent by the United States would predicate an abandonment of its independence in this respect by Great Britain, and the recognition by the latter of a concurrent right of regulation in the United States. But the Treaty conveys only a liberty to take fish in common, and neither directly nor indirectly conveys a joint right of regulation;

(*c*) Because the Treaty does not convey a common right of fishery, but a liberty to fish in common. This is evidenced by the attitude of the United States Government in 1823, with respect to the relations of Great Britain and France in regard to the fishery;

(*d*) Because if the consent of the United States were requisite for the fishery a general veto would be accorded them, the full exercise of which would be socially subversive and would lead to the consequence of an unregulatable fishery;

(*e*) Because the United States cannot by assent give legal force and validity to British legislation;

(*f*) Because the liberties to take fish in British territorial waters and to dry and cure fish on land in British territory are in principle on the same footing; but in practice a right of coöperation in the elaboration and enforcement of regulations in regard to the latter liberty (drying and curing fish on land) is unrealisable.

In any event, Great Britain, as the local sovereign, has the duty of preserving and protecting the fisheries. In so far as it is necessary for that purpose, Great Britain is not only entitled, but obliged, to provide for the protection and preservation of the fisheries; always remembering that the exercise of this right of legislation is limited by the obligation to execute the Treaty in good faith. This has been admitted by counsel and recognized by Great Britain in limiting the right of regulation to that of reasonable regulation. The inherent defect of this limitation of reasonableness, without any sanction except in diplomatic remonstrance, has been supplied by the submission to arbitral award as to existing regulations in accordance with Arts. II and III of the Special Agreement, and as to further regulation by the obligation to submit their reasonableness to an arbitral test in accordance with Art. IV of the Agreement.

It is finally contended by the United States:

That the United States did not expressly agree that the liberty granted to them could be subjected to any restriction that the grantor might choose to impose on the ground that in her judgment such restriction was reasonable. And that while admitting that all laws of a general character, controlling the conduct of men within the territory of Great Britain, are effective, binding and beyond objection by the United States, and competent to be made upon the sole determination of Great Britain or her colony, without accountability to anyone whomsoever; yet there is somewhere a line, beyond which it is not competent for Great Britain to go, or beyond which she cannot rightfully go, because to go beyond it would be an invasion of the right granted to the United States in 1818. That the legal effect of the grant of 1818 was not to leave the determination as to where that line is to be drawn to the uncontrolled judgment of the grantor, either upon the grantor's consideration as to what would be a reasonable exercise of its sovereignty over the British Empire, or upon the grantor's consideration of what would be a reasonable exercise thereof towards the grantee.

But this contention is founded on assumptions, which this Tribunal cannot accept for the following reasons in addition to those already set forth:

(*a*) Because the line by which the respective rights of both Parties accruing out of the Treaty are to be circumscribed, can refer only to the right granted by the Treaty; that is to say to the liberty of taking, drying and curing fish by American inhabitants in certain British waters in common with British subjects, and not to the exercise of rights of legislation by Great Britain not referred to in the Treaty;

(*b*) Because a line which would limit the exercise of sovereignty of a State within the limits of its own territory can be drawn only on the ground of express stipulation, and not by implication from stipulations concerning a different subject-matter;

(*c*) Because the line in question is drawn according to the principle of international law that treaty obligations are to be executed in perfect good faith, therefore excluding the right to legislate *at will* concerning the subject-matter of the Treaty, and limiting the exercise of sovereignty of the States bound by a treaty with respect to that subject-matter to such acts as are consistent with the treaty;

(*d*) Because on a true construction of the Treaty the question does not arise whether the United States agreed that Great Britain should retain the right to legislate with regard to the fisheries in her own territory; but whether the Treaty contains an abdication by Great Britain of the right which Great Britain, as the sovereign power, undoubtedly possessed when the Treaty was made, to regulate those fisheries;

(*e*) Because the right to make reasonable regulations, not inconsistent with the obligations of the Treaty, which is all that is claimed by Great Britain, for a fishery which both Parties admit requires regulation for its preservation, is not a restriction of or an invasion of the liberty granted to the inhabitants of the United States. This grant does not contain words to justify the assumption that the sovereignty of Great Britain upon its own territory was in any way affected; nor can words be found in the treaty transferring any part of that sovereignty to the United States. Great Britain assumed only duties with regard to the exercise of its sovereignty. The sovereignty of Great Britain over the coastal waters and territory of Newfoundland remains after the Treaty as unimpaired as it was before. But from the Treaty results an obligatory relation whereby the right of Great Britain to exercise its right of sovereignty by making regulations is limited to such regulations as are made in good faith, and are not in violation of the Treaty;

(*f*) Finally to hold that the United States, the grantee of the fishing right, has a voice in the preparation of fishery legislation involves the recognition of a right in that country to participate in the internal legislation of Great Britain and her Colonies, and to that extent would reduce these countries to a state of dependence.

While therefore unable to concede the claim of the United States as based on the Treaty, this Tribunal considers that such claim has been and is to some extent, conceded in the relations now existing between the two Parties. Whatever may have been the situation under the Treaty of 1818 standing alone, the exercise of the right of regulation inherent in Great Britain has been, and is, limited by the repeated recognition of the obligations already referred to, by the limitations and liabilities accepted in the Special Agreement, by the unequivocal position assumed by Great Britain in the presentation of its case before this Tribunal, and by the consequent view of this Tribunal that it would be consistent with all the circumstances, as revealed by this record, as to the duty of Great Britain, that she should submit the reasonableness of any future regulation to such an impartial arbitral test, affording full opportunity therefor, as is hereafter recommended under the authority of Article IV of the Special Agreement, whenever the reasonableness of any regulation is objected to or challenged by the United States in the manner, and within the time hereinafter specified in the said recommendation.

Now therefore this Tribunal decides and awards as follows:

The right of Great Britain to make regulations without the consent of the United States, as to the exercise of the liberty to take fish referred to in Article I of the Treaty of October 20th, 1818, in the form of municipal laws, ordinances or rules of Great Britain, Canada or Newfoundland is inherent to the sovereignty of Great Britain.

The exercise of that right by Great Britain is, however, limited by the said Treaty in respect of the said liberties therein granted to the inhabitants of the United States in that such regulations must be made bona fide and must not be in violation of the said Treaty.

Regulations which are (1) appropriate or necessary for the protection and preservation of such fisheries, or (2) desirable or necessary on grounds of public order and morals without unnecessarily interfering with the fishery itself, and in both cases equitable and fair as between local and American fishermen, and not so framed as to give unfairly an advantage to the former over the latter class, are not inconsistent with the obligation to execute the Treaty in good faith, and are therefore reasonable and not in violation of the Treaty.

For the decision of the question whether a regulation is or is not reasonable, as being or not in accordance with the dispositions of the Treaty and not in violation thereof, the Treaty of 1818 contains no special provision. The settlement of differences in this respect that might arise thereafter was left to the ordinary means of diplomatic intercourse. By reason, however, of the form in which Question I is put, and by further reason of the admission of Great Britain by her counsel before this Tribunal that it is not now for either of the Parties to the Treaty to determine the reasonableness of any regulation made by Great Britain, Canada or Newfoundland, the reasonableness of any such regulation, if contested, must be decided not by either of the Parties, but by an impartial authority in accordance with the principles hereinabove laid down, and in the manner proposed in the recommendations made by the Tribunal in virtue of Article IV of the Agreement.

The Tribunal further decides that Article IV of the Agreement is, as stated by counsel of the respective Parties at the argument, permanent in its effect, and not terminable by the expiration of the General Arbitration Treaty of 1908, between Great Britain and the United States.

In execution, therefore, of the responsibilities imposed upon this Tribunal in regard to Articles II, III and IV of the Special Agreement, we hereby pronounce in their regard as follows:

AS TO ARTICLE II.

Pursuant to the provisions of this Article, hereinbefore cited, either Party has called the attention of this Tribunal to acts of the other claimed to be inconsistent with the true interpretation of the Treaty of 1818.

But in response to a request from the Tribunal, recorded in Protocol No. XXVI of 19th July, for an exposition of the grounds of such objections, the Parties replied as reported in Protocol No. XXX of 28th July to the following effect:

His Majesty's Government considered that it would be unnecessary to call upon the Tribunal for an opinion under the second clause of Article II, in regard to the executive act of the United States of America in sending warships to the territorial waters in question, in view of the recognized motives of the United States of America in taking this action and of the relations maintained by their representatives with the local authorities. And this being the sole act to which the attention of this Tribunal has been called by His Majesty's Government, no further action in their behalf is required from this Tribunal under Article II.

The United States of America presented a statement in which their claim that specific provisions of certain legislative and executive acts of the Governments of Canada and Newfoundland were inconsistent with the true interpretation of the Treaty of 1818 was based on the contention that these provisions were not "reasonable" within the meaning of Question I.

After calling upon this Tribunal to express an opinion on these acts, pursuant to the second clause of Article II, the United States of America pointed out in that statement that under Article III any question regarding the reasonableness of any regulation might be referred by the Tribunal to a Commission of expert specialists, and expressed an intention of asking for such reference under certain circumstances.

The Tribunal having carefully considered the counter-statement presented on behalf of Great Britain at the session of August 2nd, is of opinion that the decision on the reasonableness of these regulations requires expert information about the fisheries themselves and an examination of the practical effect of a great number of these provisions in relation to the conditions surrounding the exercise of the liberty of fishery enjoyed by the inhabitants of the United States, as contemplated by Article III. No further action on behalf of the United States is therefore required from this Tribunal under Article II.

AS TO ARTICLE III.

As provided in Article III, hereinbefore cited and above referred to, "any question regarding the reasonableness of any regulation, or otherwise, which requires an examination of the practical effect of any provisions surrounding the exercise of the liberty of fishery enjoyed by the inhabitants of the United States, or which requires expert information about the fisheries themselves, may be referred by this Tribunal to a Commission of expert specialists; one to be designated by each of the Parties hereto and the third, who shall not be a national of either Party, to be designated by the Tribunal."

The Tribunal now therefore calls upon the Parties to designate within one month their national Commissioners for the expert examination of the questions submitted.

As the third non-national Commissioner this Tribunal designates Doctor P. P. C. Hoek, Scientific Adviser for the fisheries of the Netherlands and if any necessity arises therefor a substitute may be appointed by the President of this Tribunal.

After a reasonable time, to be agreed on by the Parties, for the expert Commission to arrive at a conclusion, by conference, or, if necessary, by local inspection, the Tribunal shall, if convoked by the President at the request of either Party, thereupon at the earliest convenient date, reconvene to consider the report of the Commission, and if it be on the whole unanimous shall incorporate it in the award. If not on the whole unanimous, i. e., on all points which in the opinion of the Tribunal are of essential importance, the Tribunal shall make its award as to the regulations concerned after consideration of the conclusions of the expert Commissioners and after hearing argument by counsel.

But while recognizing its responsibilities to meet the obligations imposed on it under Article III of the Special Agreement, the Tribunal hereby recommends as an alternative to having recourse to a reconvention of this Tribunal, that the Parties should accept the unanimous opinion of the Commission or the opinion of the non-national Commissioner on any points in dispute as an arbitral award rendered under the provisions of Chapter IV of the Hague Convention of 1907.

AS TO ARTICLE IV.

Pursuant to the provisions of this Article, hereinbefore cited, this Tribunal recommends for the consideration of the Parties the following rules and method of procedure under which all questions which may arise in the future regarding the exercise of the liberties above referred to may be determined in accordance with the principles laid down in this award.

1.

All future municipal laws, ordinances or rules for the regulation of the fishery by Great Britain in respect of (1) the hours, days or seasons when fish may be taken on the Treaty coasts; (2) the method, means and implements used in the taking of fish or in carrying on fishing operations; (3) any other regulation of a similar character shall be published in the London Gazette two months before going into operation.

Similar regulations by Canada or Newfoundland shall be similarly published in the Canada Gazette and the Newfoundland Gazette respectively.

2.

If the Government of the United States considers any such laws or regulations inconsistent with the Treaty of 1818, it is entitled to so notify the Government of Great Britain within the two months referred to in Rule No. 1.

3.

Any law or regulation so notified shall not come into effect with respect to inhabitants of the United States until the Permanent Mixed Fishery Commission has decided that the regulation is reasonable within the meaning of this award.

4.

Permanent Mixed Fishery Commissions for Canada and Newfoundland respectively shall be established for the decision of such questions as to the reasonableness of future regulations, as contemplated by Article IV of the Special Agreement; these Commissions shall consist of an expert national appointed by either Party for five years. The third member shall not be a national of either Party; he shall be nominated for five years by agreement of the Parties, or failing such agreement within two months, he shall be nominated by Her Majesty the Queen of the Netherlands. The two national members shall be convoked by the Government of Great Britain within one month from the date of notification by the Government of the United States.

5.

The two national members having failed to agree within one month, within another month the full Commission, under the presidency of the umpire, is to be convoked by Great Britain. It must deliver its decision, if the two Governments do not agree otherwise, at the latest in three months. The Umpire shall conduct the procedure in accordance with that provided in Chapter IV of the Convention for the Pacific Settlement of International Disputes, except in so far as herein otherwise provided.

6.

The form of convocation of the Commission including the terms of reference of the question at issue shall be as follows: "The provision hereinafter fully set forth of an Act dated ———, published in the ——— has been notified to the Government of Great Britain by the Government of the United States, under date of ————, as provided by the award of the Hague Tribunal of September 7th, 1910.

"Pursuant to the provisions of that award the Government of Great Britain hereby convokes the Permanent Mixed Fishery Commission for (Canada)/(Newfoundland), composed of ————— Commissioner for the United States of America, and of ————— Commissioner for (Canada)/(Newfoundland), which shall meet at ———— and render a decision within one month as to whether the provision so notified is reasonable and consistent with the Treaty of 1818, as interpreted by the award of the Hague Tribunal of September 7th, 1910, and if not, in what respect it is unreasonable and inconsistent therewith.

"Failing an agreement on this question within one month the Commission shall so notify the Government of Great

Britain in order that the further action required by that award may be taken for the decision of the above question.

"The provision is as follows:——

7.

The unanimous decision of the two national Commissioners, or the majority decision of the Umpire and one Commissioner, shall be final and binding.

QUESTION II.

Have the inhabitants of the United States, while exercising the liberties referred to in said Article, a right to employ as members of the fishing crews of their vessels persons not inhabitants of the United States?

In regard to this question the United States claim in substance:

1. That the liberty assured to their inhabitants by the Treaty plainly includes the right to use all the means customary or appropriate for fishing upon the sea, not only ships and nets and boats, but crews to handle the ships and the nets and the boats;
2. That no right to control or limit the means which these inhabitants shall use in fishing can be admitted unless it is provided in the terms of the Treaty and no right to question the nationality or inhabitancy of the crews employed is contained in the terms of the Treaty.

And Great Britain claims:

1. That the Treaty confers the liberty to inhabitants of the United States exclusively;

2. That the Governments of Great Britain, Canada or Newfoundland may, without infraction of the Treaty, prohibit persons from engaging as fishermen in American vessels.

Now considering (1) that the liberty to take fish is an economic right attributed by the Treaty; (2) that it is attributed to inhabitants of the United States, without any mention of their nationality; (3) that the exercise of an economic right includes the right to employ servants; (4) that the right of employing servants has not been limited by the Treaty to the employment of persons of a distinct nationality or inhabitancy; (5) that the liberty to take fish as an economic liberty refers not only to the individuals doing the manual act of fishing, but also to those for whose profit the fish are taken.

But considering, that the Treaty does not intend to grant to individual persons or to a class of persons the liberty to take fish in certain waters "in common," that is to say in company, with individual British subjects, in the sense that no law could forbid British subjects to take service on American fishing ships; (2) that the Treaty intends to secure to the United States a share of the fisheries designated therein, not only in the interest of a certain class of individuals, but also in the interest of both the United States and Great Britain, as appears from the evidence and notably from the correspondence between Mr. Adams and Lord Bathurst in 1815; (3) that the inhabitants of the United States do not derive the liberty to take fish directly from the Treaty, but from the United States Government as party to the Treaty with Great Britain and moreover exercising the right to regulate the conditions under which its inhabitants may enjoy the granted liberty; (4) that it is in the interest of the inhabitants of the United States that the fishing liberty granted to them be restricted to exercise by them and removed from the enjoyment of other aliens not entitled by this Treaty to participate in the fisheries; (5) that such restrictions have been throughout enacted in the British Statute of June 15, 1819, and that of June 3, 1824, to this effect, that no alien or stranger whatsoever shall fish in the waters designated therein, except in so far as by treaty thereto entitled, and that this exception will, in virtue of the Treaty of 1818, as hereinabove interpreted by this award, exempt from these statutes American fishermen fishing by the agency of non-inhabitant aliens employed in their service; (6) that the Treaty does not affect the sovereign right of Great Britain as to aliens, non-inhabitants of the United States, nor the right of Great Britain to regulate the engagement of British subjects, while these aliens or British subjects are on British territory.

Now therefore, in view of the preceding considerations this Tribunal is of opinion that the inhabitants of the United States while exercising the liberties referred to in the said article have a right to employ, as members of the fishing crews of their vessels, persons not inhabitants of the United States.

But in view of the preceding considerations the Tribunal, to prevent any misunderstanding as to the effect of its award, expresses the opinion that non-inhabitants employed as members of the fishing crews of the United States vessels derive no benefit or immunity from the Treaty and it is so decided and awarded.

QUESTION III.

Can the exercise by the inhabitants of the United States of the liberties referred to in the said Article be subjected, without the consent of the United States, to the requirements of entry or report at custom-houses or the payment of light or harbour or other dues, or to any other similar requirement or condition or exaction?

The Tribunal is of opinion as follows:

It is obvious that the liberties referred to in this question are those that relate to taking fish and to drying and curing fish on certain coasts as prescribed in the Treaty of October 20, 1818. The exercise of these liberties by the inhabitants of the United States in the prescribed waters to which they relate, has no reference to any commercial privileges which may or may not attach to such vessels by reason of any supposed authority outside the Treaty, which itself confers no commercial privileges whatever upon the inhabitants of the United States or the vessels in which they may exercise the fishing liberty. It follows, therefore, that when the inhabitants of the United States are not seeking to exercise the commercial privileges accorded to trading vessels for the vessels in which they are exercising the granted liberty of fishing, they ought not to be subjected to requirements as to report and entry at custom houses that are only appropriate to the exercise of commercial privileges. The exercise of the fishing liberty is distinct from the exercise of commercial or trading privileges and it is not competent for Great Britain or her colonies to impose upon the former exactions only appropriate to the latter. The reasons for the requirements enumerated in the case of commercial vessels, have no relation to the case of fishing vessels.

We think, however, that the requirement that American fishing vessels should report, if proper conveniences and an opportunity for doing so are provided, is not unreasonable or inappropriate. Such a report, while serving the purpose of a notification of the presence of a fishing vessel in the treaty waters for the purpose of exercising the treaty liberty, while it gives an opportunity for a proper surveillance of such vessel by revenue officers, may also serve to afford to such fishing vessel protection from interference in the exercise of the fishing liberty. There should be no such requirement, however, unless reasonably convenient opportunity therefor be afforded in person or by telegraph, at a custom-house or to a customs official.

The Tribunal is also of opinion that light and harbor dues, if not imposed on Newfoundland fishermen, should not be imposed on American fishermen while exercising the liberty granted by the Treaty. To impose such dues on American fishermen only would constitute an unfair discrimination between them and Newfoundland fishermen and one inconsistent with the liberty granted to American fishermen to take fish, etc., "in common with the subjects of His Britannic Majesty."

Further, the Tribunal considers that the fulfilment of the requirement as to report by fishing vessels on arrival at the fishery would be greatly facilitated in the interests of both parties by the adoption of a system of registration, and distinctive marking of the fishing boats of both parties, analogous to that established by Articles V to XIII, inclusive, of the International Convention signed at The Hague, 8 May, 1882, for the regulation of the North Sea Fisheries.

The Tribunal therefore decides and awards as follows:

The requirement that an American fishing vessel should report, if proper conveniences for doing so are at hand, is not unreasonable, for the reasons stated in the foregoing opinion. There should be no such requirement, however, unless there be reasonably convenient opportunity afforded to report in person or by telegraph, either at a custom-house or to a customs official.

But the exercise of the fishing liberty by the inhabitants

of the United States should not be subjected to the purely commercial formalities of report, entry and clearance at a custom-house, nor to light, harbor or other dues not imposed upon Newfoundland fishermen.

QUESTION IV.

Under the provision of the said Article that the American fishermen shall be admitted to enter certain bays or harbours for shelter, repairs, wood, or water, and for no other purpose whatever, but that they shall be under such restrictions as may be necessary to prevent their taking, drying, or curing fish therein or in any other manner whatever abusing the privileges thereby reserved to them, is it permissible to impose restrictions making the exercise of such privileges conditional upon the payment of light or harbour or other dues, or entering or reporting at custom-houses or any similar conditions?

The Tribunal is of opinion that the provision in the first Article of the Treaty of October 20th, 1818, admitting American fishermen to enter certain bays or harbors for shelter, repairs, wood and water, and for no other purpose whatever, is an exercise in large measure of those duties of hospitality and humanity which all civilized nations impose upon themselves and expect the performance of from others. The enumerated purposes for which entry is permitted all relate to the exigencies in which those who pursue their perilous calling on the sea may be involved. The proviso which appears in the first article of the said Treaty immediately after the so-called renunciation clause, was doubtless due to a recognition by Great Britain of what was expected from the humanity and civilization of the then leading commercial nation of the world. To impose restrictions making the exercise of such privileges conditional upon the payment of light, harbor or other dues, or entering and reporting at custom-houses, or any similar conditions would be inconsistent with the grounds upon which such privileges rest and therefore is not permissible.

And it is decided and awarded that such restrictions are not permissible.

It seems reasonable, however, in order that these privileges accorded by Great Britain on these grounds of hospitality and humanity should not be abused, that the American fishermen entering such bays for any of the four purposes aforesaid and remaining more than 48 hours therein, should be required, if thought necessary by Great Britain or the Colonial Government, to report, either in person or by telegraph, at a custom-house or to a customs official, if reasonably convenient opportunity therefor is afforded.

And it is so decided and awarded.

QUESTION V.

From where must be measured the "three marine miles of any of the coasts, bays, creeks, or harbours" referred to in the said Article?

In regard to this question, Great Britain claims that the renunciation applies to all bays generally and

The United States contend that it applies to bays of a certain class or condition.

Now, considering that the Treaty used the general term "bays" without qualification, the Tribunal is of opinion that these words of the Treaty must be interpreted in a general sense as applying to every bay on the coast in question that might be reasonably supposed to have been considered as a bay by the negotiators of the Treaty under the general conditions then prevailing, unless the United States can adduce satisfactory proof that any restrictions or qualifications of the general use of the term were or should have been present to their minds.

And for the purpose of such proof the United States contend:

1°. That while a State may renounce the treaty right to fish in foreign territorial waters, it cannot renounce the natural right to fish on the High Seas.

But the Tribunal is unable to agree with this contention. Because though a State cannot grant rights on the High Seas it certainly can abandon the exercise of its right to fish on the High Seas within certain definite limits. Such an abandonment was made with respect to their fishing rights in the waters in question by France and Spain in 1763. By a convention between the United Kingdom and the United States in 1846, the two countries assumed ownership over waters in Fuca Straits at distances from the shore as great as 17 miles.

The United States contend moreover:

2°. That by the use of the term "liberty to fish" the United States manifested the intention to renounce the liberty in the waters referred to only in so far as that liberty was dependent upon or derived from a concession on the part of Great Britain, and not to renounce the right to fish in those waters where it was enjoyed by virtue of their natural right as an independent State.

But the Tribunal is unable to agree with this contention:

(*a*) Because the term "liberty to fish" was used in the renunciatory clause of the Treaty of 1818 because the same term had been previously used in the Treaty of 1783 which gave the liberty; and it was proper to use in the renunciation clause the same term that was used in the grant with respect to the object of the grant; and, in view of the terms of the grant, it would have been improper to use the term "right" in the renunciation. Therefore the conclusion drawn from the use of the term "liberty" instead of the term "right" is not justified;

(*b*) Because the term "liberty" was a term properly applicable to the renunciation which referred not only to fishing in the territorial waters but also to drying and curing on the shore. This latter right was undoubtedly held under the provisions of the Treaty and was not a right accruing to the United States by virtue of any principle of International law.

3°. The United States also contend that the term "bays of His Britannic Majesty's Dominions" in the renunciatory clause must be read as including only those bays which were under the territorial sovereignty of Great Britain.

But the Tribunal is unable to accept this contention:

(*a*) Because the description of the coast on which the fishery is to be exercised by the inhabitants of the United States is expressed throughout the Treaty of 1818 in geographical terms and not by reference to political control; the Treaty describes the coast as contained between capes;

(*b*) Because to express the political concept of dominion as equivalent to sovereignty, the word "dominion" in the singular would have been an adequate term and not "dominions" in the plural; this latter term having a recognized and well settled meaning as descriptive of those portions of the Earth which owe political allegiance to His Majesty; e. g. "His Britannic Majesty's Dominions beyond the Seas."

4°. It has been further contended by the United States that the renunciation applies only to bays six miles or less in width "inter fauces terrae," those bays only being territorial bays, because the three mile rule is, as shown by this Treaty, a principle of international law applicable to coasts and should be strictly and systematically applied to bays.

But the Tribunal is unable to agree with this contention:

(*a*) Because admittedly the geographical character of a bay contains conditions which concern the interests of the territorial sovereign to a more intimate and important extent than do those connected with the open coast. Thus conditions of national and territorial integrity, of defence, of commerce and of industry are all vitally concerned with the control of the bays penetrating the national coast line. This interest varies, speaking generally in proportion to the penetration inland of the bay; but as no principle of international law recognizes any specified relation between the concavity of the bay and the requirements for control by the territorial sovereignty, this Tribunal is unable to qualify by the application of any new principle its interpretation of the Treaty of 1818 as excluding bays in general from the strict and systematic application of the three mile rule; nor can this Tribunal take cognizance in this connection of other principles concerning the territorial sovereignty over bays such as ten mile or twelve mile limits of exclusion based on international acts subsequent to the treaty of 1818 and relating to coasts of a different configuration and conditions of a different character;

(*b*) Because the opinion of jurists and publicists quoted in the proceedings conduce to the opinion that speaking generally the three mile rule should not be strictly and systematically applied to bays;

(*c*) Because the treaties referring to these coasts, antedating the treaty of 1818, made special provisions as to bays, such as the Treaties of 1686 and 1713 between Great Britain and France, and especially the Treaty of 1778 between the United States and France. Likewise JAY's Treaty of 1794 Art. 25, distinguished bays from the space "within cannon-shot of the coast" in regard to the right of seizure in times of war. If the proposed treaty of 1806 and the treaty of 1818 contained no disposition to that effect, the explanation may be found in the fact that the first extended the marginal belt to five miles, and also in the circumstance that the American proposition of 1818 in that respect was not limited to "bays," but extended to "chambers formed by headlands" and to "five marine miles from a right line from one headland to another," a proposition which in the times of the Napoleonic wars would have affected to a very large extent the operations of the British navy;

(*d*) Because it has not been shown by the documents and correspondence in evidence here that the application of the three mile rule to bays was present to the minds of the negotiators in 1818 and they could not reasonably have been expected either to presume it or to provide against its persumption;

(*e*) Because it is difficult to explain the words in art. III of the Treaty under interpretation "country . . . together with its bays, harbours and creeks" otherwise than that all bays without distinction as to their width were, in the opinion of the negotiators, part of the territory;

(*f*) Because from the information before this Tribunal

it is evident that the three mile rule is not applied to bays strictly or systematically either by the United States or by any other Power;

(*g*) It has been recognized by the United States that bays stand apart, and that in respect of them territorial jurisdiction may be exercised farther than the marginal belt in the case of Delaware bay by the report of the United States Attorney General of May 19th 1793; and the letter of Mr. JEFFERSON to Mr. GENET of Nov. 8th 1793 declares the bays of the United States generally to be, "as being landlocked, within the body of the United States."

5°. In this latter regard it is further contended by the United States, that such exceptions only should be made from the application of the three mile rule to bays as are sanctioned by conventions and established usage; that all exceptions for which the United States of America were responsible are so sanctioned; and that His Majesty's Government are unable to provide evidence to show that the bays concerned by the Treaty of 1818 could be claimed as exceptions on these grounds either generally, or except possibly in one or two cases, specifically.

But the Tribunal while recognizing that conventions and established usage might be considered as the basis for claiming as territorial those bays which on this ground might be called historic bays, and that such claim should be held valid in the absence of any principle of international law on the subject; nevertheless is unable to apply this, *a contrario,* so as to subject the bays in question to the three mile rule, as desired by the United States:

(*a*) Because Great Britain has during this controversy asserted a claim to these bays generally, and has enforced such claim specifically in statutes or otherwise, in regard to the more important bays such as Chaleurs, Conception and Miramichi;

(*b*) Because neither should such relaxations of this claim, as are in evidence, be construed as renunciations of it; nor should omissions to enforce the claim in regard to bays as to which no controversy arose, be so construed. Such a construction by this Tribunal would not only be intrinsically inequitable but internationally injurious; in that it would discourage conciliatory diplomatic transactions and encourage the assertion of extreme claims in their fullest extent;

(*c*) Because any such relaxations in the extreme claim of Great Britain in its international relations are compensated by recognitions of it in the same sphere by the United States; notably in relations with France for instance in 1823 when they applied to Great Britain for the protection of their fishery in the bays on the western coast of Newfoundland, whence they had been driven by French war vessels on the ground of the pretended exclusive right of the French. Though they never asserted that their fishermen had been disturbed within the three mile zone, only alleging that the disturbance had taken place in the bays, they claimed to be protected by Great Britain for having been molested in waters which were, as Mr. RUSH stated "clearly within the jurisdiction and sovereignty of Great Britain."

6°. It has been contended by the United States that the words "coasts, bays, creeks or harbours" are here used only to express different parts of the coast and are intended to express and be equivalent to the word "coast," whereby the three marine miles would be measured from the sinuosities of the coast and the renunciation would apply only to the waters of bays within three miles.

But the Tribunal is unable to agree with this contention:

(*a*) Because it is a principle of interpretation that words in a document ought not to be considered as being without any meaning if there is not specific evidence to that purpose and the interpretation referred to would lead to the consequence, practically, of reading the words "bays, coasts and harbours" out of the Treaty; so that it would read "within three miles of any of the coasts" including therein the coasts of the bays and harbours;

(*b*) Because the word "therein" in the proviso—"restrictions necessary to prevent their taking, drying or curing fish therein" can refer only to "bays," and not to the belt of three miles along the coast; and can be explained only on the supposition that the words "bays, creeks and harbours" are to be understood in their usual ordinary sense and not in an artificially restricted sense of bays within the three mile belt;

(*c*) Because the practical distinction for the purpose of this fishery between coasts and bays and the exceptional conditions pertaining to the latter has been shown from the correspondence and the documents in evidence, especially the Treaty of 1783, to have been in all probability present to the minds of the negotiators of the Treaty of 1818;

(*d*) Because the existence of this distinction is confirmed in the same article of the Treaty by the proviso permitting the United States fishermen to enter bays for certain purposes;

(*e*) Because the word "coasts" is used in the plural form whereas the contention would require its use in the singular;

(*f*) Because the Tribunal is unable to understand the term "bays" in the renunciatory clause in other than its geographical sense, by which a bay is to be considered as an indentation of the coast, bearing a configuration of a particular character easy to determine specifically, but difficult to describe generally.

The negotiators of the Treaty of 1818 did probably not trouble themselves with subtle theories concerning the notion of "bays"; they most probably thought everybody would know what was a bay. In this popular sense the term must be interpreted in the Treaty. The interpretation must take into account all the individual circumstances which for any one of the different bays are to be appreciated, the relation of its width to the length of penetration inland, the possibility and the necessity of its being defended by the State in whose territory it is indented; the special value which it has for the industry of the inhabitants of its shores; the distance which it is secluded from the highways of nations on the open sea and other circumstances not possible to enumerate in general.

For these reasons the Tribunal decides and awards:

In the case of bays the three marine miles are to be measured from a straight line drawn across the body of water at the place where it ceases to have the configuration and characteristics of a bay. At all other places the three marine miles are to be measured following the sinuosities of the coast.

But considering the Tribunal cannot overlook that this answer to Question V, although correct in principle and the only one possible in view of the want of a sufficient basis for a more concrete answer, is not entirely satisfactory as to its practical applicability, and that it leaves room for doubts and differences in practice. Therefore the Tribunal considers it its duty to render the decision more practicable and to remove the danger of future differences by adjoining to it, a recommendation in virtue of the responsibilities imposed by Art. IV of the Special Agreement.

Considering, moreover, that in treaties with France, with the North German Confederation and the German Empire and likewise in the North Sea Convention, Great Britain has adopted for similar cases the rule that only bays of ten miles width should be considered as those wherein the fishing is reserved to nationals. And that in the course of the negotiations between Great Britain and the United States a similar rule has been on various occasions proposed and adopted by Great Britain in instructions to the naval officers stationed on these coasts. And that though these circumstances are not sufficient to constitute this a principle of international law, it seems reasonable to propose this rule with certain exceptions, all the more that this rule with such exceptions has already formed the basis of an agreement between the two Powers.

Now therefore this Tribunal in pursuance of the provisions of art. IV hereby recommends for the consideration and acceptance of the High Contracting Parties the following rules and method of procedure for determining the limits of the bays hereinbefore enumerated.

1.

In every bay not hereinafter specifically provided for the limits of exclusion shall be drawn three miles seaward from a straight line across the bay in the part nearest the entrance at the first point where the width does not exceed ten miles.

2.

In the following bays where the configuration of the coast and the local climatic conditions are such that foreign fishermen when within the geographic headlands might reasonably and bona fide believe themselves on the high seas, the limits of exclusion shall be drawn in each case between the headlands hereinafter specified as being those at and within which such fishermen might be reasonably expected to recognize the bay under average conditions.

For the Baie des Chaleurs the line from the Light at Birch Point on Miscou Island to Macquereau Point Light: for the Bay of Miramichi, the line from the Light at Point Escuminac to the Light on the Eastern Point of Tabisintac Gully; for Egmont Bay, in Prince Edward Island, the line from the light at Cape Egmont to the Light at West Point; and off St. Ann's Bay, in the Province of Nova Scotia, the line from the Light at Point Anconi to the nearest point on the opposite shore of the mainland.

For Fortune Bay, in Newfoundland, the line from Connaigre Head to the Light on the Southeasterly end of Brunet Island, thence to Fortune Head.

For or near the following bays the limits of exclusion shall be three marine miles seawards from the following lines, namely:

For or near Barrington Bay, in Nova Scotia, the line from the Light on Stoddart Island to the Light on the south point of Cape Sable, thence to the light at Baccaro Point; at Chedabucto and St. Peter's Bays, the line from Cranberry Island Light to Green Island Light, thence to Point Rouge; for Mira Bay, the line from the Light on the East Point of Scatari Island to the Northeasterly Point of Cape Morien; and at Placentia Bay, in Newfoundland, the line from Latine Point, on the Eastern mainland shore, to the most Southerly Point of Red Island, thence by the most Southerly Point of Merasheen Island to the mainland.

Long Island and Bryer Island, on St. Mary's Bay, in Nova Scotia, shall, for the purpose of delimitation, be taken as the coasts of such bays.

It is understood that nothing in these rules refers either

to the Bay of Fundy considered as a whole apart from its bays and creeks or as to the innocent passage through the Gut of Canso, which were excluded by the agreement made by exchange of notes between Mr. Bacon and Mr. Bryce dated February 21st 1909 and March 4th 1909; or to Conception Bay, which was provided for by the decision of the Privy Council in the case of the Direct United States Cable Company v. The Anglo American Telegraph Company, in which decision the United States have acquiesced.

QUESTION VI.

Have the inhabitants of the United States the liberty under the said Article or otherwise, to take fish in the bays, harbours, and creeks on that part of the southern coast of Newfoundland which extends from Cape Ray to Rameau Islands, or on the western and northern coasts of Newfoundland from Cape Ray to Quirpon Islands, or on the Magdalen Islands?

In regard to this question, it is contended by the United States that the inhabitants of the United States have the liberty under Art. I of the Treaty of taking fish in the bays, harbours and creeks on that part of the Southern Coast of Newfoundland which extends from Cape Ray to Rameau Islands or on the western and northern coasts of Newfoundland from Cape Ray to Quirpon Islands and on the Magdalen Islands. It is contended by Great Britain that they have no such liberty.

Now considering that the evidence seems to show that the intention of the Parties to the Treaty of 1818, as indicated by the records of the negotiations and by the subsequent attitude of the Governments was to admit the United

States to such fishery, this Tribunal is of opinion that it is incumbent on Great Britain to produce satisfactory proof that the United States are not so entitled under the Treaty.

For this purpose Great Britain points to the fact that whereas the Treaty grants to American fishermen liberty to take fish "on the coasts, bays, harbours, and creeks from Mount Joly on the Southern coast of Labrador" the liberty is granted to the "coast" only of Newfoundland and to the "shore" only of the Magdalen Islands; and argues that evidence can be found in the correspondence submitted indicating an intention to exclude Americans from Newfoundland bays on the Treaty Coast, and that no value would have been attached at that time by the United States Government to the liberty of fishing in such bays because there was no cod fishery there as there was in the bays of Labrador.

But the Tribunal is unable to agree with this contention:

(*a*) Because the words "part of the southern coast . . . from . . . to" and the words "Western and Northern Coast . . . from . . . to," clearly indicate one uninterrupted coast-line; and there is no reason to read into the words "coasts" a contradistinction to bays, in order to exclude bays. On the contrary, as already held in the answer to Question V, the words "liberty, forever, to dry and cure fish in any of the unsettled bays, harbours and creeks of the Southern part of the Coast of Newfoundland hereabove described," indicate that in the meaning of the Treaty, as in all the preceding treaties relating to the same territories, the words coast, coasts, harbours, bays, etc., are used, without attaching to the word "coast" the specific meaning of excluding bays. Thus in the provision of the Treaty of 1783 giving liberty "to take fish on such part of the coast of Newfoundland as British fishermen shall use"; the word "coast" necessarily includes bays, because if the intention had been to prohibit the entering of the bays for fishing the following words "but not to dry or cure the same on that island," would have no meaning. The contention that in the Treaty of 1783 the word "bays" is inserted lest otherwise Great Britain would have had the right to exclude the Americans to the three mile line, is inadmissible, because in that Treaty that line is not mentioned;

(*b*) Because the correspondence between Mr. ADAMS and Lord BATHURST also shows that during the negotiations for the Treaty the United States demanded the former rights enjoyed under the Treaty of 1783, and that Lord BATHURST in the letter of 30th October 1815 made no objection to granting those "former rights" "placed under some modifications," which latter did not relate to the right of fishing in bays, but only to the "preoccupation of British harbours and creeks by the fishing vessels of the United States and the forcible exclusion of British subjects where the fishery might be most advantageously conducted," and "to the clandestine introduction of prohibited goods into the British colonies." It may be therefore assumed that the word "coast" is used in both Treaties in the same sense, including bays;

(*c*) Because the Treaty expressly allows the liberty to dry and cure in the unsettled bays, etc. of the southern part of the coast of Newfoundland, and this shows that, a fortiori, the taking of fish in those bays is also allowed; because the fishing liberty was a lesser burden than the grant to cure and dry, and the restrictive clauses never refer to fishing in contradistinction to drying, but always to drying in contradistinction to fishing. Fishing is granted without drying, never drying without fishing;

(*d*) Because there is not sufficient evidence to show that the enumeration of the component parts of the coast of Labrador was made in order to discriminate between the coast of Labrador and the coast of Newfoundland.

(*e*) Because the statement that there is no codfish in the bays of Newfoundland and that the Americans only took interest in the codfishery is not proved; and evidence to the contrary is to be found in Mr. JOHN ADAMS Journal of Peace Negotiations of November 25, 1782;

(*f*) Because the Treaty grants the right to take fish of every kind, and not only codfish;

(*g*) Because the evidence shows that, in 1823, the Americans were fishing in Newfoundland bays and that Great Britain when summoned to protect them against expulsion therefrom by the French did not deny their right to enter such bays.

Therefore this Tribunal is of opinion that American inhabitants are entitled to fish in the bays, creeks and harbours of the Treaty coasts of Newfoundland and the Magdalen Islands and it is so decided and awarded.

QUESTION VII.

Are the inhabitants of the United States whose vessels resort to the Treaty coasts for the purpose of exercising the liberties referred to in Article I of the Treaty of 1818 entitled to have for those vessels, when duly authorized by the United States in that behalf, the commercial privileges on the Treaty coasts accorded by agreement or otherwise to United States trading vessels generally?

Now assuming that commercial privileges on the Treaty coasts are accorded by agreement or otherwise to United States trading vessels generally, without any exception, the inhabitants of the United States, whose vessels resort to the same coasts for the purpose of exercising the liberties referred to in Article I of the Treaty of 1818, are entitled to have for those vessels when duly authorized by the United

States in that behalf, the above mentioned commercial privileges, the Treaty containing nothing to the contrary. But they cannot at the same time and during the same voyage exercise their Treaty rights and enjoy their commercial privileges, because Treaty rights and commercial privileges are submitted to different rules, regulations and restraints.

For these reasons this Tribunal is of opinion that the inhabitants of the United States are so entitled in so far as concerns this Treaty, there being nothing in its provisions to disentitle them provided the Treaty liberty of fishing and the commercial privileges are not exercised concurrently and it is so decided and awarded.

Done at the Hague, in the Permanent Court of Arbitration, in triplicate original, September 7th, 1910.

H. LAMMASCH.
A. F. DE SAVORNIN LOHMAN.
GEORGE GRAY.
C. FITZPATRICK.
LUIS M. DRAGO.

Signing the Award, I state pursuant to Article IX clause 2 of the Special Agreement my dissent from the majority of the Tribunal in respect to the considerations and enacting part of the Award as to Question V.

Grounds for this dissent have been filed at the International Bureau of the Permanent Court of Arbitration.

LUIS M. DRAGO.

GROUNDS FOR THE DISSENT

TO

the Award on Question V

BY

DR. LUIS M. DRAGO.

Counsel for Great Britain have very clearly stated that according to their contention the territoriality of the bays referred to in the Treaty of 1818 is immaterial because whether they are or are not territorial, the United States should be excluded from fishing in them by the terms of the renunciatory clause, which simply refers to "bays, creeks or harbours of His Britannic Majesty's Dominions" without any other qualification or description. If that were so, the necessity might arise of discussing whether or not a nation has the right to exclude another by contract or otherwise from any portion or portions of the high seas. But in my opinion the Tribunal need not concern itself with such general question, the wording of the treaty being clear enough to decide the point at issue.

Article I begins with the statement that differences have arisen respecting the liberty claimed by the United States for the inhabitants thereof to take, dry and cure fish on "certain coasts, bays, harbours and creeks of His Britannic Majesty's Dominions in America," and then proceeds to locate the specific portions of the coast with its corresponding indentations, in which the liberty of taking, drying and curing fish should be exercised. The renunciatory clause, which the Tribunal is called upon to construe, runs thus: "And the United States hereby renounce, forever, any liberty heretofore enjoyed or claimed by the inhabitants thereof, to take, dry or cure fish on, or within three marine miles of any of the Coasts, Bays, Creeks or Harbours of His Britannic Majesty's Dominions in America not included within the above mentioned limits.'' This language does not lend itself to different constructions. If the bays in which the liberty has been renounced are those ''of His Britannic Majesty's Dominions in America,'' they must necessarily be territorial bays, because in so far as they are not so considered they should belong to the high seas and consequently form no part of His Britannic Majesty's Dominions, which, by definition, do not extend to the high seas. It cannot be said, as has been suggested, that the use of the word ''dominions,'' in the plural, implies a different meaning than would be conveyed by the same term as used in the singular, so that in the present case, ''the British dominions in America'' ought to be considered as a mere geographical expression, without reference to any right of sovereignty or ''*dominion.*'' It seems to me, on the contrary, that ''dominions,'' or ''possessions,'' or ''estates,'' or such other equivalent terms, simply designate the places over which the ''dominion'' or property rights are exercised. Where there is no possibility of appropriation or dominion, as on the high seas, we cannot speak of dominions. The ''dominions'' extend exactly to the point which the ''dominion'' reaches; they are simply the actual or physical thing over which the abstract power or authority, the *right*, as given to the proprietor or the ruler, applies. The interpretation as to the territoriality of the bays as mentioned in the renunciatory clause of the treaty appears stronger when considering that the United States specifically renounced the ''liberty,'' not the ''right'' to fish or to cure and dry fish. ''The United States renounce, forever, any *liberty* heretofore enjoyed or claimed, to take, cure or dry fish on, or within three marine miles of any of the coasts, bays, creeks or harbours of His Britannic Majesty's Dominions in America." It is well known that the negotiators of the Treaty of 1783 gave a very different meaning to the terms *liberty* and *right*, as distinguished from each other. In this connection Mr. ADAMS' Journal may be recited. To this Journal the British Counter Case refers in the following terms: "From an entry in Mr. ADAMS' Journal it appears that he drafted an article by which he distinguished the *right* to take fish (both on the high seas and on the shores) and the *liberty* to take and cure fish on the land. But on the following day he presented to the British negotiators a draft in which he distinguishes between the "*right*" to take fish on the high seas, and the "*liberty*" to take fish on the "*coasts*," and to dry and cure fish on the land. . . . The British Commissioner called attention to the distinction thus suggested by Mr. ADAMS and proposed that the word *liberty* should be applied to the privileges both on the water and on the land. Mr. ADAMS thereupon rose up and made a vehement protest, as is recorded in his Diary, against the suggestion that the United States enjoyed the fishing on the banks of Newfoundland by any other title than that of *right*. " . . . The application of the word *liberty* to the coast fishery was left as Mr. ADAMS proposed." "The incident, proceeds the British Case, is of importance, since it shows that the difference between the two phrases was intentional." (British Counter Case, page 17). And the British Argument emphasizes again the difference. "More cogent still is the distinction between the words *right* and *liberty*. The word *right* is applied to the sea fisheries, and the word *liberty* to the shore fisheries. The history of the negotiations shows that this distinction was advisedly adopted." If then a *liberty* is a grant and not the recognition of a *right;* if, as the British Case, Counter Case and Argument recognize, the United States had the right to fish in the open sea in contradistinction with the *liberty* to fish near the shores or portions of the shores, and if what has been renounced in the words of the treaty is the *"liberty"* to fish on, or within three miles of the bays, creeks and harbours of His Britannic Majesty's Dominions, it clearly follows that such *liberty* and the corresponding renunciation refers only to such portions of the bays which were under the sovereignty of Great Britain and not to such other portions, if any, as form part of the high seas.

And thus it appears that far from being immaterial the territoriality of bays is of the utmost importance. The treaty not containing any rule or indication upon the subject, the Tribunal cannot help a decision as to this point, which involves the second branch of the British contention that all so-called bays are not only geographical but wholly territorial as well, and subject to the jurisdiction of Great Britain. The situation was very accurately described on almost the same lines as above stated by the British Memorandum sent in 1870 by the Earl of Kimberly to Governor SIR JOHN YOUNG: "The right of Great Britain to exclude American fishermen from waters within three miles of the coasts is unambiguous, and, it is believed, uncontested. But there appears to be some doubt what are the waters described as within three miles of bays, creeks or harbors. When a bay is less than six miles broad its waters are within the three mile limit, and therefore clearly within the meaning of the treaty; *but when it is more than that breadth, the question arises whether it is a bay of Her Britannic Majesty's Dominions.* This is a question which has to be considered in each particular case with regard to international law and usage. When such a bay is not a bay of Her Majesty's dominions, the American fishermen shall be entitled to fish in it, except within three marine miles of the 'coast'; when it is a bay of Her Majesty's dominions they will not be entitled to fish within three miles of it, that is to say (it is presumed) within three miles of a line drawn from headland to headland." (American Case Appendix, page 629).

Now, it must be stated in the first place that there does not seem to exist any general rule of international law which may be considered final, even in what refers to the marginal belt of territorial waters. The old rule of the cannon-shot, crystallized into the present three marine miles measured from low water mark, may be modified at a later period inasmuch as certain nations claim a wider jurisdiction and an extension has already been recommended by the Institute of International Law. There is an obvious reason for that. The marginal strip of territorial waters based originally on the cannon-shot, was founded on the necessity of the riparian State to protect itself from outward attack, by providing something in the nature of an insulating zone, which very reasonably should be extended with the accrued possibility of offense due to the wider range of modern ordnance. In what refers to bays, it has been proposed as a general rule (subject to certain important exceptions) that the marginal belt of territorial waters should follow the sinuosities of the coast more or less in the manner held by the United States in the present contention, so that the marginal belt being of three miles, as in the Treaty under consideration, only such bays should be held as territorial as have an entrance not wider than six miles. (See Sir THOMAS BARCLAY'S Report to Institute of International Law, 1894, page 129, in which he also strongly recommends these limits). This is the doctrine which WESTLAKE, the eminent English writer on International Law, has summed up in very few words: "As to bays," he says, "if the entrance to one of them is not more than twice the width of the littoral sea enjoyed by the country in question,—that is, not more than six sea miles in the ordinary case, eight in that of Norway, and so forth—there is no access from the open sea to the bay except through the territorial water of that country, and the inner part of the bay will belong to that country no matter how widely it may expand. The line drawn from shore to shore at the part where, in approaching from the open sea, the width first contracts to that mentioned, will take the place of the line of low water, and the littoral sea belonging to the State will be measured outwards from that line to the distance of three miles or more, proper to the State''; (WESTLAKE, Vol. 1, page 187). But the learned author takes care to add: ''But although this is the general rule it often meets with an exception in the case of bays which penetrate deep into the land and are called gulfs. Many of these are recognized by immemorial usage as territorial sea of the States into which they penetrate, notwithstanding that their entrance is wider than the general rule for bays would give as a limit for such appropriation.'' And he proceeds to quote as examples of this kind the Bay of Conception in Newfoundland, which he considers as wholly British, Chesapeake and Delaware Bays, which belong to the United States, and others. (*Ibid*, page 188). The Institute of International Law, in its Annual Meeting of 1894, recommended a marginal belt of six miles for the general line of the coast and as a consequence established that for bays the line should be drawn up across at the nearest portion of the entrance toward the sea where the distance between the two sides do not exceed twelve miles. But the learned association very wisely added a proviso to the effect, ''that bays should be so considered and measured *unless a continuous and established usage* has sanctioned a greater breadth.'' Many great authorities are agreed as to that. Counsel for the United States proclaimed the right to the exclusive jurisdiction of certain bays, no matter what the width of their entrance should be, when the littoral nation has asserted its right to take it into their jurisdiction upon reasons which go always back to the doctrine of protection. Lord BLACKBURN, one of the most eminent of English Judges, in delivering the opinion of the Privy Council about Conception Bay in Newfoundland, adhered to the same doctrine when he asserted the territoriality of that branch of the sea, giving as a reason for such finding "that the British Government for a long period had exercised dominion over this bay and its claim had been acquiesced in by other nations, so as to show that the bay had been for a long time occupied exclusively by Great Britain, a circumstance which, in the tribunals of any country, would be very important." "And moreover," he added, "the British Legislature has, by Acts of Parliament, declared it to be part of the British territory, and part of the country made subject to the legislation of Newfoundland." (Direct U. S. Cable Co. v. The Anglo-American Telegraph Co., Law Reports, 2 Appeal Cases, 374).

So it may be safely asserted that a certain class of bays, which might be properly called the historical bays such as Chesapeake Bay and Delaware Bay in North America and the great estuary of the River Plate in South America, form a class distinct and apart and undoubtedly belong to the littoral country, whatever be their depth of penetration and the width of their mouths, when such country has asserted its sovereignty over them, and particular circumstances such as geographical configuration, immemorial usage and above all, the requirements of self-defense, justify such a pretension. The right of Great Britain over the bays of Conception, Chaleur and Miramichi are of this description. In what refers to the other bays, as might be termed the common, ordinary bays, indenting the coasts, over which no special claim or assertion of sovereignty has been made, there does not seem to be any other general principle to be applied than the one resulting from the custom and usage of each individual nation as shown by their Treaties and their general and time honored practice.

The well known words of BYNKERSHOEK might be very appropriately recalled in this connection when so many and divergent opinions and authorities have been recited: "The common law of nations," he says, "can only be learnt from reason and custom. I do not deny that authority may add weight to reason, but I prefer to seek it in a constant custom of concluding treaties in one sense or another and in examples that have occurred in one country or another. (Questiones Jure Publici, Vol. 1, Cap. 3).

It is to be borne in mind in this respect that the Tribunal has been called upon to decide as the subject matter of this controversy, the construction to be given to the fishery Treaty of 1818 between Great Britain and the United States. And so it is that from the usage and the practice of Great Britain in this and other like fisheries and from Treaties entered into by them with other nations as to fisheries, may be evolved the right interpretation to be given to the particular convention which has been submitted. In this connection the following Treaties may be recited:

Treaty between Great Britain and France. 2nd August, 1839. It reads as follows:

Article IX. The subjects of Her Britannic Majesty shall enjoy the exclusive right of fishery within the distance of 3 miles from low water mark along the whole extent of the coasts of the British Islands.

It is agreed that the distance of three miles fixed as the general limit for the exclusive right of fishery upon the coasts of the two countries, shall, with respect to bays, the mouths of which do not exceed ten miles in width, be measured from a straight line drawn from headland to headland.

Article X. It is agreed and understood, that the miles mentioned in the present Convention are geographical miles, whereof 60 make a degree of latitude.

(HERTSLETT'S Treaties and Conventions, Vol. V. p. 89).

Regulations between Great Britain and France 24th May, 1843.

Art. II. The limits, within which the general right of fishery is exclusively reserved to the subjects of the two kingdoms respectively, are fixed (with the exception of those in Granville Bay) at 3 miles distance from low water mark.

With respect to bays, the mouths of which do not exceed ten miles in width, the 3 mile distance is measured from a straight line drawn from headland to headland.

Art. III. The miles mentioned in the present regulations are geographical miles, of which 60 makes a degree of latitude.

(HERTSLETT, Vol. VI, p. 416.)

Treaty between Great Britain and France. November 11, 1867.

Art. I. British fishermen shall enjoy the exclusive right of fishery within the distance of 3 miles from low water mark, along the whole extent of the coasts of the British Islands.

The distance of 3 miles fixed as the general limit for the exclusive right of fishery upon the coasts of the two countries shall, with respect to bays, the mouths of which do not exceed ten miles in width be measured from a straight line drawn from headland to headland.

The miles mentioned in the present convention are geographical miles whereof 60 make a degree of latitude.

(HERTSLETT'S Treaties, Vol. XII, p. 1126, British Case App. p. 38).

Great Britain and North German Confederation. British notice to fishermen by the Board of Trade. Board of Trade, November 1868.

Her Majesty's Government and the North German Confederation having come to an agreement respecting the regulations to be observed by British fishermen fishing off the coasts of the North German Confederation, the following notice is issued for the guidance and warning of British fishermen:

1. The exclusive fishery limits of the German Empire are designated by the Imperial Government as follows: that tract of the sea which extends to a distance of 3 sea miles from the extremest limits which the ebb leaves dry of the German North Sea Coast of the German Islands or flats lying before it, as well as those bays and incurvations of the coast which are ten sea miles or less in breadth reckoned from the extremest points of the land and the flats, must be considered as under the territorial sovereignty of North Germany.

(HERTSLETT'S Treaties, Vol. XIV, p. 1055).

Great Britain and German Empire. British Board of Trade, December, 1874.

(Same recital referring to an arrangement entered into between Her Britannic Majesty and the German Government).

Then the same articles follow with the alteration of the words "German Empire" for "North Germany."

(HERTSLETT'S, Vol. XIV, p. 1058).

Treaty between Great Britain, Belgium, Denmark, France, Germany and the *Netherlands* for regulating the police of the North Sea Fisheries, May 6, 1882.

II. Les pêcheurs nationaux jouiront du droit exclusif de pêche dans le rayon de 3 milles, à partir de la laisse de basse mer, le long de toute l'étendue des côtes de leurs pays respectifs, ainsi que des îles et des bancs qui en dépendent.

Pour les baies le rayon de 3 milles sera mesuré à partir d'une ligne droite, tirée, en travers de la baie, dans la partie la plus rapprochée de l'entrée, au premier point où l'ouverture n'excédera pas 10 milles.

(HERTSLETT, Vol. XV, p. 794).

British Order in Council, October 23rd, 1877.

Prescribes the obligation of not concealing or effacing numbers or marks on boats, employed in fishing or dredging for purposes of sale on the coasts of England, Wales, Scotland and the Islands of Guernsey, Jersey, Alderney, Sark and Man, and not going outside;

(*a*) The distance of 3 miles from low water mark along the whole extent of the said coasts;

(*b*) In cases of bays less than 10 miles wide the line joining the headlands of said bays.

(HERTSLETT'S Vol. XIV, p. 1032).

To this list may be added the unratified Treaty of 1888 between Great Britain and the United States which is so familiar to the Tribunal. Such unratified Treaty contains an authoritative interpretation of the Convention of October 20th, 1818, *sub-judice:* "The three marine miles mentioned in Article I of the Convention of October 20th, 1818, shall be measured seaward from low-water mark; but at every bay, creek or harbor, not otherwise specifically provided for in this Treaty, such three marine miles shall be measured seaward from a straight line drawn across the bay, creek or harbor, in the part nearest the entrance at the first point where the width does not exceed ten marine miles," which is recognizing the exceptional bays as aforesaid and laying the rule for the general and common bays.

It has been suggested that the Treaty of 1818 ought not to be studied as hereabove in the light of any Treaties of a later date, but rather be referred to such British international Conventions as preceded it and clearly illustrate, according to this view, what were, at the time, the principles maintained by Great Britain as to their sovereignty over the sea and over the coast and the adjacent territorial waters. In this connection the Treaties of 1686 and 1713 with France and of 1763 with France and Spain have been recited and offered as examples also of exclusion of nations by agreement from fishery rights on the high seas. I cannot partake of such a view. The treaties of 1686, 1713 and 1763 can hardly be understood with respect to this, otherwise than as examples of the wild, obsolete claims over the common ocean which all nations have of old abandoned with the progress of an enlightened civilization. And if certain nations accepted long ago to be excluded by convention from fishing on what is to-day considered a common sea, it is precisely because it was then understood that such tracts of water, now free and open to all, were the exclusive property of a particular power, who, being the owners, admitted or excluded others from their use. The Treaty of 1818 is in the meantime one of the few which mark an era in the diplomacy of the world. As a matter of fact it is the very first which commuted the rule of the cannon-shot into the three marine miles of coastal jurisdiction. And it really would appear unjustified to explain such historic document, by referring it to international Agreements of a hundred and two hundred years before when the doctrine of SELDEN's *Mare Clausum* was at its height and when the coastal waters were fixed at such distances as sixty miles, or a hundred miles, or two days' journey from the shore and the like. It seems very appropriate, on the contrary, to explain the meaning of the Treaty of

1818 by comparing it with those which immediately followed and established the same limit of coastal jurisdiction. As a general rule a Treaty of a former date may be very safely construed by referring it to the provisions of like Treaties made by the same nation on the same matter at a later time. Much more so when, as occurs in the present case, the later Conventions, with no exception, starting from the same premise of the three miles coastal jurisdiction arrive always to an uniform policy and line of action in what refers to bays. As a matter of fact all authorities approach and connect the modern fishery Treaties of Great Britain and refer them to the Treaty of 1818. The second edition of KLUBER, for instance, quotes in the same sentence the Treaties of October 20th, 1818, and August 2, 1839, as fixing a distance of three miles from low water mark for coastal jurisdiction. And FIORI, the well-known Italian jurist, referring to the same marine miles of coastal jurisdiction, says: "This rule recognized as early as the Treaty of 1818 between the United States and Great Britain, and that between Great Britain and France in 1839, has again been admitted in the treaty of 1867." (Nouveau Droit International Public, Paris, 1885, Section 803).

This is only a recognition of the permanency and the continuity of States. The Treaty of 1818 is not a separate fact unconnected with the later policy of Great Britain. Its negotiators were not parties to such international Convention and their powers disappeared as soon as they signed the document on behalf of their countries. The parties to the Treaty of 1818 were the United States and Great Britain, and what Great Britain meant in 1818 about bays and fisheries, when they for the first time fixed a marginal jurisdiction of three miles, can be very well explained by what Great Britain, the same permanent political entity, understood in 1839, 1843, 1867, 1874,

1878 and 1882, when fixing the very same zone of territorial waters. That a bay in Europe should be considered as different from a bay in America and subject to other principles of international law cannot be admitted in the face of it. What the practice of Great Britain has been outside the Treaties is very well known to the Tribunal, and the examples might be multiplied of the cases in which that nation has ordered its subordinates to apply to the bays on these fisheries the ten mile entrance rule or the six miles according to the occasion. It has been repeatedly said that such have been only relaxations of the strict right, assented to by Great Britain in order to avoid friction on certain special occasions. That may be. But it may also be asserted that such relaxations have been very many and that the constant, uniform, never contradicted, practice of concluding fishery Treaties from 1839 down to the present day, in all of which the ten miles entrance bays are recognized, is the clear sign of a policy. This policy has but very lately found a most public, solemn and unequivocal expression. "On a question asked in Parliament on the 21st of February, 1907, says PITT COBBETT, a distinguished English writer, with respect to the Moray Firth Case, it was stated that, according to the view of the Foreign Office, the Admiralty, the Colonial Office, the Board of Trade and the Board of Agriculture and Fisheries, the term "territorial waters" was deemed to include waters extending from the coast line of any part of the territory of a State to three miles from the low-water mark of such coast line and the waters of all bays, the entrance to which is not more than *six miles*, and of which the entire land boundary forms part of the territory of the same state. (PITT COBBETT Cases and Opinions on International Law, Vol. 1, p. 143).

Is there a contradiction between these six miles and the ten miles of the treaties just referred to? Not at all.

The six miles are the consequence of the three miles marginal belt of territorial waters in their coincidence from both sides at the inlets of the coast and the ten miles far from being an arbitrary measure are simply an extension, a margin given for convenience to the strict six miles with fishery purposes. Where the miles represent sixty to a degree in latitude the ten miles are besides the sixth part of the same degree. The American Government in reply to the observations made to Secretary BAYARD'S Memorandum of 1888, said very precisely: "The width of ten miles was proposed not only because it had been followed in Conventions between many other powers, but also because it was deemed reasonable and just in the present case; this Government recognizing the fact that while it might have claimed a width of six miles as a basis of settlement, fishing within bays and harbors only slightly wider would be confined to areas so narrow as to render it practically valueless and almost necessarily expose the fishermen to constant danger of carrying their operations into forbidden waters." (British Case Appendix, page 416). And Professor JOHN BASSET MOORE, a recognized authority on International law, in a communication addressed to the Institute of International law, said very forcibly: "Since you observe that there does not appear to be any convincing reason to prefer the ten mile line in such a case to that of double three miles, I may say that there have been supposed to exist reasons both of convenience and of safety. The ten mile line has been adopted in the cases referred to as a practical rule. The transgression of an encroachment upon territorial waters by fishing vessels is generally a grave offense, involving in many instances the forfeiture of the offending vessel, and it is obvious that the narrower the space in which it is permissible to fish the more likely the offense is to be committed. In order, therefore, that fishing may be practicable and safe and not constantly attended with the risk of violating territorial waters, it has been thought to be expedient not to allow it where the extent of free waters between the three miles drawn on each side of the bay is less than four miles. This is the reason of the ten mile line. Its intention is not to hamper or restrict the right to fish, but to render its exercise practicable and safe. When fishermen fall in with a shoal of fish, the impulse to follow it is so strong as to make the possibilities of transgression very serious within narrow limits of free waters. Hence it has been deemed wiser to exclude them from space less than four miles each way from the forbidden lines. In spaces less than this operations are not only hazardous, but so circumscribed as to render them of little practical value." (Annuaire de l'Institut de Droit International, 1894, p. 146).

So the use of the ten mile bays so constantly put into practice by Great Britain in its fishery Treaties has its root and connection with the marginal belt of three miles for the territorial waters. So much so that the Tribunal having decided not to adjudicate in this case the ten miles entrance to the bays of the treaty of 1818, this will be the only one exception in which the ten miles of the bays do not follow as a consequence the strip of three miles of territorial waters, the historical bays and estuaries always excepted.

And it is for that reason that an usage so firmly and for so long a time established ought, in my opinion, be applied to the construction of the Treaty under consideration, much more so, when custom, one of the recognized sources of law, international as well as municipal, is supported in this case by reason and by the acquiescence and the practice of many nations.

The Tribunal has decided that: "In case of bays the 3 miles (of the Treaty) are to be measured from a straight line drawn across the body of water at the place where it ceases to have the configuration characteristic of a bay. At all other places the three miles are to be measured following the sinuosities of the coast." But no rule is laid out or general principle evolved for the parties to know what the nature of such configuration is or by what methods the points should be ascertained from which the bay should lose the characteristics of such. There lies the whole contention and the whole difficulty, not satisfactorily solved, to my mind, by simply recommending, without the scope of the award and as a system of procedure for resolving future contestations under Article IV of the Treaty of Arbitration, a series of lines, which practical as they may be supposed to be, cannot be adopted by the Parties without concluding a new Treaty.

These are the reasons for my dissent, which I much regret, on Question Five.

Done at the Hague, September 7th, 1910.

LUIS M. DRAGO.

INDEX

Printed in Dunstable, United Kingdom

91220127R00268